CELL DEATH

The Role of Poly (ADP-ribose) polymerase

Pharmacology and Toxicology: Basic and Clinical Aspects

Mannfred A. Hollinger, Series Editor
University of California, Davis

Published Titles

Manual of Immunological Methods, 1999, Pauline Brousseau, Yves Payette, Helen Tryphonas, Barry Blakley, Herman Boermans, Denis Flipo, Michel Fournier

CNS Injuries: Cellular Responses and Pharmacological Strategies, 1999, Martin Berry and Ann Logan

Infectious Diseases in Immunocompromised Hosts, 1998, Vassil St. Georgiev

Pharmacology of Antimuscarinic Agents, 1998, Laszlo Gyermek

Basis of Toxicity Testing, Second Edition, 1997, Donald J. Ecobichon

Anabolic Treatments for Osteoporosis, 1997, James F. Whitfield and Paul Morley

Antibody Therapeutics, 1997, William J. Harris and John R. Adair

Muscarinic Receptor Subtypes in Smooth Muscle, 1997, Richard M. Eglen

Antisense Oligodeonucleotides as Novel Pharmacological Therapeutic Agents, 1997, Benjamin Weiss

Airway Wall Remodelling in Asthma, 1996, A.G. Stewart

Drug Delivery Systems, 1996, Vasant V. Ranade and Mannfred A. Hollinger

Brain Mechanisms and Psychotropic Drugs, 1996, Andrius Baskys and Gary Remington

Receptor Dynamics in Neural Development, 1996, Christopher A. Shaw

Ryanodine Receptors, 1996, Vincenzo Sorrentino

Therapeutic Modulation of Cytokines, 1996, M.W. Bodmer and Brian Henderson

Pharmacology in Exercise and Sport, 1996, Satu M. Somani

Placental Pharmacology, 1996, B. V. Rama Sastry

Pharmacological Effects of Ethanol on the Nervous System, 1996, Richard A. Deitrich

Immunopharmaceuticals, 1996, Edward S. Kimball

Chemoattractant Ligands and Their Receptors, 1996, Richard Horuk

Pharmacological Regulation of Gene Expression in the CNS, 1996, Kalpana Merchant

Experimental Models of Mucosal Inflammation, 1995, Timothy S. Gaginella

Human Growth Hormone Pharmacology: Basic and Clinical Aspects, 1995, Kathleen T. Shiverick and Arlan Rosenbloom

Placental Toxicology, 1995, B. V. Rama Sastry

Stealth Liposomes, 1995, Danilo Lasic and Frank Martin

TAXOL®: Science and Applications, 1995, Matthew Suffness

Endothelin Receptors: From the Gene to the Human, 1995, Robert R. Ruffolo, Jr.

Alternative Methodologies for the Safety Evaluation of Chemicals in the Cosmetic Industry, 1995, Nicola Loprieno

Phospholipase A_2 in Clinical Inflammation: Molecular Approaches to Pathophysiology, 1995, Keith B. Glaser and Peter Vadas

Serotonin and Gastrointestinal Function, 1995, Timothy S. Gaginella and James J. Galligan

Chemical and Structural Approaches to Rational Drug Design, 1994, David B. Weiner and William V. Williams

Biological Approaches to Rational Drug Design, 1994, David B. Weiner and William V. Williams

Direct and Allosteric Control of Glutamate Receptors, 1994, M. Palfreyman, I. Reynolds, and P. Skolnick

Genomic and Non-Genomic Effects of Aldosterone, 1994, Martin Wehling

Peroxisome Proliferators: Unique Inducers of Drug-Metabolizing Enzymes, 1994, David E. Moody

Pharmacology and Toxicology: Basic and Clinical Aspects

Published Titles (*Continued*)

Angiotensin II Receptors, Volume I: Molecular Biology, Biochemistry, Pharmacology, and Clinical Perspectives, 1994, Robert R. Ruffolo, Jr.
Angiotensin II Receptors, Volume II: Medicinal Chemistry, 1994, Robert R. Ruffolo, Jr.
Beneficial and Toxic Effects of Aspirin, 1993, Susan E. Feinman
Preclinical and Clinical Modulation of Anticancer Drugs, 1993, Kenneth D. Tew, Peter Houghton, and Janet Houghton
In Vitro *Methods of Toxicology,* 1992, Ronald R. Watson
Basis of Toxicity Testing, 1992, Donald J. Ecobichon
Human Drug Metabolism from Molecular Biology to Man, 1992, Elizabeth Jeffreys
Platelet Activating Factor Receptor: Signal Mechanisms and Molecular Biology, 1992, Shivendra D. Shukla
Biopharmaceutics of Ocular Drug Delivery, 1992, Peter Edman
Pharmacology of the Skin, 1991, Hasan Mukhtar
Inflammatory Cells and Mediators in Bronchial Asthma, 1990, Devendra K. Agrawal and Robert G. Townley

CELL DEATH

The Role of Poly (ADP-ribose) polymerase

Edited by
Csaba Szabó, M.D., Ph.D.

CRC Press
Boca Raton   London   New York   Washington, D.C.

Library of Congress Cataloging-in-Publication Data
Cell death : the role of PARP / edited by Csaba Szabó.
    p. cm. — (Pharmacology and toxicology)
Includes bibliographical references and index.
ISBN 0-8493-2267-7 (alk. paper)
    1. Cell death. 2. NAD-ADP-ribosyltransferase. 3. DNA damage. 4. DNA repair. I. Szabó, Csaba, M.D. II. Pharmacology & toxicology (Boca Raton, Fla.)
        [DNLM: 1. Cell Death—physiology. 2. DNA Damage. 3. DNA Repair. 4. NAD+ ADP-Ribosyltransferase—physiology. 5. Oxidative Stress—physiology.
QH 671 C39385 2000]
571.9'39—dc21
                                                                                                                00-025646
                                                                                                                       CIP

This book contains information obtained from authentic and highly regarded sources. Reprinted material is quoted with permission, and sources are indicated. A wide variety of references are listed. Reasonable efforts have been made to publish reliable data and information, but the author and the publisher cannot assume responsibility for the validity of all materials or for the consequences of their use.

Neither this book nor any part may be reproduced or transmitted in any form or by any means, electronic or mechanical, including photocopying, microfilming, and recording, or by any information storage or retrieval system, without prior permission in writing from the publisher.

All rights reserved. Authorization to photocopy items for internal or personal use, or the personal or internal use of specific clients, may be granted by CRC Press LLC, provided that $.50 per page photocopied is paid directly to Copyright Clearance Center, 222 Rosewood Drive, Danvers, MA 01923 USA. The fee code for users of the Transactional Reporting Service is ISBN 0-8493-2267-7/00/$0.00+$.50. The fee is subject to change without notice. For organizations that have been granted a photocopy license by the CCC, a separate system of payment has been arranged.

The consent of CRC Press LLC does not extend to copying for general distribution, for promotion, for creating new works, or for resale. Specific permission must be obtained in writing from CRC Press LLC for such copying.

Direct all inquiries to CRC Press LLC, 2000 N.W. Corporate Blvd., Boca Raton, Florida 33431.

Trademark Notice: Product or corporate names may be trademarks or registered trademarks, and are used only for identification and explanation, without intent to infringe.

© 2000 by CRC Press LLC

No claim to original U.S. Government works
International Standard Book Number 0-8493-2267-7
Library of Congress Card Number 00-025646
Printed in the United States of America 1 2 3 4 5 6 7 8 9 0
Printed on acid-free paper

The Editor

Csaba Szabó, M.D., Ph.D., Vice President for Research, Inotek Corporation, is a pharmacologist, physiologist, and cell biologist. He received his M.D. degree at the Semmelweis University Medical School in Hungary, and obtained a Ph.D. degree in physiology at the Hungarian Academy of Sciences. He next completed a 3-year tenure with Nobel Laureate Sir John Vane at the William Harvey Research Institute in London, England, and has many major publications in the pathophysiology of shock and inflammation to his credit, including highly cited papers on the role of nitric oxide in endotoxic and hemorrhagic shock. He also identified several classes of nitric oxide synthase inhibitors selective toward the inducible isoform of nitric oxide synthase.

He was only 26 when he was initially recruited by Cincinnati's Children's Hospital to lead the research program of the Division of Critical Care, where he worked between 1994 and 1999 as an assistant and, later, as an associate professor and established one of the country's preeminent research programs for pediatric critical care.

Currently, Dr. Szabó holds an M.D. and two Ph.D. degrees, and, in addition to his position as Vice President for Research at Inotek Corporation, he has an adjunct full professorship (surgery) at the University of Medicine and Dentistry of New Jersey. He is recognized internationally as a leader in the field of the pathophysiological roles of nitric oxide, and has pioneered the field of the role of poly(ADP-ribose) synthetase activation in various forms of shock, inflammation, and reperfusion injury. His expertise ranges from reperfusion injury of the brain and heart, through local and systemic inflammatory conditions, septic and hemorrhagic shock, diabetes mellitus, to neuroimmunology of the inflammatory response. He is author of over 200 peer-reviewed original publications, author of numerous invited reviews and book chapters, a frequent invited lecturer at national and international meetings, a reviewer for over 30 major journals, and an invited grant reviewer for the National Institutes of Health, the Welcome Foundation, and other organizations. He has numerous issued and pending patents and patent applications in the areas of nitric oxide synthase inhibitors, nitric oxide donors, and poly(ADP-ribose) synthetase.

Dr. Szabó is spearheading diverse in-house basic research and preclinical development programs at Inotek, and collaborating with over 30 research groups worldwide. Over the last 5 years, he has attracted substantial extramural research funding from the National Institutes of Health, the American Heart Association, the American Lung Association, the Juvenile Diabetes Foundation, and other institutions. He is a member of the editorial boards of the *British Journal of Pharmacology, Shock,* and the *International Journal of Molecular Medicine.* He has been awarded a number of scientific awards, including the Pro Scientia Award (1989), the Lloyd's of London Tercentenary Foundation Award (1992), the Servier Young Investigator Award

(1994), the Hans Selye Award (1997), and the Peroxynitrite Award (1998). At the age of 31, he was elected the youngest ever Doctor of the Academy of Sciences in his native Hungary.

Dr. Szabó is married and has a 1-year-old daughter, Lili Szabó.

Contributors

El Bachir Affar
Health and Environment Unit
Laval University Medical Research
  Center, CHUQ, and
Faculty of Medicine
Laval University
Ste-Foy, Québec, Canada

Betty J. Benton
U.S. Army Medical Institute of
  Chemical Defense
Aberdeen Proving Ground, Aberdeen,
Maryland

Sándor Bernáth
N-Gene Research Laboratories, Inc.
Budapest, Hungary

Claudia Boucher
Health and Environment Unit
Laval University Medical Research
  Center, CHUQ, and
Faculty of Medicine
Laval University
Ste-Foy, Québec, Canada

Sylvie Bourassa
Health and Environment Unit
Laval University Medical Research
  Center, CHUQ, and
Faculty of Medicine
Laval University
Ste-Foy, Québec, Canada

Volker Burkart
Diabetes Research Institute
Heinrich-Heine University
Düsseldorf, Germany

Ted M. Dawson
Departments of Neurology and
  Neuroscience
Johns Hopkins University School of
  Medicine
Baltimore, Maryland

Valina L. Dawson
Departments of Neurology,
  Neuroscience, and Physiology
Johns Hopkins University School of
  Medicine
Baltimore, Maryland

Gilbert de Murcia
Laboratoire Conventionné avec le
  Commissariat à l'Energie
  Atomique
Centre National de la Recherche
  Scientifique
Ecole Supérieure de Biotechnologie
  de Strasbourg
Illkirch-Graffenstaden, France

Ferenc Gallyas, Jr.
Department of Biochemistry
University Medical School Pécs
Pécs, Hungary

Marc Germain
Health and Environment Unit
Laval University Medical Research
  Center, CHUQ, and
Faculty of Medicine
Laval University
Ste-Foy, Québec, Canada

Stéphane Gobeil
Health and Environment Unit
Laval University Medical Research
  Center, CHUQ, and
Faculty of Medicine
Laval University
Ste-Foy, Québec, Canada

Elaine L. Jacobson
Department of Clinical Sciences
Center for Nutritional Sciences
Lucille P. Markey Cancer Center
Advanced Science and Technology
  Commercialization Center
University of Kentucky
Lexington, Kentucky

Myron K. Jacobson
College of Pharmacy
Lucille P. Markey Cancer Center
Advanced Science and Technology
  Commercialization Center
University of Kentucky
Lexington, Kentucky

Jia-He Li
Guilford Pharmaceuticals, Inc.
Baltimore, Maryland

Josiane Ménissier-de Murcia
Laboratoire Conventionné avec le
  Commissariat à l'Energie
  Atomique
Centre National de la Recherche
  Scientifique
Ecole Supérieure de Biotechnologie
  de Strasbourg
Illkirch-Graffenstaden, France

Péter Literati Nagy
N-Gene Research Laboratories, Inc.
Budapest, Hungary

Claude Niedergang
Laboratoire Conventionné avec le
  Commissariat à l'Energie Atomique
Centre National de la Recherche
  Scientifique
Ecole Supérieure de Biotechnologie de
  Strasbourg
Illkirch-Graffenstaden, France

F. Javier Oliver
Laboratoire Conventionné avec le
  Commissariat à l'Energie Atomique
Centre National de la Recherche
  Scientifique
Ecole Supérieure de Biotechnologie de
  Strasbourg
Illkirch-Graffenstaden, France

Guy G. Poirier
Health and Environment Unit
Laval University Medical Research
  Center, CHUQ, and
Faculty of Medicine
Laval University
Ste-Foy, Québec, Canada

György Rablóczky
N-Gene Research Laboratories, Inc.
Budapest, Hungary

Ildikó Rácz
N-Gene Research Laboratories, Inc.
Budapest, Hungary

Radharaman Ray
U.S. Army Medical Institute of
  Chemical Defense
Aberdeen Proving Ground, Aberdeen,
  Maryland

Dean S. Rosenthal
Department of Biochemistry and
  Molecular Biology
Georgetown University School of
  Medicine
Washington, D.C.

Jérôme St-Cyr
Health and Environment Unit
Laval University Medical Research
  Center, CHUQ, and
  Faculty of Medicine
Laval University
Ste-Foy, Québec, Canada

Andrew L. Salzman
Inotek Corporation
Beverly, Massachusetts

A. Ivana Scovassi
Instituto di Genetica Biochimica
  Evoluzionistica CNR
Pavia, Italy

Rashmi Shah
Health and Environment Unit
Laval University Medical Research
  Center, CHUQ, and
  Faculty of Medicine
Laval University
Ste-Foy, Québec, Canada

Cynthia M. Simbulan-Rosenthal
Department of Biochemistry and
  Molecular Biology
Georgetown University School of
  Medicine
Washington, D.C.

William J. Smith
U.S. Army Medical Institute of
  Chemical Defense
Aberdeen Proving Ground, Aberdeen,
  Maryland

Mark E. Smulson
Department of Biochemistry and
  Molecular Biology
Georgetown University School of
  Medicine
Washington, D.C.

Solomon H. Snyder
Department of Neuroscience,
  Pharmacology and Molecular
  Sciences, and Psychiatry
Johns Hopkins University School of
  Medicine
Baltimore, Maryland

Balázs Sümegi
Department of Biochemistry
University Medical School Pécs
Pécs, Hungary

Csaba Szabó
Inotek Corporation
Beverly, Massachusetts

Kálmán Tory
N-Gene Research Laboratories, Inc.
Budapest, Hungary

Gábor Várbiró
Department of Biochemistry
University Medical School Pécs
Pécs, Hungary

László Virág
Department of Medical Chemistry
University Medical School of
  Debrecen
Debrecen, Hungary

Eric Winstall
Health and Environment Unit
Laval University Medical Research
  Center, CHUQ, and
  Faculty of Medicine
Laval University
Ste-Foy, Québec, Canada

Jie Zhang
Guilford Pharmaceuticals, Inc.
Baltimore, Maryland

Basilia Zingarelli
Division of Critical Care Medicine
Children's Hospital Medical Center
Cincinnati, Ohio

Table of Contents

Introduction
Cell Death by Energy Depletion: A Reemerging Concept via the Nitric Oxide Connection 1
Csaba Szabó and Solomon H. Snyder

Chapter 1
Nitric Oxide and Poly(ADP-Ribose) Polymerase in Neuronal Cell Death 7
Jie Zhang and Solomon H. Snyder

Chapter 2
Role of Poly(ADP-Ribose) Polymerase Mediating Acute and Chronic Neuronal Injury and Neurodegeneration 23
Valina L. Dawson and Ted M. Dawson

Chapter 3
Importance of Poly(ADP-Ribose) Polymerase Activation in Myocardial Reperfusion Injury 41
Basilia Zingarelli

Chapter 4
Role of Poly(ADP-Ribose) Polymerase Activation in Inflammation 61
Andrew L. Salzman

Chapter 5
Poly(ADP-Ribose) Polymerase Activation and the Pathogenesis of Circulatory Shock 81
Csaba Szabó

Chapter 6
Role of Poly(ADP-Ribose) Polymerase in the Pathogenesis of Pancreatic Islet Cell Death and Type 1 Diabetes 103
Volker Burkart

Chapter 7
Poly(ADP-Ribose) Polymerase in the Immune System: Focus on Reactive Nitrogen Intermediates and Cell Death 131
László Virág

Chapter 8
Protective Effect of Poly(ADP-Ribose) Polymerase Inhibitors against
Cell Damage Induced by Antiviral and Anticancer Drugs 167
*Balázs Sümegi, György Rablóczky, Ildikó Rácz, Kálmán Tory, Sándor Bernáth,
Gábor Várbiró, Ferenc Gallyas, Jr., and Péter Literati Nagy*

Chapter 9
Involvement of Poly(ADP-Ribose) Polymerase in the Cellular Response
to DNA Damage 183
*Claude Niedergang, F. Javier Oliver, Josiane Ménissier-de Murcia, and
Gilbert de Murcia*

Chapter 10
Role of Poly(ADP-Ribose) Polymerase in Apoptosis 209
Marc Germain, A. Ivana Scovassi, and Guy G. Poirier

Chapter 11
Poly(ADP-Ribose) Polymerase is an Active Participant in Programmed
Cell Death and Maintenance of Genomic Stability 227
*Dean S. Rosenthal, Cynthia M. Simbulan-Rosenthal, William J. Smith,
Betty J. Benton, Radharaman Ray, and Mark E. Smulson*

Chapter 12
Pleiotropic Roles of Poly(ADP-Ribosyl)ation of DNA-Binding Proteins 251
Cynthia M. Simbulan-Rosenthal, Dean S. Rosenthal, and Mark E. Smulson

Chapter 13
Poly(ADP-Ribose) Polymerase Inhibition by Genetic and Pharmacological
Means 279
Jie Zhang and Jia-He Li

Chapter 14
Determination of Poly(ADP-Ribose) Polymerase Activation and Cleavage
during Cell Death 305
*El Bachir Affar, Stéphane Gobeil, Marc Germain, Eric Winstall, Rashmi Shah,
Sylvie Bourassa, Jérôme St-Cyr, Claudia Boucher, and Guy G. Poirier*

Chapter 15
Isoforms of Poly(ADP-Ribose) Polymerase: Potential Roles in Cell Death 323
Elaine L. Jacobson and Myron K. Jacobson

Index 331

Introduction

Cell Death by Energy Depletion: A Reemerging Concept via the Nitric Oxide Connection

Csaba Szabó and Solomon H. Snyder

Poly(ADP-ribose) polymerase (PARP), also termed poly(ADP-ribose) synthetase (PARS), is a nuclear enzyme with a wide range of functions including regulation of DNA repair, cell differentiation, and gene expression.[1,2] More than a decade after the identification of PARP-like enzymatic activities in mammalian cells, Dr. Nathan Berger and colleagues[3,4] proposed a novel role for this enzyme, mediating a suicidal mechanism triggered by DNA strand breakage. As summarized in 1995 by Berger, "when DNA strand breaks are extensive or when breaks fail to be repaired, the stimulus for activation of poly(ADP-ribose) persists and the activated enzyme is capable of totally consuming cellular pools of NAD. Depletion of NAD and consequent lowering of cellular ATP pools, due to activation of poly(ADP-ribose) polymerase, may account for rapid cell death before DNA repair takes place and before the genetic effects of DNA damage become manifest."[5] This hypothesis has since become a controversial centerpiece of the PARP field. In many experimental systems, the PARP suicide theory has been confirmed and extended. For instance, in the mid 1980s, Dr. Charles Cochrane and his group carried out a detailed characterization of the PARP suicide pathway in white blood cells stimulated with hydrogen peroxide.[6-8] Furthermore, in the early 1980s, a role of PARP was proposed in conjunction with streptozotocin-induced islet cell death and diabetes mellitus.[9,10]

The theoretical and practical implications of the PARP suicide pathway were not exploited extensively until the 1990s; in the 1980s only a handful of pathophysiologically relevant studies, almost exclusively focusing on diabetes mellitus, had been completed in conjunction with the pathway (e.g., References 11 and 12). What accounts for this "lag phase"? One factor was a shift of attention toward investigations into molecular mechanisms of PARP cleavage and its role in apoptosis. Another reason may have been the limited availability of potent and specific PARP inhibitors, and the lack of PARP-deficient cells and animals. The fact that the field was lacking a pathophysiologically relevant stimulus of DNA single-strand breakage and PARP activation may also have played an important role.

In the early 1990s, one of our laboratories[5,15] and a group of diabetes researchers in Düsseldorf, Germany, led by Dr. Hubert Kolb, provided evidence that nitric oxide (NO), a labile free radical with multiple roles in physiology and pathophysiology, can activate a pathway leading to cell death (neuronal cell death and pancreatic cell death, respectively).[13,14] Peroxynitrite, a reactive oxidant species produced from the reaction of NO and superoxide free radicals, has subsequently been established by one of the authors[15] as an endogenously produced trigger of DNA single-strand breakage and PARP activation (overviewed in Reference 16). Peroxynitrite, rather than NO is the likely trigger of cell death in stroke,[17] as well as in various other forms of reperfusion and inflammation.[18]

Appreciation of the NO connection introduced new investigators (mainly from the NO field) to the field of PARP, many of them with interest in pathophysiology and *in vivo* models of disease. About this time, the first line of knockout mice, lacking functional PARP, was generated.[19] Cells from the PARP knockout mice were found to be resistant to various forms of oxidant injury.[20-22] Using the PARP-deficient mice, it was also demonstrated that lack of PARP confers protection against a wide range of insults including animal models of stroke,[21-23] brain trauma,[24] myocardial ischemia–reperfusion injury,[25-27] colitis,[28] endotoxic shock,[29] diabetes,[30-32] and Parkinson's disease.[33] The findings in knockout animals have also been reproduced in animals treated with pharmacological inhibitors of PARP (e.g., References 21 through 36). PARP rapidly became a target for drug development for the experimental therapy of various forms of inflammation and reperfusion injury.

Recent investigations have now clarified the mode of oxidant-induced cell death affected by PARP. Exposure to massive amounts of DNA-damaging mediators leads to correspondingly massive levels of DNA damage, with consequent massive synthesis of poly(ADP-ribose) polymer and loss of cellular NAD. It is now clear that the PARP suicide pathway, the process of "DNA single-strand breakage → PARP activation → energy depletion" does not directly induce apoptosis, but rather triggers necrotic cell death characterized by rapid cell injury, changes in membrane integrity, and release of intracellular content evidenced by increases in the plasma levels of various necrotic markers, such as lactate dehydrogenase.[14,20,21,37-39] Pharmacological inhibition or genetic inactivation of PARP protects against cell necrosis. In some systems, e.g., in cells challenged with high levels of peroxynitrite or hydrogen peroxide, inhibition of PARP can divert the mode of cell death from the necrotic toward the apoptotic mode by providing energy to apoptosis (apoptosis being an energy-dependent, active process). The fact that cell necrosis can be influenced by pharmacological means can be considered a key revelation, because necrotic cell death is not generally considered to be amenable to pharmacological or therapeutic interventions. Necrotic cell death is a key component of many pathophysiological conditions. For example, in the reperfused myocardium, or the failing liver, necrotic cells lose the integrity of the plasma membrane and release their intracellular contents. Indeed, the release of intracellular enzymes from specific tissues has long been used as a clinical diagnostic tool; examples are the measurement of the plasma levels of creatine kinase isoforms specific for the heart in the diagnosis of myocardial infarction and the detection of various hepatic transaminase enzymes in the plasma in the diagnosis of liver injury. Similarly, in various forms of stroke, neuronal necrosis is a dominant part of the histological picture. Inhibitors of PARP may represent one of the first pharmacological means to prevent cell necrosis. The "PARP activation → cell necrosis" pathway has also transformed the conventional wisdom regarding the role of PARP cleavage in the process of cell death: according to recent studies, PARP cleavage can be considered a protective mechanism, which prevents cell necrosis (because cleaved PARP is catalytically inactive and cannot trigger NAD$^+$ and ATP depletion) and thereby helps the completion of the apoptotic process.[40]

The current status of the field of PARP and cell death is well represented in this book. Investigators from the diverse fields of neuroinjury, myocardial injury, diabetes, shock, and inflammation have contributed. In addition, the area of PARP and apoptosis, PARP and DNA repair, PARP and regulation of gene expression are well represented. Furthermore, separate chapters focus on developments in the respective areas of pharmacological inhibition of PARP, and on novel ways of measuring PARP activation and PARP cleavage. Finally, the novel, emerging field of PARP isoforms is presented, with their potential role in cell death.

Tremendous progress has been made in the area of PARP and cell death. It is also clear that the area of PARP research will grow further in the future, and may ultimately enter the clinical arena. Nevertheless, many controversies remain to be clarified, and many recent discoveries and observations need to be further developed and explored. We hope that the current volume not only will present a state-of-the art overview of the field, but also will provide some help for this quest.

REFERENCES

1. de Murcia, G., Schreiber, V., Molinete, M., Saulier, B., Poch, O., Masson, M., Niedergang, C., and Menissier-de Murcia, J., Structure and function of Poly(ADP-ribose) polymerase. *Mol. Cell. Biochem.,* 138, 15–24, 1994.
2. Le Rhun, Y., Kirkland, J.B., and Shah, G.M., Cellular responses to DNA damage in the absence of Poly(ADP-ribose) polymerase. *Biochem. Biophys. Res. Commun.,* 245, 1–10, 1998.
3. Berger, N.A., Catino, D.M., and Vietti, T.J., Synergistic antileukemic effect of 6-aminonicotinamide and 1,3-bis(2-chloroethyl)-1-nitrosourea on L1210 cells *in vitro* and *in vivo. Cancer Res.,* 42, 4382–4386, 1982.
4. Sims, J.L., Berger, S.J., and Berger, N.A., Poly(ADP-ribose) Polymerase inhibitors preserve nicotinamide adenine dinucleotide and adenosine 5'-triphosphate pools in DNA-damaged cells: mechanism of stimulation of unscheduled DNA synthesis. *Biochemistry,* 22, 5188–5194, 1983.
5. Berger, N.A., Poly(ADP-ribose) in the cellular response to DNA damage. *Radiat. Res.,* 101, 4–15, 1985.
6. Schraufstatter, I.U., Hinshaw, D.B., Hyslop, P.A., Spragg, R.G., and Cochrane, C.G., Oxidant injury of cells. DNA strand-breaks activate polyadenosine diphosphate-ribose polymerase and lead to depletion of nicotinamide adenine dinucleotide. *J. Clin. Invest.,* 77, 1312–1320, 1986.

7. Schraufstatter, I.U., Hyslop, P.A., Hinshaw, D.B., Spragg, R.G., Sklar, L.A., and Cochrane, C.G., Hydrogen peroxide-induced injury of cells and its prevention by inhibitors of poly(ADP-ribose) polymerase. *Proc. Natl. Acad. Sci. U.S.A.,* 83, 4908–4912, 1986.
8. Cochrane, C.G., Cellular injury by oxidants. *Am. J. Med.,* 91, 23S–30S, 1991.
9. Yamamoto, H. and Okamoto, H., Protection by picolinamide, a novel inhibitor of poly (ADP-ribose) synthetase, against both streptozotocin-induced depression of proinsulin synthesis and reduction of NAD content in pancreatic islets. *Biochem. Biophys. Res. Commun.,* 95, 474–481, 1980.
10. Uchigata, Y., Yamamoto, H., Kawamura, A., and Okamoto, H., Protection by superoxide dismutase, catalase, and poly(ADP-ribose) synthetase inhibitors against alloxan- and streptozotocin-induced islet DNA strand breaks and against the inhibition of proinsulin synthesis. *J. Biol. Chem.,* 257, 6084–6088, 1982.
11. Yonemura, Y., Takashima, T., Miwa, K., Miyazaki, I., Yamamoto, H., and Okamoto, H., Amelioration of diabetes mellitus in partially depancreatized rats by poly(ADP-ribose) synthetase inhibitors. Evidence of islet B-cell regeneration. *Diabetes,* 33, 401–404, 1984.
12. Shima, K., Hirota, M., Sato, M., Numoto, S., and Oshima, I., Effect of poly(ADP-ribose) synthetase inhibitor administration to streptozotocin-induced diabetic rats on insulin and glucagon contents in their pancreas. *Diabetes Res. Clin. Pract.,* 3, 135–142, 1987.
13. Zhang, J., Dawson, V.L., Dawson, T.M., and Snyder, S.H., Nitric oxide activation of poly (ADP-ribose) synthetase in neurotoxicity. *Science,* 263, 687–689, 1994.
14. Radons, J., Heller, B., Burkle, A., Hartmann, B., Rodriguez, M.L., Kroncke, K.D., Burkart, V., and Kolb, H., Nitric oxide toxicity in islet cells involves poly (ADP-ribose) polymerase activation and concomitant NAD depletion. *Biochem. Biophys. Res. Commun.,* 199, 1270–1277, 1994.
15. Szabó, C., Zingarelli, B., O'Connor, M., and Salzman, A.L., DNA strand breakage, activation of poly-ADP ribosyl synthetase, and cellular energy depletion are involved in the cytotoxicity in macrophages and smooth muscle cells exposed to peroxynitrite. *Proc. Natl. Acad. Sci. U.S.A.,* 93, 1753–1758, 1996.
16. Szabó, C., DNA strand breakage and activation of poly-ADP ribosyltransferase: a cytotoxic pathway triggered by peroxynitrite. *Free Radical Biol. Med.,* 21, 855–869, 1996.
17. Eliasson, M.J., Huang, Z., Ferrante, R.J., Sasamata, M., Molliver, M.E., Snyder, S.H., and Moskowitz, M.A., Neuronal nitric oxide synthase activation and peroxynitrite formation in ischemic stroke linked to neural damage. *J. Neurosci.,* 19, 5910–5918, 1999.
18. Szabó, C. and Dawson, V.L., Role of poly (ADP-ribose) synthetase activation in inflammation and reperfusion injury. *Trends Pharmacol. Sci.,* 19, 287–298, 1998.
19. Wang, Z.Q., Auer, B., Stingl, L., Berghammer, H., Haidacher, D., Schweiger, M., and Wagner, E.F., Mice lacking ADPRT and poly(ADP-ribosyl)ation develop normally but are susceptible to skin disease. *Genes Dev.,* 9, 510–520, 1995.
20. Heller, B., Wang, Z.Q., Wagner, E.F., Radons, J., Burkle, A., Fehsel, K., Burkart, V., and Kolb, H., Inactivation of the poly(ADP-ribose) polymerase gene affects oxygen radical and nitric oxide toxicity in islet cells. *J. Biol. Chem.,* 270, 11176–11180, 1985.
21. Eliasson, M.J.L, Sampei, K., Mandir, A.S., Hurn, P.D., Traystman, R.J., Bao, J., Pieper, A., Wang, Z.Q., Dawson, T.M., Snyder, S.H., and Dawson, V.L., Poly (ADP-ribose) polymerase gene disruption renders mice resistant to cerebral ischaemia. *Nat. Med.,* 3, 1089–1095, 1997.

22. Szabó, C., Virág, L., Cuzzocrea, S., Scott, G.J., Hake, P., O'Connor, M.P, Zingarelli, B., Salzman, A.L., and Kun, E., Protection against peroxynitrite-induced fibroblast injury and arthritis development by inhibition of poly (ADP-ribose) synthetase. *Proc. Natl. Acad. Sci. U.S.A.*, 95, 3867–3872, 1998.
23. Szabó, C., Lim, L.H., Cuzzocrea, S., Getting, S.J., Zingarelli, B., Flower, R.J., Salzman, A.L., and Perretti, M., Inhibition of poly (ADP-ribose) synthetase exerts anti-inflammatory effects and inhibits neutrophil recruitment. *J. Exp. Med.*, 186, 1041–1049, 1997.
24. Whalen, M.J., Clark, R.S.B., Dixon, C.E., Robichaud, P., Marion, D.W., Vagni, V., Graham, S., Virág, L., Haskó, G., Stachlewitz, R., Szabó, C., and Kochanek, P., Reduction in deficits of memory and motor function after traumatic brain injury in mice deficient in poly (ADP-ribose) polymerase. *J. Cereb. Blood Flow Metabol.*, 19, 835–842, 1999.
25. Zingarelli, B., Salzman. A.L., and Szabó, C., Genetic disruption of poly (ADP ribose) synthetase inhibits the expression of P-selectin and intercellular adhesion molecule-1 in myocardial ischemia-reperfusion injury. *Circ. Res.*, 83, 85–94, 1998.
26. Yang, Z., Zingarelli, B., and Szabó, C., Role of poly (ADP-ribose) synthetase in the delayed myocardial ischemia-reperfusion injury. *Shock*, 13, 60–66, 2000.
27. Grupp, I.L., Jackson, T.M., Hake, P., Grupp, G., and Szabó, C., Protection against hypoxia-reoxygenation in the absence of poly (ADP-ribose) synthetase in isolated working hearts. *J. Mol. Cell. Cardiol.*, 31, 297–303, 1999.
28. Zingarelli, B., Szabó, C., and Salzman, A.L., Poly (ADP-ribose) synthetase triggers a positive feedback cycle of neutrophil recruitment, oxidant generation, and mucosal injury in colitis. *Gastroenterology*, 116, 335–345, 1999.
29. Oliver, F.J., Menissier-de Murcia, J., Nacci, C., Decker, P., Andriantsitohaina, R., Muller, S., De la Rubia, G., Stoclet, J.C., and de Murcia, G., Resistance to endotoxic shock as a consequence of defective NF-kappaB activation in poly (ADP-ribose) polymerase-1 deficient mice. *EMBO J.*, 18, 4446–4454, 1999.
30. Burkart, V., Wang, Z.Q., Radons, J., Heller, B., Herceg, Z., Stingl, L., Wagner, E.F., and Kolb, H., Mice lacking the poly(ADP-ribose) polymerase gene are resistant to pancreatic beta-cell destruction and diabetes development induced by streptozocin. *Nat. Med.*, 5, 314–319, 1999.
31. Masutani, M., Suzuki, H., Kamada, N., Watanabe, M., Ueda, O., Nozaki, T., Jishage, K., Watanabe, T., Sugimoto, T., Nakagama, H., Ochiya, T., and Sugimura, T.P., Poly(ADP-ribose) polymerase gene disruption conferred mice resistant to streptozotocin-induced diabetes. *Proc. Natl. Acad. Sci. U.S.A.*, 96, 2301–2304, 1999.
32. Pieper, A.A., Brat, D.J., Krug, D.K., Watkins, C.C., Gupta, A., Blackshaw, S., Verma, A., Wang, Z.Q., and Snyder, S.H., Poly(ADP-ribose) polymerase-deficient mice are protected from streptozotocin-induced diabetes. *Proc. Natl. Acad. Sci. U.S.A.*, 96, 3059–3064, 1999.
33. Mandir, A.S., Przedborski, S., Jackson-Lewis, V., Wang, Z.Q., Simbulan-Rosenthal, D., Smulson, M.E., Hoffman, B.E., Guastella, D.B., Dawson, V.L., and Dawson, T.M., Poly(ADP-ribose) polymerase activation mediates 1-methyl-4-phenyl-1,2,3,6-tetrahydropyridine (MPTP)-induced parkinsonism. *Proc. Natl. Acad. Sci. U.S.A.*, 96, 5774-5779, 1999.
34. Takahashi, K., Pieper, A.A., Croul, S.E., Zhang, J., Snyder, S.H., and Greenberg, J.H., Post-treatment with an inhibitor of poly(ADP-ribose) polymerase attenuates cerebral damage in focal ischemia. *Brain Res.*, 829, 46–54, 1999.

35. Szabó, C., Zingarelli, B., and Salzman, A.L., Role of poly-ADP ribosyltransferase activation in the nitric oxide- and peroxynitrite-induced vascular failure. *Circ. Res.*, 78, 1051–1063, 1996.
36. Thiemermann, C., Bowes, J., Myint, F.P., and Vane, J.R., Inhibition of the activity of poly (ADP ribose) synthetase reduces ischemia-reperfusion injury in the heart and skeletal muscle. *Proc. Natl. Acad. Sci. U.S.A.,* 94, 679–683, 1997.
37. Virág, L., Scott, G.S., Salzman, A.L., and Szabó, C., Peroxynitrite-induced thymocyte apoptosis: the role of caspases and poly-(ADP-ribose) synthetase (PARS) activation. *Immunology,* 94, 345–355, 1998.
38. Virág, L., Salzman, A.L., and Szabó, C., Poly (ADP-ribose) synthetase activation mediates mitochondrial injury during oxidant-induced cell death. *J. Immunol.,* 161, 3753–3759, 1998.
39. Ha, H.C. and Snyder, S.H., Poly(ADP-ribose) polymerase is a mediator of necrotic cell death by ATP depletion. *Proc. Natl. Acad. Sci. U.S.A.,* 96, 13978–13982, 1999.
40. Herceg, Z. and Wang, Z.Q., Failure of poly(ADP-ribose) polymerase cleavage by caspases leads to induction of necrosis and enhanced apoptosis. *Mol. Cell. Biol.,* 19, 5124–5133, 1999.

1 Nitric Oxide and Poly(ADP-Ribose) Polymerase in Neuronal Cell Death

Jie Zhang and Solomon H. Snyder

CONTENTS

1.1 Introduction 7
1.2 Nitric Oxide 8
1.3 Relationship of Nitric Oxide to Poly(ADP-Ribose) Polymerase (PARP) 10
1.4 Use of PARP Knockout Mice to Clarify Neurotoxicity 11
1.5 Conclusions: Therapeutic Relevance 16
References 17

1.1 INTRODUCTION

Mechanisms of neural damage have been of considerable clinical concern long before the augmented interest in programmed cell death, apoptosis, as a sequence of defined steps that can be approached through selective modulation of individual genes. Much of this interest involved efforts to deal therapeutically with stroke and neurodegenerative diseases such as Alzheimer's disease, Huntington's disease, and Parkinson's disease. As most strokes are caused by occlusion of cerebral blood vessels, it was thought for many years that neural damage simply involved infarction of tissue due to absence of blood flow so that no therapeutic intervention would be feasible. In the case of neurodegenerative diseases, whose specific causes remain largely unknown, there was little to be done about the inexorable process of cell death. This line of thinking began to change with evidence that, following stroke, only a limited amount of tissue is infarcted due to total ischemia, while the major neural damage proceeds gradually over a day or two following toxic insult to tissue that was only partially hypoxic. Numerous toxic chemicals are released as a result of ischemia, including products of lipid peroxidation and assorted oxygen free radicals. A particularly large augmentation of glutamate release into the extracellular space occurs following vascular occlusion with 50-fold or greater increases detected by microdialysis.[1]

Glutamate is of interest because of evidence from the late 1970s that glutamate can be a neurotoxin. Indeed, the considerable potency of various rigid glutamate analogues has led to their use as the preferred means of producing brain lesions that selectively destroy neurons while leaving glia and other supporting cells intact.[2]

Glutamate neurotoxicity is particularly notable, because glutamate is a physiological excitatory neurotransmitter that acts at specific receptors so that pharmacological agents can be developed to perturb its effects. Glutamate acts at two major types of receptors, metabotropic receptors, which are coupled to G proteins and act through cyclic AMP or inositol 1,4,5-trisphosphate (IP_3), and ionotropic receptors. There are two major classes of ionotropic receptors, those selectively stimulated by N-methyl-D-aspartate (NMDA) and those activated by (α-amino-3-hydroxy-5-methylisoxazole-4-propionic acid (AMPA). All of the subtypes of glutamate appear to play some role in neurotoxicity. However, the greatest attention has been focused upon NMDA receptors. The NMDA receptor–associated ion channel admits sodium and calcium. It is also voltage activated so that a neuron must be partially depolarized, usually by AMPA receptor activation, before a physiological block of the NMDA channel by magnesium is relieved. The NMDA receptor is also unique in that it requires two agonists for physiological activation. A so-called glycine site of the receptor must be activated in addition to the glutamate recognition site. Recent evidence indicates that the predominant endogenous ligand for the glycine site is in fact D-serine formed by an enzyme, serine racemase colocalized in astrocytes adjacent to NMDA receptors, which converts L-serine to D-serine.[3]

Drugs acting at all the various subtypes of glutamate receptors influence neurotoxicity. Most extensively studied have been agents affecting NMDA receptors. Drugs blocking either the glutamate or "glycine" sites of the receptor prevent neurotoxicity in multiple models of neuronal culture and diminish stroke damage in intact animals. Agents blocking AMPA receptors are also therapeutic, as are drugs influencing metabotropic sites. These pharmacological data have provided the most persuasive evidence for a role of glutamate excess in neurotoxicity and stroke. Models of neurodegenerative disease are difficult to assess. However, the destruction of dopamine neurons by 1-methyl-4-phenyl-1,2,3,6-tetrahydropyridine (MPTP) is regarded as a meaningful model mimicking salient features of Parkinson's disease. Various glutamate antagonists prevent MPTP neurotoxicity (see Chapter 2 by Dawson and Dawson).

1.2 NITRIC OXIDE

Nitric oxide (NO) was first identified as a physiological modulator of cellular activity through its actions as endothelial-derived relaxing factor in blood vessels and its role in mediating the ability of activated macrophages to kill tumor cells and bacteria. Garthwaite and associates[4] detected a substance with activity resembling NO in cerebellar cultures, suggesting that NO might also be a messenger molecule in the brain. In blood vessels, NO relaxes smooth muscle by stimulating the activity of guanylyl cyclase to synthesize cyclic GMP. In the brain, the highest levels of cyclic GMP occur in the cerebellum of young rats. NMDA receptor activation elicits a rapid ten-fold increase in cyclic GMP levels. This increase was found to be associated with a concomitant tripling of NO synthase (NOS) activity that occurs within a matter of a few minutes.[5] Both the augmentation of NOS activity and of cyclic GMP levels could be blocked by arginine derivatives that are inhibitors of NOS.[6] Purification of neuronal NOS (nNOS) showed that it requires calmodulin and associated calcium for activity.[7] This explains the abrupt augmentation of nNOS activity following NMDA receptor activation, as the calcium entry into cells through the NMDA channel stimulates nNOS. This mode of calcium entry is selective as general increases in calcium elicited by augmentation of intracellular IP_3 do not activate nNOS. Instead, NMDA receptors are physically linked to nNOS by the postsynaptic density protein, PSD-95, which binds both to the NMDA receptor and to nNOS.[8,9]

Purification of nNOS permitted the development of antisera and immunohistochemical mapping of NO neurons as well as cloning of nNOS and, subsequently, separate forms of the enzyme derived from distinct genes, endothelial NOS (eNOS), and macrophage or inducible NOS (iNOS).[10-17] nNOS neurons comprise only about 1% of neuronal cell bodies in the cerebral cortex and most other parts of the brain. However, their extensively branched processes ramify to so great an extent that it seems likely that virtually every neuron in the brain is in proximity to an nNOS neuronal process.[18]

The possibility that NO might influence a wide range of neuronal cells in the brain suggests that, just as NO mediates effects of glutamate on cyclic GMP, perhaps it mediates the neurotoxic actions of glutamate. Cerebral cortical cultures have been a valuable system for exploring the neurotoxicity, especially by NMDA activation. Exposure of cortical cultures to glutamate or NMDA itself for a brief period is followed 24 h later by death of the majority of neurons. The ability of drugs to block neurotoxicity in this model correlates well with their antistroke actions in intact animals. NOS inhibitors block NMDA neurotoxicity in these cultures.[19] Subsequently, cultures from mice with targeted deletion of nNOS were shown to be markedly protected from NMDA neurotoxicity.[19]

Evidence for NO as a mediator of neurotoxicity has been translated into direct demonstrations for a major role of NO in stroke damage. Numerous laboratories showed that NOS inhibitors substantially reduce stroke damage even when administered after ligation of the middle cerebral artery.[21] The extent of protection of stroke damage with NOS inhibitors is comparable to what one obtains with glutamate receptor antagonists. $nNOS^{-/-}$ mice are also protected against stroke damage although the protection is less than with NOS inhibitors.[22] This discrepancy may be related to the fact that the $nNOS^{-/-}$ animals employed in these studies are deficient only in the major form of the enzyme, nNOS-α, while two alternatively spliced forms, nNOS-β and nNOS-γ, are retained in the knockout mice.[23] Evidence that these alternatively spliced forms of the enzyme are functional in intact animals comes from immunohistochemical staining with antibodies generated against citrulline, which is formed stoichiometrically with NO by NOS.[24] The staining pattern for citrulline in intact brain resembles that for nNOS. In $nNOS^{-/-}$ mice, citrulline staining is abolished in some areas such as the cerebellum. However, in other brain regions such as the cerebral cortex and corpus striatum, citrulline staining levels are 30 to 50% of control values. This fits with studies monitoring catalytic activity of nNOS in homogenates from various brain regions of $nNOS^{-/-}$ animals.[25]

The citrulline staining technique provided an approach to direct examination of the hypothesis that strokes do elicit a major augmentation in NO production, which had been hypothesized but never directly demonstrated, except in very preliminary efforts. Occluding the middle cerebral artery in rats does indeed markedly stimulate citrulline staining. However, the area stained does not correspond to infarcted, dead brain tissue. Instead, citrulline staining is most intense in penumbra areas, which are partially hypoxic but in which neurons are still alive.[26] By contrast, nitrotyrosine staining is most evident in infarcted zones. Nitrotyrosine reflects the actions of peroxynitrite, formed by the combination of NO and superoxide. These findings thus establish that NO alone is insufficient for eliciting stroke damage. Rather, it must combine with superoxide, leading to the formation of peroxynitrite, which then degenerates to the very toxic hydroxyl free radical.

The citrulline staining studies provide insight to the cellular dynamics of NO turnover in the brain.[26] In control brains, citrulline staining cells represent only a small portion of the total number of nNOS staining cells, generally only about 20%. However, following cerebral artery occlusion, all nNOS cells stain for citrulline. Thus, about 80% of the nNOS neurons in the brain can be regarded as "quiescent," in that they presumably are not actively forming NO under normal circumstances. In this way, one might regard NO somewhat as a "stress neurotransmitter" formed and released in large amounts only in response to marked perturbations.

1.3 RELATIONSHIP OF NITRIC OXIDE TO POLY(ADP-RIBOSE) POLYMERASE (PARP)

A variety of evidence suggests that cyclic GMP, while mediating actions of physiological levels of NO, is not responsible for the neurotoxic effects of large amounts of released NO. Cyclic GMP derivatives that penetrate into cells fail to produce neurotoxicity, and inhibitors of guanylyl cyclase fail to prevent toxicity. This led to a search for other targets of NO action. As a free radical molecule, NO is highly reactive and so can influence a wide range of potential targets. Extensive attention has been devoted to the ability of NO to nitrosylate cysteine groups in proteins, and there is good evidence that this mechanism may account for the influence of NO in activating Ras and thereby influencing nuclear events in cells.[27] NO can bind to the glycolytic enzyme *cis*-aconitase, which also functions as an iron regulatory protein altering the stability of the messenger RNA for the iron regulatory proteins ferritin and the transferrin receptor.[28] This mechanism physiologically mediates NMDA receptor activation upon iron dynamics in cells. NO binds to a variety of mitochondrial proteins as well.

Our focus on PARP arose out of an interest in identifying targets of NO in addition to soluble guanylyl cyclase. We had noted a report that NO stimulates the ADP-ribosylation of a 36 to 38-kDa protein.[29] Because the molecular weight was similar to that of G proteins, which are well known to be ADP-ribosylated, the authors had suggested that G proteins may be a target of NO stimulated ADP-ribosylation. Because a substantial number of different G proteins exists, we wondered what might be the exact identity of the ADP-ribosylated protein. To isolate it, we developed a biotin-linked derivative of NAD, the substrate for ADP-ribosylation. We then purified and sequenced the protein, which turned out not to be a G protein at all but instead glyceraldehyde-3-phosphate dehydrogenase (GAPDH).[30,31] Other groups independently identified the apparent ADP-ribosylation of GAPDH.[32,33] Certain investigators developed evidence that instead of attaching only an ADP-ribose group to GAPDH, what might be involved is the attachment of the entire NAD molecule to GAPDH.[34,35] The exact nature of the modified GAPDH has not been resolved, and it may be that both NAD attachment and ADP-ribosylation can take place. The ADP-ribosylation is an auto-ribosylation in that NO mixed with pure GAPDH results in ADP-ribosylation, which involves a single cysteine at the active catalytic site of the enzyme.[31,36,37] The exact function of the auto-ADP-ribosylation of GAPDH is not altogether clear. Some studies suggest that GAPDH catalytic activity in cells is diminished by NO effects on the enzyme so that this process may be a way of decreasing glycolytic metabolism. However, GAPDH has functions other than glycolysis. GAPDH can translocate to the nucleus[38,39] and bind to DNA and RNA as well as other proteins.[40-42] Indeed, during several modes of cell death in neuronal and non-neuronal systems GAPDH synthesis is markedly augmented[38,39,43] and the "new" GAPDH appears in the nucleus.[38,44] Antisense to GAPDH, which blocks this augmented synthesis, also prevents cell death.[38,39,45,46] Conceivably, NO and auto-ADP-ribosylation of GAPDH influence this process.

Utilizing the biotin-linked NAD, we sought other proteins whose ADP-ribosylation might be augmented by NO. We isolated a 110-kDa protein, which we showed to be poly(ADP-ribose) polymerase (PARP).[47] PARP is an abundant nuclear enzyme activated by DNA strand breaks and implicated in facilitating the DNA repair process. PARP utilizes NAD as its substrate and attaches branched chains of 50 to 200 ADP-ribose groups to a variety of nuclear proteins including histones, topoisomerases, and PARP itself. In brain extracts exposed to NO, we found that PARP itself is virtually the sole target for NO-activated ADP-ribosylation.

Berger and associates[48] hypothesized that radiation might exert a major portion of its cell-killing activity by DNA damage and PARP overactivation. The overactivation of PARP leads to depletion of its substrate NAD, and, in efforts to resynthesize NAD, cellular ATP is depleted and the cell dies from energy loss (Figure 1.1). We wondered whether such a mechanism might account for NMDA neurotoxicity in cortical cultures. In studies in collaboration with the Dawsons, we found that benzamide-related PARP inhibitors blocked NMDA neurotoxicity in cultures in rough proportion to their potencies as PARP inhibitors, while agents that inhibited only mono-ADP-ribosylation and not PARP were ineffective.[47] Independently, Cosi et al.[49] explored the role of PARP in glutamate toxicity in cerebellar granule cells in culture. These workers observed an increase in levels of the poly(ADP-ribose) polymer following glutamate treatment and showed that benzamide-related PARP inhibitors diminished toxicity.

1.4 USE OF PARP KNOCKOUT MICE TO CLARIFY NEUROTOXICITY

Benzamide-related inhibitors of PARP have provided hints of a role for PARP in neurotoxicity. However, one must be extremely cautious about interpreting results

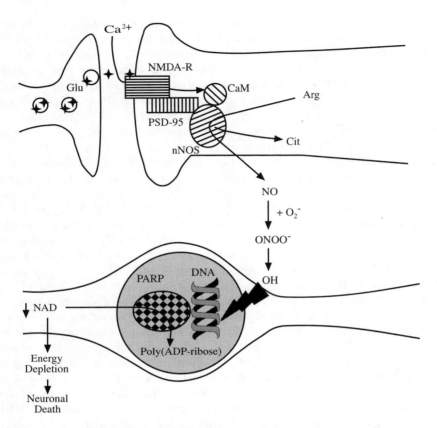

FIGURE 1.1 The mechanism of PARP activation in mediating excitotoxicity. Excess release of glutamate (Glu) binds to and opens the postsynaptic NMDA receptor (NMDA-R). The influx of calcium (Ca^{2+}) combines with calmodulin (CaM) to activate nNOS, which is coupled to the NMDA receptor through PSD-95 (postsynaptic density protein 95 kDa). Large amounts of NO are produced by nNOS using arginine (Arg) as a substrate with citrulline (Cit) as a by-product. The released NO and superoxide (O^{2-}) form peroxynitrite ($ONOO^-$), which damages DNA either by itself or through its decomposition to the reactive hydroxy radicals (OH·). Overactivation of PARP by extensive DNA damage causes depletion of NAD which contributes to neuronal death.

with such agents, as they are weak inhibitors of PARP with IC_{50} values around 10 to 100 μM. Most studies of their effects on neurotoxicity have employed much higher concentrations. Virtually any compound used in millimolar concentrations will exert many different effects. In the case of benzamide derivatives, a number of these effects have been characterized and might account for inhibition of neurotoxicity in various systems. For instance, benzamides at high concentration were found to affect glucose metabolism and DNA synthesis.[50] They also inhibit the expression of iNOS as well as cell adhesion receptors that could be relevant to neurotoxicity. Such effects may relate to their ability to inhibit gene expression for many agents including various transcription factors. Nicotinamide has also been employed as a PARP inhibitor and is even less potent than benzamide analogues.[51] Nicotinamide is also an oxygen free radical scavenger. Because it is the precursor of NAD, any neuroprotective effects of nicotinamide might derive from augmenting NAD levels that were being depleted by neurotoxicity. More recently, more potent agents have been developed with greater selectivity but with various problems. For instance, 3,4-dihydro-5-[4-1(1-piperidynil)buthoxy]-1(2H)-isoquinolinone (DPQ) inhibits PARP in low micromolar concentrations. It provides stroke protection in rats, but at higher doses loses neuroprotection, which suggests that this drug exerts other actions besides inhibition to PARP.[52]

Much greater selectivity can be obtained with targeted deletion of genes. Three separate groups have developed mutant mice lacking PARP (PARP-/-).[53-55] In brain tissue of these mice, PARP activity appears to be abolished. However, some residual PARP activity and protein can be demonstrated in fibroblasts from PARP knockout mice.[56] These findings led to the suggestion that there might exist genes encoding other forms of PARP. Quite recently, two separate PARP genes have been identified, and they are distinct from the original form of PARP, which is designated PARP$_1$.[57-60] The two new forms of PARP are only about half the molecular weight of PARP$_1$. In terms of amino acid sequence, PARP$_2$ and PARP$_3$ display 40 and 31% sequence identity to the catalytic C-terminal domain of PARP$_1$. Although they both lack the zinc-finger DNA-binding domain that occurs in the N-terminus of PARP$_1$ and they also lack the central automodification domain which is where ADP-ribose groups are attached to PARP, at least PARP$_2$ still displays auto-ADP-ribosylation activity in a DNA-dependent manner.[60] The PARP$_2$ DNA-binding domain resides in the first 64 amino acids at the N-terminus. Unlike PARP$_1$, PARP$_2$ does not modify purified histones by ADP-ribosylation.[60] About 90% of DNA damage–induced poly(ADP-ribose) synthesis can be accounted for by PARP$_1$ activity, with PARP$_2$, and perhaps PARP$_3$, contributing the rest. In PARP$_1$-/- cells, only 5 to 10% residual activity was detected, which might reflect lesser catalytic activity for PARP$_2$ and PARP$_3$.[56] Differences between PARP$_1$ and PARP$_2$ can be explained by their intrinsic catalytic properties. The K_m for PARP$_1$ is 2.6-fold lower than that for PARP$_2$ and the k_{cat}/K_m ratio for PARP$_1$ is 18 times higher than that for PARP$_2$.[60] It appears that severe DNA damage is required to activate PARP$_2$, since only high doses of DNA damaging agents stimulate poly(ADP-ribose) synthesis in PARP$_1$-/- cells, while lower doses, known to activate PARP$_1$, do not influence PARP$_2$.[60] Taken together, these results suggest that PARP$_1$ is the dominant isoform responsible for the majority of poly(ADP-ribose) synthesis and the depletion of NAD. Like PARP$_1$, PARP$_2$ and PARP$_3$ occur in a wide range of peripheral tissues and in the brain. The three forms of PARP derive from distinct genes and occur on different chromosomes, with PARP$_1$ being on human chromosome 1q42, while PARP$_2$ and PARP$_3$ are on chromosomes 14q11.2-q12 and 3p21.1-p22.2, respectively.

The existence of PARP$_2$ and PARP$_3$ raises a number of questions about studies utilizing PARP-/- animals, which in fact are only deficient in PARP$_1$. For instance, because of the role of PARP in DNA repair, most researchers assumed that PARP-/- mice would be extremely susceptible to the development of tumors. However, there appears to be no increased incidence of tumors in any of the three strains of PARP-/- animals, even though some of them have been followed for up to 2 years. Definite abnormalities in sister chromatid exchange have been observed as well as susceptibility to lethal dose of X-ray irradiation. Otherwise, PARP$^{-/-}$ animals appear to be healthy with normal tissue morphology and locomotor activity. Whether their apparently normal physical and behavioral function reflects compensation by PARP$_2$ and PARP$_3$ is not clear. No augmented expression of PARP$_2$ was detected in PARP$_1^{-/-}$ mice.[58,60] Creating PARP$_2$ and PARP$_3$ knockout mice should help elucidate the physiological roles of PARP isozymes.

Although the existence of PARP$_2$ and PARP$_3$ makes it necessary to be cautious in interpreting any results with PARP$_1^{-/-}$ animals, nonetheless these animals have been of considerable help in elucidating the biological functions of PARP. In evaluating energy depletion following overactivation of PARP, the PARP$_1$ knockout mice should be meaningful models because of the dominant role of PARP$_1$ in poly(ADP-ribose) synthesis and NAD depletion. In the brain, evidence is quite strong that PARP$_1$ accounts for essentially all PARP catalytic activity.

Initial studies focused on the role of PARP in neurotoxicity in cerebral cortical cultures.[61] Toxicity elicited by oxygen–glucose depletion, NMDA stimulation, or NO are all abolished in cultures from PARP$^{-/-}$ animals. Cultures from heterozygote PARP$^{+/-}$ animals display about a 70% reduction in neurotoxicity in all of these conditions, corresponding to a 50 to 70% reduction in PARP protein and catalytic activity. These findings indicate that 30 to 50% of normal PARP activity is not sufficient to provide the NAD depletion that would be neurotoxic. This suggests that the PARP overactivation that kills cells requires stimulation of almost all of the PARP that occurs normally in cells. If PARP inhibitor drugs are to be therapeutic, then presumably one would not need to inhibit all PARP activity to obtain a beneficial effect.

Studies of cerebral focal ischemia in intact animals have shown impressive therapeutic effects. In one study PARP$^{-/-}$ animals manifested an 80% reduction in the extent of neural damage following middle cerebral artery occlusion.[61] In this study PARP$^{+/-}$ animals displayed about 50% protection. In another study utilizing the same PARP$^{-/-}$ mice, only about 50% reduction of stroke damage was observed.[62] Discrepancies between those two studies may derive from the genetic background of the knockout mice, one of which is of Sv129 after several generations of backcrossing,[61] and the other of Sv129×C58Bl6 mixed background.[62] It is well known that susceptibility to stroke damage varies considerably among different strains of mice.[63]

PARP inhibitors also reduce stroke damage in both permanent and transient models of cerebral ischemia in rodents. DPQ inhibits neural damage following middle cerebral artery occlusion in rats with significant effects at as little as 5 mg/kg when administered prior to the occlusion.[52] At 10 mg/kg stroke damage is reduced by more than 50%. Interestingly, at 20 and 40 mg/kg, protection against damage is lessened, suggesting that DPQ itself elicits some sort of toxic effect distinct from its inhibition of PARP. The adverse effect appears to be compound specific rather than PARP related, since another PARP inhibitor, GPI 6150, offers substantial neuroprotection against stroke in rats without exhibiting diminished effect at higher dose.[64] In the transient cerebral ischemia model, the level of protection afforded by GPI 6150 treatment approaches the maximal protective level in PARP$^{-/-}$ mice.[61] In

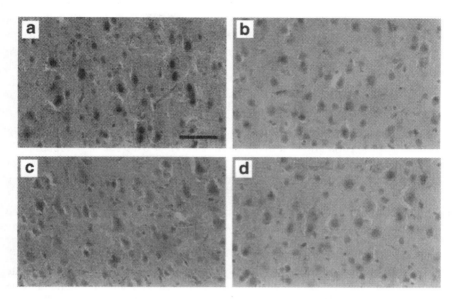

FIGURE 1.2 PARP inhibitor diminishes poly(ADP-ribose) accumulation in the ischemic region of rat brain. Immunohistochemistry staining of poly(ADP-ribose) reveals a substantial increase of nuclear poly(ADP-ribose) synthesis in the ischemic tissues (a), compared with the contralateral side (b). Administration of DPQ, a PARP inhibitor, to the rats strongly inhibits the formation of poly(ADP-ribose) in the ischemic region (c) without affecting the contralateral area (d). All photographs are from layer V of the frontoparietal cortex. (From Takahashi, K. et al., *Brain Res.*, 46, © 1999. With permission from Elsevier Science.)

other studies to mimic clinical situations, DPQ was administered to rats 30 min after blood vessel occlusion and still produced substantial reduction in stroke damage.[65] Similarly, GPI 6150 given at 1 h postischemia achieves the same level of infarct reduction as in pretreatment. Both DPQ and GPI 6150 diminish poly(ADP-ribose) accumulation in the ischemic area, suggesting that the neuroprotective effect of the compounds is due to PARP inhibition (Figure 1.2).[65,66]

Stroke damage is not the only neural insult linked to PARP. MPTP destruction of dopamine neurons is virtually abolished in PARP-/- mice,[67] fitting with the protective effects against MPTP damage observed previously with PARP inhibitors.[68,69] In an animal model of traumatic brain injury, neural lesion in the cortex is associated with an acute PARP activation.[70] PARP inhibition by GPI 6150 treatment in the rats significantly attenuates the extent of the trauma-induced lesion area.[71] In a controlled cortical impact model of traumatic brain injury, PARP-/- mice demonstrated improved motor and memory function after the impact, compared with litter mates, although there was no difference of the contusion volumes between PARP-/- mice and their littermates.[72]

The possibility that overactivity of PARP is involved in the pathophysiology of stroke and neurodegenerative diseases comes from recent neuropathological studies. In one study, patients with global cerebral ischemia following cardiac arrest demonstrated augmented levels of PARP protein and the poly(ADP-ribose) polymer compared with individuals who died by mechanisms that did not appear to involve cerebral ischemia.[73] The same investigators similarly found increased poly(ADP-ribose) polymer and PARP protein in brains of patients with Alzheimer's disease compared with the controls.[74]

1.5 CONCLUSIONS: THERAPEUTIC RELEVANCE

Possible clinical application for PARP inhibitors is especially tantalizing in light of the dramatic protection afforded in animal models of stroke. The 80% reduction in neural damage in PARP$^{-/-}$ mice in one study substantially exceeds the level of protection that has been provided by any other therapeutic intervention, including a wide range of glutamate receptor antagonists and calcium channel blockers. This level of protection may not be surprising if one regards PARP overactivation as a final common pathway for eliciting stroke damage. Glutamate overflow triggering the release of NO, which combines with supreroxide to form peroxynitrite, is one way in which free radicals can lead to DNA damage and PARP activation. However, free radicals can be formed independently of glutamate and NO and also activate PARP. Massive calcium entry into cells activates proteases, which indirectly damage DNA, as well as calcium-dependent DNA cleaving enzymes.

One might ask why DNA damage, PARP activation, and energy depletion seem to be a final downstream mediator of cell death as contrasted to other forms of cellular insult, such as lipid peroxidation of cellular membranes, the destruction of numerous proteins by calcium-activated proteases, and mitochondrial damage by calcium and other substances. One answer might relate to the time course of events. DNA damage provides almost instant activation of PARP. PARP is highly abundant, accounting for 2% of nuclear protein, and poly(ADP-ribose) has a high turnover rate. Accordingly, PARP activation and consequent NAD and ATP depletion occur quite early. Thus, while other molecular alterations may, in principle, be serious enough to kill cells, the PARP-associated energy depletion takes place so rapidly that it suffices to provide a rate-limiting lethal insult and so is an ideal target for therapeutic intervention.

The fact that PARP is downstream of a number of toxic events, such as glutamate release, might provide an augmented therapeutic window for administering PARP inhibitors. At present, the only widely used antistroke drugs are clot-dissolving agents such as tissue plasminogen activator (TPA). To achieve therapeutic benefit, TPA must be administered within 3 h following the initial stroke symptoms. Unfortunately, most patients with clinical stroke are slow to appreciate the significance of initial symptoms, such as dizziness. They reach emergency rooms of hospitals much later than patients suffering from myocardial infarction. If PARP inhibitors could be employed as long as 12 to 16 h following the initial symptoms, the number of patients with stroke who are candidates for treatment would be greatly augmented.

In the therapy of neurodegenerative diseases one would want to administer PARP inhibitors chronically, perhaps for years. Such chronic use would require evidence of safety. One might be leery about administering on a chronic basis a drug that inhibits an enzyme that is associated with DNA repair. However, thus far there does not appear to be major damage in animals with chronic total deletion of PARP. As already indicated, one might need only about 50 to 60% inhibition to produce major clinical benefit. Conceivably, almost total inhibition might be required to elicit adverse effects associated with DNA repair.

Because PARP is ubiquitous, one might anticipate that its overactivation would be involved in other clinical situations. Indeed, myocardial infarction is associated with PARP overactivation and PARP$^{-/-}$ mice are protected from PARP damage, while PARP inhibitors also provide similar protection.[75,76] Additionally, PARP gene knockout[55,77,78] and PARP inhibitors[79] protect from diabetic damage associated with pancreatic insults.

REFERENCES

1. Benveniste, H., Drejer, J., Schousboe, A., and Diemer, N.H., Elevation of the extracellular concentrations of glutamate and aspartate in rat hippocampus during transient cerebral ischemia monitored by intracerebral microdialysis. *J. Neurochem.*, 43:1369–1374, 1984.
2. Choi, D.W., Koh, J.Y., and Peters, S., Pharmacology of glutamate neurotoxicity in cortical cell culture: attenuation by NMDA antagonists. *J. Neurosci.*, 8:185–196, 1988.
3. Wolosker, H., Sheth, K.N., Takahashi, M., Mothet, J.P., Brady, R.O., Jr., Ferris, C.D., and Snyder, S.H., Purification of serine racemase: biosynthesis of the neuromodulator D-serine. *Proc. Natl. Acad. Sci. U.S.A.*, 96:721–725, 1999.
4. Garthwaite, J., Charles, S.L., and Chess-Williams, R., Endothelium-derived relaxing factor release on activation of NMDA receptors suggests role as intercellular messenger in the brain. *Nature*, 336:385–388, 1988.
5. Bredt, D.S. and Snyder, S.H., Nitric oxide mediates glutamate-linked enhancement of cGMP levels in the cerebellum. *Proc. Natl. Acad. Sci. U.S.A.*, 86:9030–9033, 1989.
6. Garthwaite, J., Garthwaite, G., Palmer, R.M., and Moncada, S., NMDA receptor activation induces nitric oxide synthesis from arginine in rat brain slices. *Eur. J. Pharmacol.*, 17:413–416, 1989.
7. Bredt, D.S. and Snyder, S.H., Isolation of nitric oxide synthetase, a calmodulin-requiring enzyme. *Proc. Natl. Acad. Sci. U.S.A.*, 87:682–685, 1990.
8. Brenman, J.E., Chao, D.S., Gee, S.H., McGee, A.W., Craven, S.E., Santillano, D.R., Wu, Z., Huang, F., Xia, H., Peters, M.F., Froehner, S.C., and Bredt, D.S., Interaction of nitric oxide synthase with the postsynaptic density protein PSD-95 and alpha1-syntrophin mediated by PDZ domains. *Cell*, 84:757–767, 1996.
9. Kornau, H.C., Schenker, L.T., Kennedy, M.B., and Seeburg, P.H., Domain interaction between NMDA receptor subunits and the postsynaptic density protein PSD-95. *Science*, 269:1737–1740, 1995.
10. Bredt, D.S., Hwang, P.M., and Snyder, S.H., Localization of nitric oxide synthase indicating a neural role for nitric oxide. *Nature*, 347:768–770, 1990.
11. Bredt, D.S., Hwang, P.M., Glatt, C.E., Lowenstein, C., Reed, R.R., and Snyder, S.H., Cloned and expressed nitric oxide synthase structurally resembles cytochrome P-450 reductase. *Nature*, 351:714–718, 1991.
12. Xie, Q.W., Cho, H.J., Calaycay, J., Mumford, R.A., Swiderek, K.M., Lee, T.D., Ding, A., Troso, T., and Nathan, C., Cloning and characterization of inducible nitric oxide synthase from mouse macrophages. *Science*, 256:225–228, 1992.

13. Lowenstein, C.J., Glatt, C.S., Bredt, D.S., and Snyder, S.H., Cloned and expressed macrophage nitric oxide synthase contrasts with the brain enzyme. *Proc. Natl. Acad. Sci. U.S.A.*, 89:6711–6715, 1992.
14. Lyons, C.R., Orloff, G.J., and Cunningham, J.M., Molecular cloning and functional expression of an inducible nitric oxide synthase from a murine macrophage cell line. *J. Biol. Chem.*, 267:6370–6374, 1992.
15. Lamas, S., Marsden, P.A., Li, G.K., Tempst, P., and Michel, T., Endothelial nitric oxide synthase: molecular cloning and characterization of a distinct constitutive enzyme isoform. *Proc. Natl. Acad. Sci. U.S.A.*, 89:6348–6352, 1992.
16. Sessa, W.C., Harrison, J.K., Barber, C.M., Zeng, D., Durieux, M.E., D'Angelo, D.D., Lynch, K.R., and Peach, M.J., Molecular cloning and expression of a cDNA encoding endothelial cell nitric oxide synthase. *J. Biol. Chem.*, 267:15274–1526, 1992.
17. Janssens, S.P., Simouchi, A., Quertermous, T., Bloch, D.B., and Bloch, K.D., Cloning and expression of a cDNA encoding human endothelium-derived relating factor/nitric oxide synthase. *J. Biol. Chem.*, 267:14519–14522, 1992.
18. Dawson, T.M., Bredt, D.S., Fotuhi, M., Hwang, P.M., and Snyder, S.H., Nitric oxide synthase and neuronal NADPH diaphorase are identical in brain and peripheral tissues. *Proc. Natl. Acad. Sci. U.S.A.*, 88:7797–7801, 1991.
19. Dawson, V.L., Dawson, T.M., London, E.D., Bredt, D.S., and Snyder, S.H., Nitric oxide mediates glutamate neurotoxicity in primary cortical cultures. *Proc. Natl. Acad. Sci. U.S.A.*, 88:6368–6371, 1991.
20. Dawson, V.L., Kizushi, V.M., Huang, P.L., Snyder, S.H., and Dawson, T.M., Resistance to neurotoxicity in cortical cultures from neuronal nitric oxide synthase-deficient mice. *J. Neurosci.*, 16:2479–2487, 1996.
21. Dawson, V.L. and Dawson, T.M., Nitric oxide in neurodegeneration. *Prog. Brain Res.*, 118:215–229, 1998.
22. Huang, Z., Huang, P.L., Panahian, N., Dalkara, T., Fishman, M.C., and Moskowitz, M.A., Effects of cerebral ischemia in mice deficient in neuronal nitric oxide synthase. *Science*, 265:1883–1885, 1994.
23. Brenman, J.E., Xia, H., Chao, D.S., Black, S.M., and Bredt, D.S., Regulation of neuronal nitric oxide synthase through alternative transcripts. *Dev. Neurosci.*, 19:224–231, 1997.
24. Eliasson, M.J., Blackshaw, S., Schell, M.J., and Snyder, S.H., Neuronal nitric oxide synthase alternatively spliced forms: prominent functional localizations in the brain. *Proc. Natl. Acad. Sci. U.S.A.*, 94:3396–3401, 1997.
25. Huang, P.L., Dawson, T.M., Bredt, D.S., Snyder, S.H., and Fishman, M.C., Targeted disruption of the neuronal nitric oxide synthase gene. *Cell*, 75:1273–1286, 1993.
26. Eliasson, M.J., Huang, Z., Ferrante, R.J., Sasamata, M., Molliver, M.E., Snyder, S.H., and Moskowitz, M.A., Neuronal nitric oxide synthase activation and peroxynitrite formation in ischemic stroke linked to neural damage. *J. Neurosci.*, 19:5910–5918, 1999.
27. Lander, H.M., Ogiste, J.S., Pearce, S.F., Levi, R., and Novogrodsky, A., Nitric oxide-stimulated guanine nucleotide exchange on p21ras. *J. Biol. Chem.*, 270:7017–7020, 1995.
28. Jaffrey, S.R., Cohen, N.A., Rouault, T.A., Klausner, R.D., and Snyder, S.H., The iron-responsive element binding protein: a target for synaptic actions of nitric oxide. *Proc. Natl. Acad. Sci. U.S.A.*, 91:12994–12998, 1994.
29. Brune, B. and Lapetina, E.G., Activation of a cytosolic ADP-ribosyltransferase by nitric oxide-generating agents. *J. Biol. Chem.*, 264:8455–9458, 1989.

30. Zhang, J. and Snyder, S.H., Purification of a nitric oxide-stimulated ADP-ribosylated protein using biotinylated beta-nicotinamide adenine dinucleotide. *Biochemistry,* 32:2228–2233, 1993.
31. Zhang, J. and Snyder, S.H., Nitric oxide stimulates auto-ADP-ribosylation of glyceraldehyde-3-phosphate dehydrogenase. *Proc. Natl. Acad. Sci. U.S.A.,* 89:9382–9385, 1992.
32. Dimmeler, S., Lottspeich, F., and Brune, B., Nitric oxide causes ADP-ribosylation and inhibition of glyceraldehyde-3-phosphate dehydrogenase. *J. Biol. Chem.,* 267:16771–16774, 1992.
33. Kots, A.Y.A., Skurat, A.V., Sergienko, E.A., Bulargina, T.V., and Severin, E.S., Nitroprusside stimulates the cysteine-specific mono(ADP-ribosylation) of glyceraldehyde-3-phosphate dehydrogenase from human erythrocytes. *FEBS Lett.,* 300:9–12, 1992.
34. McDonald, L.J. and Moss, J., Stimulation by nitric oxide of an NAD linkage to glyceraldehyde-3-phosphate dehydrogenase. *Proc. Natl. Acad. Sci. U.S.A.,* 90:6238–6241, 1993.
35. Minetti, M., Pietraforte, D., Di Stasi, A.M., and Mallozzi, C., Nitric oxide-dependent NAD linkage to glyceraldehyde-3-phosphate dehydrogenase: possible involvement of a cysteine thiol radical intermediate. *Biochem. J.,* 319:369–375, 1996.
36. Kots, A.Y.A., Sergienko, E.A., Bulargina, T.V., and Severin, E.S., Glyceraldehyde-3-phosphate activates auto-ADP-ribosylation of glyceraldehyde-3-phosphate dehydrogenase. *FEBS Lett.,* 324:33–36, 1993.
37. Dimmeler, S. and Brune, B., Nitric oxide preferentially stimulates auto-ADP-ribosylation of glyceraldehyde-3-phosphate dehydrogenase compared to alcohol or lactate dehydrogenase. *FEBS Lett.,* 315:21–24, 1993.
38. Sawa, A., Khan, A.A., Hester, L.D., and Snyder, S.H., Glyceraldehyde-3-phosphate dehydrogenase: nuclear translocation participates in neuronal and nonneuronal cell death. *Proc. Natl. Acad. Sci. U.S.A.,* 94:11669–11674, 1997.
39. Saunders, P.A., Chen, R.W., and Chuang, D.M., Nuclear translocation of glyceraldehyde-3-phosphate dehydrogenase isoforms during neuronal apoptosis. *J. Neurochem.,* 72:925–932, 1997.
40. Sirover, M.A., Role of the glycolytic protein, glyceraldehyde-3-phosphate dehydrogenase, in normal cell function and in cell pathology. *J. Cell. Biochem.,* 66:133–140, 1997.
41. Singh, R. and Green, M.R., Sequence-specific binding of transfer RNA by glyceraldehyde-3-phosphate dehydrogenase. *Science,* 259:365–368, 1993.
42. Baxi, M.D. and Vishwanatha, J.K., Uracil DNA-glycosylase/glyceraldehyde-3-phosphate dehydrogenase is an Ap4A binding protein. *Biochemistry,* 34:9700–9707, 1995.
43. Tokunaga, K., Nakamura, Y., Sakata, K., Fujimori, K., Ohkubo, M., Sawada, K., and Sakiyama, S., Enhanced expression of a glyceraldehyde-3-phosphate dehydrogenase gene in human lung cancers. *Cancer Res.,* 47:5616–5619, 1987.
44. Ishitani, R., Sunaga, K., Hirano, A., Saunders, P., Katsube, N., and Chuang, D.M., Evidence that glyceraldehyde-3-phosphate dehydrogenase is involved in age-induced apoptosis in mature cerebellar neurons in culture. *J. Neurochem.,* 66:928–935, 1996.
45. Shashidharan, P., Chalmers-Redman, R.M., Carlile, G.W., Rodic, V., Gurvich, N., Yuen, T., Tatton, W.G., and Sealfon, S.C., Nuclear translocation of GAPDH-GFP fusion protein during apoptosis. *Neuroreport,* 10:1149–1153, 1999.
46. Ishitani, R. and Chuang, D.M., Glyceraldehyde-3-phosphate dehydrogenase antisense oligodeoxynucleotides protect against cytosine arabinonucleoside-induced apoptosis in cultured cerebellar neurons. *Proc. Natl. Acad. Sci. U.S.A.,* 93:9937–9941, 1996.

47. Zhang, J., Dawson, V.L., Dawson, T.M., and Snyder, S.H., Nitric oxide activation of poly(ADP-ribose) synthetase in neurotoxicity. *Science*, 263:687–689, 1994.
48. Berger, N.A., Poly(ADP-ribose) in the cellular response to DNA damage. *Radiat. Res.*, 101:4–15, 1985.
49. Cosi, C., Suzuki, H., Milani, D., Facci, L., Menegazzi, M., Vantini, G., Kanai, Y., and Skaper, S.D., Poly(ADP-ribose) polymerase: early involvement in glutamate-induced neurotoxicity in cultured cerebellar granule cells. *J. Neurosci. Res.*, 39:38–46, 1994.
50. Milam, K.M. and Cleaver, J.E., Inhibitors of poly(adenosine diphosphate-ribose) synthesis: effect on other metabolic processes. *Science*, 223:589–591, 1984.
51. Pellat-Deceunynck, C., Wietzerbin, J., and Drapier, J.C., Nicotinamide inhibits nitric oxide synthase mRNA induction in activated macrophages. *Biochem. J.*, 297:53–58, 1994.
52. Takahashi, K., Greenberg, J.H., Jackson, P., Maclin, K., and Zhang, J., Neuroprotective effects of inhibiting poly(ADP-ribose) synthetase on focal cerebral ischemia in rats. *J. Cereb. Blood Flow Metab.*, 17:1137–1142, 1997.
53. Wang, Z.Q., Auer, B., Stingl, L., Berghammer, H., Haidacher, D., Schweiger, M., and Wagner, E.F., Mice lacking ADPRT and poly(ADP-ribosyl)ation develop normally but are susceptible to skin disease. *Genes Dev.*, 9:509–520, 1995.
54. de Murcia, J.M., Niedergang, C., Trucco, C., Ricoul, M., Dutrillaux, B., Mark, M., Oliver, F.J., Masson, M., Dierich, A., LeMeur, M., Walztinger, C., Chambon, P., and de Murcia, G., Requirement of poly(ADP-ribose) polymerase in recovery from DNA damage in mice and in cells. *Proc. Natl. Acad. Sci. U.S.A.*, 94:7303–7307, 1997.
55. Masutani, M., Suzuki, H., Kamada, N., Watanabe, M., Ueda, O., Nozaki, T., Jishage, K., Watanabe, T., Sugimoto, T., Nakagama, H., Ochiya, T., and Sugimura, T., Poly(ADP-ribose) polymerase gene disruption conferred mice resistant to streptozotocin-induced diabetes. *Proc. Natl. Acad. Sci. U.S.A.*, 96:2301–2304, 1999.
56. Shieh, W.M., Ame, J.C., Wilson, M.V., Wang, Z.Q., Koh, D.W., Jacobson, M.K., and Jacobson, E.L., Poly(ADP-ribose) polymerase null mouse cells synthesize ADP-ribose polymers. *J. Biol. Chem.*, 273:30069–30072, 1998.
57. Babiychuk, E., Cottrill, P.B., Storozhenko, S., Fuangthong, M., Chen, Y., O'Farrell, M.K., Van Montagu, M., Inze, D., and Kushnir, S., Higher plants possess two structurally different poly(ADP-ribose) polymerases. *Plant J.*, 15:635–645, 1998.
58. Berghammer, H., Ebner, M., Marksteiner, R., and Auer, B., pADPRT-2: a novel mammalian polymerizing(ADP-ribosyl)transferase gene related to truncated pADPRT homologues in plants and *Caenorhabditis elegans. FEBS Lett.*, 449:259–263, 1999.
59. Johansson, M., A human poly(ADP-ribose) polymerase gene family (ADPRTL): cDNA cloning of two novel poly(ADP-ribose) polymerase homologues. *Genomics*, 57:442–445, 1999.
60. Ame, J.C., Rolli, V., Schreiber, V., Niedergang, C., Apiou, F., Decker, P., Muller, S., Hoger, T., Menissier-de Murcia, J., and de Murcia, G., PARP-2, a novel mammalian DNA damage-dependent poly(ADP-ribose) polymerase. *J. Biol. Chem.*, 274:17860–17868, 1999.
61. Eliasson, M.J., Sampei, K., Mandir, A.S., Hurn, P.D., Traystman, R.J., Bao, J., Pieper, A., Wang, Z.Q., Dawson, T.M., Snyder, S.H., and Dawson, V.L., Poly(ADP-ribose) polymerase gene disruption renders mice resistant to cerebral ischemia. *Nat. Med.*, 3:1089–1095, 1997.

62. Endres, M., Wang, Z.Q., Namura, S., Waeber, C., and Moskowitz, M.A., Ischemic brain injury is mediated by the activation of poly(ADP-ribose)polymerase. *J. Cereb. Blood Flow Metab.*, 17:1143–1151, 1997.
63. Choi, D.W., At the scene of ischemic brain injury: is PARP a perp? *Nat. Med.*, 3:1073–1074, 1997.
64. Zhang, J., Li, J.-H., and Lautar, S., Post-ischemia protection by a potent PARP inhibitor in transient cerebral focal ischemia. *Soc. Neurosci. Abstr.*, 24:1226, 1998.
65. Takahashi, K., Pieper, A.A., Croul, S.E., Zhang, J., Snyder, S.H., and Greenberg, J.H., Post-treatment with an inhibitor of poly(ADP-ribose) polymerase attenuates cerebral damage in focal ischemia. *Brain Res.*, 829:46–54, 1999.
66. Lautar, S., Pieper, A.A., Verma, A., Snyder, S.H., Li, J.-H., and Zhang, J., Post-ischemia treatment with GPI 6150 diminishes poly(ADP-ribose) accumulation during ischemia-reperfusion injury in rat brain. *Soc. Neurosci. Abstr.*, 24:1226, 1998.
67. Mandir, A.S., Przedborski, S., Jackson-Lewis, V., Wang, Z.Q., Simbulan-Rosenthal, C.M., Smulson, M.E., Hoffman, B.E., Guastella, D.B., Dawson, V.L., and Dawson, T.M., Poly(ADP-ribose) polymerase activation mediates 1-methyl-4-phenyl-1, 2,3,6-tetrahydropyridine (MPTP)-induced parkinsonism. *Proc. Natl. Acad. Sci. U.S.A.*, 96:5774–5779, 1999.
68. Zhang, J., Pieper, A., and Snyder, S.H., Poly(ADP-ribose) synthetase activation: an early indicator of neurotoxic DNA damage. *J. Neurochem.*, 65:1411–1414, 1995.
69. Cosi, C. and Marien, M., Decreases in mouse brain NAD^+ and ATP induced by 1-methyl-4-phenyl-1,2,3,6-tetrahydropyridine (MPTP): prevention by the poly(ADP-ribose) polymerase inhibitor, benzamide. *Brain Res.*, 809:58–67, 1998.
70. LaPlaca, M.C., Raghupathi, R., Verma, A., Pieper, A.A., Saatman, K.E., Snyder, S.H., and McIntosh, T.K., Temporal patterns of poly(ADP-ribose) polymerase activation in the cortex following experimental brain injury in the rat. *J. Neurochem.*, 73:205–213, 1999.
71. LaPlaca, M.C., Zhang, J., Li, J.-H., Raghupathi, R., Smith, F., Bareye, F.M., Graham, D.I., and McIntosh, T.K., Pharmacologic inhibition of poly(ADP-ribose) polymerase is neuroprotective following traumatic brain injury in rats, *J. Neurotrauma*, 16:976, 1999 (abstract).
72. Whalen, M.J., Clark, R.S., Dixon, C.E., Robichaud, P., Marion, D.W., Vagni, V., Graham, S.H., Virag, L., Hasko, G., Stachlewitz, R., Szabó, C., and Kochanek, P.M., Reduction of cognitive and motor deficits after traumatic brain injury in mice deficient in poly(ADP-ribose) polymerase. *J. Cereb. Blood Flow Metab.*, 19:835–842, 1999.
73. Love, S., Barber, R., and Wilcock, G.K., Neuronal accumulation of poly(ADP-ribose) after brain ischaemia. *Neuropathol. Appl. Neurobiol.*, 25:98–103, 1999.
74. Love, S., Barber, R., and Wilcock, G.K., Increased poly(ADP-ribosyl)ation of nuclear proteins in Alzheimer's disease. *Brain*, 122:247–253, 1999.
75. Thiemermann, C., Bowes, J., Myint, F.P., and Vane, J.R., Inhibition of the activity of poly(ADP ribose) synthetase reduces ischemia-reperfusion injury in the heart and skeletal muscle. *Proc. Natl. Acad. Sci. U.S.A.*, 94:679–683, 1997.
76. Zingarelli, B., Salzman, A.L., and Szabo, C., Genetic disruption of poly(ADP-ribose) synthetase inhibits the expression of P-selectin and intercellular adhesion molecule-1 in myocardial ischemia/reperfusion injury. *Circ. Res.*, 83:85–94, 1998.
77. Pieper, A.A., Brat, D.J., Krug, D.K., Watkins, C.C., Gupta, A., Blackshaw, S., Verma, A., Wang, Z.Q., and Snyder, S.H., Poly(ADP-ribose) polymerase-deficient mice are protected from streptozotocin-induced diabetes. *Proc. Natl. Acad. Sci. U.S.A.*, 96:3059–3064, 1999.

78. Burkart, V., Wang, Z.Q., Radons, J., Heller, B., Herceg, Z., Stingl, L., Wagner, E.F., and Kolb, H., Mice lacking the poly(ADP-ribose) polymerase gene are resistant to pancreatic beta-cell destruction and diabetes development induced by streptozocin. *Nat. Med.,* 5:314–319, 1999.
79. Masiello, P., Cubeddu, T.L., Frosina, G., and Bergamini, E., Protective effect of 3-aminobenzamide, an inhibitor of poly (ADP-ribose) synthetase, against streptozotocin-induced diabetes. *Diabetologia,* 28:683–686, 1985.

2 Role of Poly(ADP-Ribose) Polymerase Mediating Acute and Chronic Neuronal Injury and Neurodegeneration

Valina L. Dawson and Ted M. Dawson

CONTENTS

2.1 Cell Death 23
2.2 Excitotoxicity 25
2.3 Ischemic Injury 27
2.4 Traumatic Neuronal Injury 30
2.5 Parkinson's Disease 30
2.6 Alzheimer's Disease 32
2.7 Conclusions 33
References 34

2.1 CELL DEATH

Cell death is the fundamental biological process that is relevant to both the steady-state kinetics of healthy adult tissue,[1] as well as the pathogenesis of tissue damage in disease.[2-4] In the central nervous system, cell death plays a major role in the normal growth and differentiation of the developing nervous system.[5-7] During development, a large number of neurons die in a process that is thought to be responsible for matching neuronal populations to target areas. Once the developing nervous system matures, neurons become postmitotic, and no longer have the ability to reenter the cell cycle and undergo mitosis and cell division. Pathological insults, which result in neuronal cell death, can have devastating consequences on the central nervous system due to the loss of neurons and the inability of the tissue to replace these cells adequately. In the human brain and spinal cord, neurons degenerate after acute insults, including stroke and trauma. Neuronal degeneration also occurs during progressive, adult-onset diseases such as Parkinson's disease and Alzheimer's disease. Glutamate receptor–mediated excitotoxicity has been implicated in many neurological conditions.[8] Nevertheless, effective approaches to prevent or limit neuronal

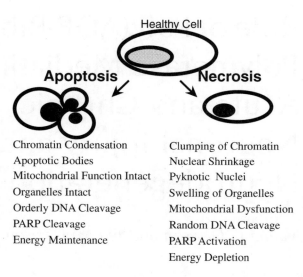

FIGURE 2.1 PARP in apoptosis and necrosis. Apoptosis and necrosis are two forms of cell death with distinct morphologic features. The biochemical pathways that mediate these forms of cell death are under intense investigation, and there may be significant overlap in the molecules that execute each of these death mechanisms. PARP may be a critical molecule that directs the cell toward apoptosis or necrosis. In apoptosis, PARP is cleaved by caspases and inactivated. Cellular energy is maintained until the end of the death program. In necrosis, PARP is not cleaved but rather activated. PARP rapidly consumes NAD and subsequently ATP. Cellular energy is impaired early in the necrotic death cycle.

damage in these disorders remain elusive, due largely to an incomplete understanding of neuronal death pathways *in vivo*. Recently, poly(ADP-ribose) polymerase (PARP) has been indentified as an important mediator of neuronal cell death.[9-12]

Cell death traditionally has been classified as being one of two distinct types: apoptotic or necrotic, each of which clearly has distinct morphologic features (Figure 2.1). The biochemical distinction between apoptosis and necrosis is becoming less distinct. As cell death is examined in detail in the central nervous system, this binary scheme is clearly not sufficient to describe the various types of cell death that occur with degeneration of neurons.

Apoptosis is a distinctive morphologic alteration in the cell[1,13] for which some biochemical pathways recently have been described.[14-17] In apoptosis, the chromatin condenses into sharply delineated, uniformly dense masses within the nucleus. Nucleolar disintegration also occurs early during this process. The cytoplasm condenses, and the cell shrinks in size while the plasma membrane and mitochondrial function remain intact. Subsequently, the nuclear and plasma membranes become convoluted and the cell undergoes a process termed *budding*.[13] During this process, the nucleus and organelles become fragmented and dissociated with bits of condensed cytoplasm forming cellular debris, termed *apoptotic bodies*. Eventually, this cellular debris is phagocytosed by resident cells, typically without generating an acute inflammatory response.[13] Biochemical studies have delineated the importance of various protein families in mediating this elegant and organized mechanism of cell death. In particular, a group of cysteine proteases termed *caspases,* in concert with mitochondrial proteins and endonucleases, mediate many components of apoptosis.[18] One of the target proteins for caspase cleavage is PARP. Proteolytic cleavage of PARP is such a common target for caspase activation that PARP cleavage has become a biochemical hallmark for identification of cells undergoing apoptosis. However, the reason for this cleavage remains unclear. The rate of apoptosis and the intensity of the signal necessary to induce apotosis are not altered between wild-type and PARP null tissues following exposure to anti-Fas, tumor neurosis factor-α, γ irradiation, dexamethasone,[19] topoisomerase I inhibitor CPT-11, or by interleukin-3 removal, but exposure of PARP null tissue to alkylating agents accelerates apoptotic injury.[20,21] A caspase uncleavable mutant PARP (D214A-PARP) introduced into PARP$^{-/-}$ fibroblasts delays the rate of apoptosis following exposure to anti-CD95 or an alkylating agent.[20] A similar point mutation into the cleavage site (DEVD) of PARP also renders PARP resistant to caspase cleavage *in vitro* and *in vivo*.[22] Treatment with tumor necrosis factor-α of fibroblasts expressing this caspase-resistant PARP mutant results in accelerated cell death. This enhanced cell death is due to the induction of necrosis and increased apoptosis.[22] Depletion of NAD$^+$ and ATP was observed in cells expressing caspase-resistant PARP. 3-Aminobenzamide, a PARP inhibitor, prevented NAD$^+$ depletion and concomitantly inhibited necrosis, but increased apoptosis. These data suggest that PARP is a key regulator, controlling the cellular switch that determines whether a cell will die by apoptosis or necrosis.

Necrosis is a form of cell death that is perhaps at the extreme opposite of the cell death spectrum from apoptosis. In contrast to the well-organized cellular dismantling that occurs in apoptosis, following a severe pathological insult, cell death can occur through necrosis with morphology reflecting in part the initial damage or dysfunction.[13] In necrosis, both the nucleus and cytoplasm show ultrastructural changes, with the main features including clumping of chromatin, swelling and degeneration of organelles, destruction of membrane integrity, immediate inhibition of mitochondrial function, and eventual dissolution of the cell.[2] Nuclear pyknosis with condensation of chromatin into irregularly shaped clumps sharply contrasts with the formation of few, uniformly dense, and regularly shaped chromatin aggregates that occurs in apoptosis. Furthermore, in cellular necrosis, the nuclei of dying cells do not bud to form discrete membrane-bound fragments, but rather the nucleus condenses.[2] These different morphologic features of nuclear structure, pyknotic condensed nuclei vs. nuclei-containing apoptotic bodies, are considered diagnostic and hallmark features to distinguish between pure apoptosis and necrosis. A further characteristic of cellular necrosis is that its rate of progression depends on the initial severity of the insult. In contrast, apoptosis is an all-or-none phenomenon and does not follow the clear "dose–response" principle. In the nervous system, excitotoxicity and ischemic injury have most of the major hallmarks of necrosis. In these models, inhibition or deletion of PARP results in neuronal survival.

2.2 EXCITOTOXICITY

In the nervous system, neuronal death can be induced by excitotoxicity (Figure 2.2), a concept coined by Olney[23-27] that subsequently has been implicated in a variety of

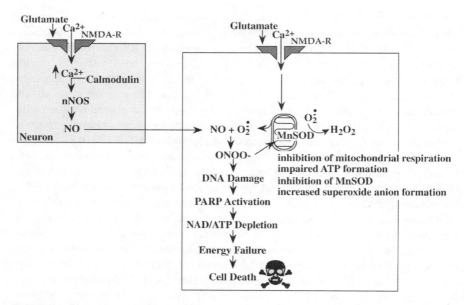

FIGURE 2.2 PARP mediated excitotoxicity. Acting via NMDA receptors, glutamate triggers an influx of Ca^{2+}, activating mitochondria and, in nNOS neurons, inducing the production of NO. Mitochondrial activation results in an increase in superoxide anion (O_2^-) production. NO is a diffusible molecule that combines with superoxide to form peroxynitrite ($ONOO^-$). Peroxynitrite damages mitochondrial enzymes, decreasing mitochondrial respiration and the production of ATP, and damages MnSOD, further increasing the production of superoxide anion. Peroxynitrite has limited diffusion across membranes and can leave the mitochondria and enter the nucleus where it damages DNA and activates PARP. Massive activation of PARP leads to ADP-ribosylation and depletion of NAD. ATP is depleted in an effort to resynthesize NAD. In the setting of impaired energy generation due to mitochondrial dysfunction, this loss of NAD and ATP leads to cell death.

neurological disorders. Excitotoxicity is mediated by excessive glutamate acting on glutamate-gated ion channel receptors with subsequent activation of voltage-dependent ion channels. Glutamate elicits its actions by acting on a series of postsynaptic receptors, including the *N*-methyl-D-aspartate receptors (NMDA), non-NMDA receptors, and metabotropic glutamate receptors. Activation of these receptors increases cytosolic free calcium and results in the activation of a host of calcium-dependent processes including calcium-sensitive proteases, protein kinases, protein phosphatases, phospholipases, and endonucleases.[28,29] Additionally, the influx of calcium results in changes in mitochondrial function and activation of neuronal nitric oxide (NO) synthase (nNOS).[30,31] The most rapid and severe excitotoxic damage is mediated largely through glutamate activation of the NMDA receptor. Calcium fluxing through the NMDA receptor results in subsequent activation of nNOS in generation of NO.[32,33] NO is an important biological messenger molecule, but when generated in the presence of superoxide anion, it can form the potent oxidant, peroxynitrite.[33] Activation of NMDA receptors often stimulates superoxide anion generation from the mitochondria, setting the stage for peroxynitrite generation.[34]

Peroxynitrite can oxidize and damage most cellular constituents and, in particular, DNA.[35] DNA damage is subsequently recognized by PARP, leading to PARP activation and consumption of NAD due to the ribosylation of nuclear proteins and PARP itself. For every mole of NAD that is consumed by PARP, four free-energy equivalents of ATP are required to restore or maintain NAD levels within the cell.[36-38] Mitochondrial proteins are also a target for peroxynitrite attack.[39-42] During excitotoxic injury, mitochondrial function is impaired and oxidative phosphorylation cannot restore cellular levels of NAD and ATP. In this setting PARP activation can contribute to a rapid cellular energy depletion. *In vitro* inhibition of PARP blocks excitotoxicity and NO-mediated toxicity. Genetic deletion of PARP results in neurons that are exquisitely resistant to excitotoxicity, as well as NO-mediated neurotoxicity.[9,10] Expression of PARP by viral-mediated transfection of PARP knockout cultures restores the sensitivity of neurons to excitotoxicity, indicating that the resistance to excitotoxicity is due to the lack of PARP activity rather than a compensatory mechanism following the genetic deletion of PARP (W.J. Herring, T.M. Dawson, and V.L. Dawson, unpublished observations). Furthermore, PARP knockout mice are resistant to excitotoxicity following direct intrastriatal injection of NMDA. NMDA lesions are reduced over 94% in comparison with wild-type mice experiencing intrastriatal injection of NMDA. The lesion size in PARP knockout mice is not significantly different from that observed in vehicle injected control mice (A.S. Mandir, M. Poitras, V.L. Dawson, and T.M. Dawson, unpublished observations). These data indicate that activation of PARP is somehow critically involved in the process of excitotoxic neuronal cell death.

2.3 ISCHEMIC INJURY

Cessation of blood flow to the brain for even a few minutes results in neuronal injury due to both oxygen and nutrient deprivation and the subsequent initiation of a cascade of secondary mechanisms (Figure 2.3). The neurotoxic cascade involves derangements in both normal metabolic and physiological functions, as well as recruitment of other cell-death processes. Ischemic events result in reduction of the resting membrane potential of both glia and neurons in the brain. Leakage of potassium depolarizes neurons leading to a massive release of neurotransmitters including glutamate.[43] A major component of neuronal injury following ischemic insult is due to glutamate acting on its receptors. Glutamate-mediated injury is largely mediated via activation of NMDA receptors and subsequent activation of nNOS and generation of NO. In animal models of experimental stroke in which the middle cerebral artery is occluded for 2 h followed by 22 h of reperfusion, the NMDA antagonist, MK801, effectively reduces between 30 and 50% of the infarct volume.[44] Selective nNOS inhibitors, as well as nNOS knockout mice, result in a similar amount of protection in these experimental stroke models.[45-47] Because PARP activation following NO production is thought to mediate in large part excitotoxic neuronal damage, a role for PARP in ischemic injury was investigated. Ischemic injury can be induced in primary neuronal cultures by removal of oxygen and glucose for up to 60 min.[48] This ischemic insult particularly results in 60 to 80% neuronal cell death. In cortical cultures derived from PARP knockout mice or in wild-type cortical cultures treated

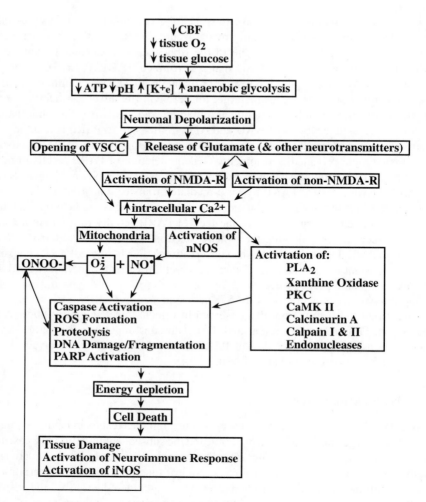

FIGURE 2.3 Ischemic activation of PARP. Ischemia induces a multitude of cellular events. PARP activation leading to cell death would fit into the known schema as follows: Ischemic events result in alterations in pH, ATP, and ionic gradients, leading to the reduction of the resting membrane potential of glia and neurons in the brain that causes a massive release of glutamate and other neurotransmitters. Acting via NMDA receptors and non-NMDA receptors, glutamate triggers rapid elevations in intracellular calcium, which activates a variety of signal cascades that can elevate intracellular free radicals and reactive oxygen species, as well as trigger apoptotic cascades. In particular, glutamate triggers a release of NO that combines with superoxide to form peroxynitrite. Peroxynitrite damages DNA, activating PARP. Massive activation of PARP leads to ADP-ribosylation and depletion of NAD. ATP is depleted in an effort to resynthesize NAD, leading to cell death. Abbreviations: CBF, cerebral blood flow; VSCC, voltage dependent calcium channels; NMDA-R, N-methyl-D-aspartate receptor; $O_2^{\cdot-}$, superoxide anion; nNOS, neuronal nitric oxide synthase; PLA_2, phospholipase A_2; PKC, protein kinase C; ROS, reactive oxygen species.

with a PARP inhibitor, ischemic damage is virtually abolished.[9] These observations have been extended to *in vivo* models of experimental stroke. Following middle cerebral artery occlusion (MCAO), infarct volume can be reduced up to 80% in the PARP knockout mice.[9,10] Immunohistochemistry in wild-type mice demonstrates dramatic activation of PARP in the ipsilateral, but not the contralateral, hemisphere, indicating that PARP is activated following ischemic insult.[9,49-52] Immunohistochemical detection of poly(ADP-ribose) was not observed in the PARP knockout mice, confirming the genotype and phenotype of these animals.[9] Interestingly, the formation of poly(ADP-ribose) observed during reperfusion is attenuated in nNOS knockout mice.[53] Similar, but not as dramatic, protection following experimental stroke has been made with the PARP inhibitors: 3-aminobenzamide, 5-iodo-6-amino-1,2-benzopyrone (INH2BP), or 3,4-dihydro-5-[4-(1- piperidinyl)butoxy]-1(2H)-isoquinolinone (DPQ).[49,52,54,55] The difference observed with inhibitors vs. genetic deletion of PARP may be due largely to the dosing paradigm, bioavailability of the inhibitors, the specificity and potency of the inhibitors, and their mechanisms of action. In particular, 3-aminobenzamide, a prototype first-generation PARP inhibitor, which is commonly used because of its solubility in biological solutions, inhibits PARP but also inhibits protein, RNA and DNA synthesis, and DNA repair.[56-60] These secondary actions of 3-aminobenzamide, independent from inhibition of PARP, can confound observations. 3-Aminobenzamide is useful for acute studies, but due to inhibition of DNA repair, it is not useful for evaluation of chronic experiments. PARP inhibition may also limit excitotoxicity during ischemic injury by attenuating glutamate release. 3-Aminobenzamide inhibition of PARP in rats prevents NMDA-induced efflux of glutamate following MCAO.[61] Lowering extracellular glutamate levels is beneficial in limiting excitotoxicity. Secondary damage to the neural tissue following ischemic insult results from neutrophil infiltration to the site of injury.[62] Neutrophil recruitment is mediated by upregulation of adhesion receptors and molecules. In PARP null mice, there is reduced endothelial expression of P-selectin and intracellular adhesion molecule-1 (ICAM-1) following myocardial ischemia and reperfusion.[63] Neutrophil infiltration is diminished in the PARP null mice or in animals treated with PARP inhibitors in experimental models of inflammation[64,65] or myocardial ischemia and reperfusion.[66] It is possible that inhibition of PARP or genetic deletion of PARP also protects the nervous system by a similar mechanism of limiting neutrophil infiltration and subsequent secondary injury. Both genetic deletion of PARP and pharmacological inhibition of PARP clearly demonstrate that PARP activation is a critical feature in the pathogenesis of ischemic injury.

One puzzling observation in this series of studies is that the protection observed with PARP inhibition or PARP knockout is much greater than the protection observed by inhibiting glutamate receptor activation or inhibiting NOS activity. This may be explained by the hypothesis that DNA damage and PARP activation compose the choke point of a final common pathway for ischemic injury. NMDA receptor activation and NOS stimulation are among several upstream activities that can subsequently damage DNA and result in neuronal injury following ischemic insult. Such a conclusion is merely conjecture and somewhat surprising because, in addition to DNA damage, increases in intracellular calcium and production of reactive oxygen species can disrupt a diverse array of metabolic and cellular processes, resulting in damage of lipid and proteins. The next step in the ischemic injury signal cascade following PARP activation that results in neuronal cell death is not known. It is thought that energy depletion may play a key role, but this is yet to be confirmed. The present data implicate PARP as the key mediator; however, its actions in the central nervous system currently are unknown, but under intensive investigation.

2.4 TRAUMATIC NEURONAL INJURY

Mechanical trauma to the brain and spinal cord triggers a series of sequential events that ultimately lead to secondary neuronal destruction. The precise pathophysiological changes in cellular signaling that occur following mechanical injury are not completely understood. However, a role for free radicals and activation of endonucleases with subsequent DNA damage has been observed.[67-71] In experimental animal models, excitotoxicity and production of DNA-damaging agents have been implicated in the pathogenesis of traumatic injury to both the brain and spinal cord.[68,72-74] In addition to excitotoxicity, traumatic injury can induce the expression of nNOS that can participate in a secondary NO/peroxynitrite-mediated neuronal injury.[75-77] In this setting, one might expect that PARP would play an important role in traumatic neuronal injury. In fact, activation of both nNOS and PARP can be detected in the spinal cords of rats following mechanical injury.[70,71,78] PARP activity is detected in lateral fluid percussion traumatic brain injury in rats within 30 min of insult.[79] Inhibitors of PARP suggest that PARP activation mediates, in part, the neuronal death following traumatic brain injury. PARP inhibition protects hippocampal slices against percussion-induced loss of CA1 pyramidal cells evoked response *in vitro*, implicating PARP activation with loss of neuronal function following mechanical injury.[80] In the controlled cortical impact model of mechanical traumatic injury, PARP knockout mice show protection from functional deficits in both motor function tests and memory acquisition tests.[81] However, in this model, contusion volume was not significantly different when comparing wild-type vs. PARP knockout mice. This may be in part due to the model chosen, as the controlled cortical impact model is a severe traumatic injury model that initiates a cascade of events that result in acute energy failure, release of excitatory amino acids, direct cellular disruption, local ischemia, and inflammation. In this model, PARP plays a key role, but it does not appear to be the choke point of the final common pathway.[81]

2.5 PARKINSON'S DISEASE

Parkinson's disease (Figure 2.4) is a chronic progressive neurological disorder that affects over 500,000 men and women in the United States alone. Typically, the clinical signs include tremor, slowness of movement, and muscular rigidity, which appear in the second half of life. Underlying the clinical symptoms of Parkinson's disease is the degeneration of the neuromelanin-containing dopaminergic neurons predominantly located in the substantia nigra.[82] It is unknown why the neurons in the substantia nigra die. Current medical therapy involves replacement of the neurotransmitter dopamine, whose levels drop due to loss of neuronal innervation of the

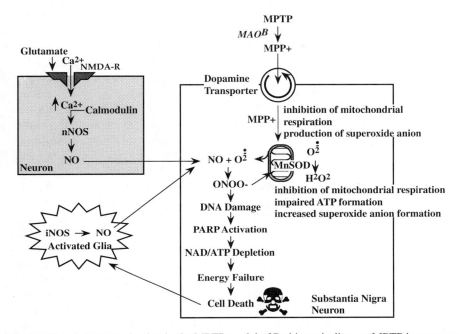

FIGURE 2.4 PARP activation in the MPTP model of Parkinson's disease. MPTP is converted to MPP+ by MAO_B. MPP+ is actively transported into dopaminergic neurons of the substantia nigra by the dopamine transporter and then transported into the mitochondria. MPP+ inhibits mitochondrial respiration, resulting in superoxide anion formation and decreased ATP formation. Decreased ATP can result in the depolarization of nigral neurons, increasing the need for ATP while further depleting ATP. Increased superoxide anion can react with locally produced NO, forming peroxynitrite ($ONOO^-$). Peroxynitrite can damage DNA, activating PARP, protein ribosylation, and NAD and ATP consumption. Peroxynitrite can also damage MnSOD and mitochondrial respiration, further compromising mitochondrial function and production of ATP, resulting ultimately in neuronal cell death. Dead and dying neurons will stimulate and activate local glia, inducing the expression of iNOS and increasing the local generation of NO, thus contributing to further neuronal damage.

striatum from the substantia nigra. Unfortunately, this therapy is palliative and does not address the underlying mechanisms of disease. The progressive neuronal degeneration proceeds, and eventually the pharmacological therapy fails.

A synthetic heroin analogue, 1-methyl-4-phenyl-1,2,3,6-tetrahydropyridine (MPTP), can selectively damage neurons in the nigrostriatal dopaminergic pathway and has provided an experimental model for parkinsonism in nonhuman primates and mice.[83-85] MPTP induces irreversible and severe motor abnormalities that are indistinguishable from those observed in Parkinson's disease. In monkeys and mice, MPTP exposure induces many of the biochemical and neuropathological features of nigrostriatal dopaminergic degeneration that are observed in postmortem studies of human Parkinson's disease. These changes include a marked reduction in striatal dopamine content and dopamine metabolites, as well as reductions in the number of dopamine cell bodies in the substantia nigra.[83-85] In the nervous system, MPTP is oxidized to MPP+ by monoamine oxidase B (MAO_B), which is then taken up by dopamine-containing neurons via the high-affinity dopamine transporter. Once inside the cell, MPP+ is actively transported into mitochondria where it blocks mitochondrial respiration, resulting in increased formation of superoxide anion.[86-90] Superoxide anion can react with NO to form the potent oxidant, peroxynitrite. While substantia nigra neurons do not express nNOS, their terminals are physically associated with nNOS-containing nerve terminals. Increased nitration of protein has been observed following MPTP exposure of mice, suggesting that peroxynitrite formation occurs.[91] Additionally, the selective nNOS inhibitor 7-nitroindazole (7-NI) provides neuroprotection against MPTP-induced neurotoxicity in a time- and dose-dependent fashion.[90,92] Further, mutant mice that lack the nNOS gene are also resistant to MPTP-induced neurotoxicity when compared with wild-type littermate controls.[92,93] MAO_B activity is unimpaired by either 7-NI or deletion of the nNOS gene, implicating a direct role for nNOS activation in MPTP-induced neurotoxicity.[92] The protection observed with 7-NI was greater than that observed in the nNOS knockout mice. 7-NI is selective for nNOS over endothelial NOS (eNOS), but has some effect on the catalytic activity of inducible NOS (iNOS). Studies performed in iNOS knockout mice show protection against MPTP toxicity, implicating a second NOS isoform, iNOS, in MPTP neurotoxicity.[94] Since peroxynitrite can effectively damage DNA, it is possible that PARP activation plays a key role in MPTP toxicity. In fact, PARP knockout mice are dramatically protected against MPTP neurotoxicity.[95] Similar results have been observed with pharmacological inhibition of PARP with 3-aminobenzamide,[96,97] but, because the pharmacokinetics of this agent are not known in the nervous system and 3-aminobenzaminde has other cellular targets, definitive conclusions could not be drawn. In wild-type mice, immunohistochemistry for poly(ADP-ribose) indicates that PARP is activated in vulnerable dopamine-containing neurons of the substantia nigra neurons following MPTP injection.[95] No PARP immunohistochemistry is observed in both the nNOS knockout mice and the PARP knockout mice, indicating in the MPTP model that NO and, likely, peroxynitrite are the major activators of PARP. On counting the number of substantia nigra neurons, there is no loss of neurons in the PARP knockout mice exposed to MPTP as compared with wild-type mice injected with saline. In contrast, there is a significant loss of substantia nigra pars compacta neurons in wild-type mice exposed to MPTP. Additionally, MPTP elicits a novel pattern of poly(ADP-ribosyl)ation of nuclear proteins that completely depends on neuronally derived NO.[95] Thus, NO, DNA damage, and PARP activation play a critical role in MPTP-induced parkinsonism and suggest that inhibitors of PARP may have protective benefit in the treatment of Parkinson's disease.

2.6 ALZHEIMER'S DISEASE

Alzheimer's disease (Figure 2.5) is the most common neurodegenerative disease worldwide. It is a chronic, progressive dementing illness and results in death in many cases. Alzheimer's disease is characterized by the progressive accumulation of neuritic plaques and neurofibrillary tangles and by selective neuronal cell death.[98] While a role for abnormal amyloid protein processing is implicated by both genetic

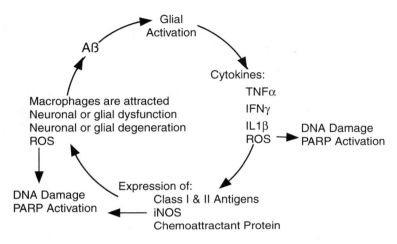

FIGURE 2.5 PARP activation in Alzheimer's disease. The role of glial inflammatory processes in Alzheimer's disease has been highlighted by recent postmortem studies and investigations of amyloid-β actions on microglia as well as the use of indomethacin to slow progression of Alzheimer's disease. Chronic activation of glial inflammatory processes by amyloid-β could set in motion a cytokine cycle of cellular and molecular events, which ultimately result in neuronal dysfunction and degeneration. The production of inflammatory and neuroactive molecules can increase reactive oxygen species (ROS) formation leading to chronic oxidation. Increased tissue oxidation could lead to DNA damage and PARP activation, resulting in the increased poly(ADP-ribose) expression observed in Alzheimer's disease brain tissue.

studies and experimental modeling,[99] the pathogenic processes that result in Alzheimer's disease are not known. In human postmortem studies, oxidative damage to DNA and other macromolecules has been reported in Alzheimer's disease.[100] PARP activation has been investigated in Alzheimer's disease by assessment of poly(ADP ribose) formation. Immunohistochemical staining for poly(ADP-ribose) was evaluated in the frontal and temporal lobe of patients with Alzheimer's disease and compared with controls. An increase in expression of PARP, as well as increased poly(ADP-ribosyl)ation was detected in Alzheimer's disease tissue when compared with control brain tissue.[101] Double labeling for poly(ADP-ribose) and markers for neuronal, astrocytic, and microglial cells indicated that most of the cells expressing increased poly(ADP ribosyl)ation were neurons. Most of these were small pyramidal neurons in cortical laminae 3 and 5. A few of the cells containing poly(ADP-ribose) were astrocytes. No poly(ADP-ribose) accumulation was detected in microglia. However, the cells with increased poly(ADP ribosyl)ation were not found within tangles and very few occurred within senile plaques.[101] While there is enhanced PARP activity observed in Alzheimer's disease, the role of PARP in Alzheimer's disease is not yet known.

2.7 CONCLUSIONS

Diverse forms of neuronal injury lead to neuronal cell death by PARP activation. Most of these neuronal cell death pathways involve increased reactive oxygen species formation, in particular, formation of NO and superoxide anion. In many situations, mitochondrial activity is impaired, preventing restoration of the NAD and ATP that is being depleted by PARP activity. In animal models of neuronal injury, inhibition of PARP or genetic deletion of PARP protects against excitotoxicity, ischemic injury, traumatic brain injury, and MPTP-induced parkinsonism. These protective effects may be due to limiting excitotoxic and free radical–induced injury, as well as to limiting secondary injury from infiltrating neutrophils. The development of selective PARP inhibitors is a new and exciting area of investigation for future clinical therapeutics.

REFERENCES

1. Kerr, J. F., Wyllie, A. H., and Currie, A. R., Apoptosis: a basic biological phenomenon with wide-ranging implications in tissue kinetics, *Br. J. Cancer*, 26, 239, 1972.
2. Martin, L. J., Al-Abdulla, N. A., Brambrink, A. M., Kirsch, J. R., Sieber, F. E., and Portera-Cailliau, C., Neurodegeneration in excitotoxicity, global cerebral ischemia, and target deprivation: a perspective on the contributions of apoptosis and necrosis, *Brain Res. Bull.*, 46, 281, 1998.
3. Schwartz, L. M. and Osborne, B. A., Programmed cell death, apoptosis and killer genes, *Immunol. Today*, 14, 582, 1993.
4. Schwartz, L. M., Smith, S. W., Jones, M. E., and Osborne, B. A., Do all programmed cell deaths occur via apoptosis? *Proc. Natl. Acad. Sci. U.S.A.*, 90, 980, 1993.
5. Boonman, Z. and Isacson, O., Apoptosis in neuronal development and transplantation: role of caspases and trophic factors, *Exp. Neurol.*, 156, 1, 1999.
6. D'Mello, S. R., Molecular regulation of neuronal apoptosis, *Curr. Top. Dev. Biol.*, 39, 187, 1998.
7. Rubin, L. L., Neuronal cell death: when, why and how, *Br. Med. Bull.*, 53, 617, 1997.
8. Meldrum, B. and Garthwaite, J., Excitatory amino acid neurotoxicity and neurodegenerative disease, *Trends Pharmacol. Sci.*, 11, 379, 1990.
9. Eliasson, M. J. L., Sampei, K., Mandir, A. S., Hurn, P. D., Traystman, R. J., Jun Bao, J., Pieper, A., Wang, Z.-Q., Dawson, T. M., Snyder, S. H., and Dawson, V. L., Poly(ADP-Ribose) polymerase gene disruption renders mice resistant to cerebral ischemia, *Nat. Med.*, 3, 1, 1997.
10. Endres, M., Wang, Z. Q., Namura, S., Waeber, C., and Moskowitz, M. A., Ischemic brain injury is mediated by the activation of poly(ADP-ribose)polymerase, *J. Cereb. Blood Flow Metab.*, 17, 1143, 1997.
11. Szabó, C. and Dawson, V. L., Role of poly (ADP-ribose) synthetase in inflammation and reperfusion injury, *Trends Pharmacol. Sci.*, 19, 287–298, 1998.
12. Zhang, J., Dawson, V. L., Dawson, T. M., and Snyder, S. H., Nitric oxide activation of poly(ADP-ribose) synthetase in neurotoxicity, *Science*, 263, 687, 1994.
13. Wyllie, A. H., Kerr, J. F., and Currie, A. R., Cell death: the significance of apoptosis, *Int. Rev. Cytol.*, 68, 251, 1980.
14. Tsujimoto, Y., Role of Bcl-2 family proteins in apoptosis: apoptosomes or mitochondria? *Genes Cells*, 3, 697, 1998.
15. Metzstein, M. M., Stanfield, G. M., and Horvitz, H. R., Genetics of programmed cell death in *C. elegans:* past, present and future, *Trends Genet.*, 14, 410, 1998.
16. Fadok, V. A. and Henson, P. M., Apoptosis: getting rid of the bodies, *Curr. Biol.*, 8, R693, 1998.

17. Peter, M. E., Heufelder, A. E., and Hengartner, M. O., Advances in apoptosis research, *Proc. Natl. Acad. Sci. U.S.A.*, 94, 12736, 1997.
18. Villa, P., Kaufmann, S. H., and Earnshaw, W. C., Caspases and caspase inhibitors, *Trends Biochem. Sci.*, 22, 388, 1997.
19. Wang, Z. Q., Stingl, L., Morrison, C., Jantsch, M., Los, M., Schulze-Osthoff, K., and Wagner, E. F., PARP is important for genomic stability but dispensable in apoptosis, *Genes Dev.*, 11, 2347, 1997.
20. Oliver, F. J., de la Rubia, G., Rolli, V., Ruiz-Ruiz, M. C., de Murcia, G., and Ménissier-de Murcia, J., Importance of poly(ADP-ribose) polymerase and its cleavage in apoptosis. Lesson from an uncleavable mutant, *J. Biol. Chem.*, 273, 33533, 1998.
21. Ménissier-de Murcia, J., Niedergang, C., Trucco, C., Ricoul, M., Dutrillaux, B., Mark, M., Oliver, F. J., Masson, M., Dierich, A., LeMeur, M., Walztinger, C., Chambon, P., and de Murcia, G., Requirement of poly(ADP-ribose) polymerase in recovery from DNA damage in mice and in cells, *Proc. Natl. Acad. Sci. U.S.A.*, 94, 7303, 1997.
22. Herceg, Z. and Wang, Z. Q., Failure of poly(ADP-ribose) polymerase cleavage by caspases leads to induction of necrosis and enhanced apoptosis, *Mol. Cell. Biol.*, 19, 5124, 1999.
23. Olney, J. W., Ho, O. L., Rhee, V., and DeGubareff, T., Neurotoxic effects of glutamate [letter], *N. Engl. J. Med.*, 289, 1374, 1973.
24. Olney, J. W. and Ho, O. L., Brain damage in infant mice following oral intake of glutamate, aspartate or cysteine, *Nature*, 227, 609, 1970.
25. Olney, J. W., Brain lesions, obesity, and other disturbances in mice treated with monosodium glutamate, *Science*, 164, 719, 1969.
26. Olney, J. W. and Sharpe, L. G., Brain lesions in an infant rhesus monkey treated with monsodium glutamate, *Science*, 166, 386, 1969.
27. Pullan, L. M., Olney, J. W., Price, M. T., Compton, R. P., Hood, W. F., Michel, J., and Monahan, J. B., Excitatory amino acid receptor potency and subclass specificity of sulfur-containing amino acids, *J. Neurochem.*, 49, 1301, 1987.
28. McBain, C. J. and Mayer, M. L., N-methyl-D-aspartic acid receptor structure and function, *Physiol. Rev.*, 74, 723, 1994.
29. Mayer, M. L., Benveniste, M., Patneau, D. K., and Vyklicky, L., Jr., Pharmacologic properties of NMDA receptors, *Ann. N.Y. Acad. Sci.*, 648, 194, 1992.
30. Bredt, D. S. and Snyder, S. H., Nitric oxide mediates glutamate-linked enhancement of cGMP levels in the cerebellum, *Proc. Natl. Acad. Sci. U.S.A.*, 86, 9030, 1989.
31. Garthwaite, J., Garthwaite, G., Palmer, R. M., and Moncada, S., NMDA receptor activation induces nitric oxide synthesis from arginine in rat brain slices, *Eur. J. Pharmacol.*, 172, 413, 1989.
32. Dawson, V. L., Dawson, T. M., Bartley, D. A., Uhl, G. R., and Snyder, S. H., Mechanisms of nitric oxide-mediated neurotoxicity in primary brain cultures, *J. Neurosci.*, 13, 2651, 1993.
33. Dawson, V. L., Dawson, T. M., London, E. D., Bredt, D. S., and Snyder, S. H., Nitric oxide mediates glutamate neurotoxicity in primary cortical cultures, *Proc. Natl. Acad. Sci. U.S.A.*, 88, 6368, 1991.
34. Dugan, L. L., Sensi, S. L., Canzoniero, L. M., Handran, S. D., Rothman, S. M., Lin, T. S., Goldberg, M. P., and Choi, D. W., Mitochondrial production of reactive oxygen species in cortical neurons following exposure to N-methyl-D-aspartate, *J. Neurosci.*, 15, 6377, 1995.
35. Beckman, J. S., The double-edged role of nitric oxide in brain function and superoxide-mediated injury, *J. Dev. Physiol.*, 15, 53, 1991.

36. de Murcia, G., Schreiber, V., Molinete, M., Saulier, B., Poch, O., Masson, M., Niedergang, C., and Ménissier-de Murcia, J., Structure and function of poly(ADP-ribose) polymerase, *Mol. Cell. Biochem.*, 138, 15, 1994.
37. de Murcia, G. and Ménissier-de Murcia, J., Poly(ADP-ribose) polymerase: a molecular nick-sensor, *Trends Biochem. Sci.*, 19, 172, 1994.
38. Lautier, D., Lagueux, J., Thibodeau, J., Menard, L., and Poirier, G. G., Molecular and biochemical features of poly(ADP-ribose) metabolism, *Mol. Cell. Biochem.*, 122, 171, 1993.
39. Brown, G. C., Foxwell, N., and Moncada, S., Transcellular regulation of cell respiration by nitric oxide generated by activated macrophages, *FEBS Lett.*, 439, 321, 1998.
40. Clementi, E., Brown, G. C., Feelisch, M., and Moncada, S., Persistent inhibition of cell respiration by nitric oxide: crucial role of S-nitrosylation of mitochondrial complex I and protective action of glutathione, *Proc. Natl. Acad. Sci. U.S.A.*, 95, 7631, 1998.
41. Lizasoain, I., Moro, M. A., Knowles, R. G., Darley-Usmar, V., and Moncada, S., Nitric oxide and peroxynitrite exert distinct effects on mitochondrial respiration which are differentially blocked by glutathione or glucose, *Biochem. J.*, 314, 877, 1996.
42. MacMillan-Crow, L. A., Crow, J. P., Kerby, J. D., Beckman, J. S., and Thompson, J. A., Nitration and inactivation of manganese superoxide dismutase in chronic rejection of human renal allografts, *Proc. Natl. Acad. Sci. U.S.A.*, 93, 11853, 1996.
43. Samdani, A. F., Dawson, T. M., and Dawson, V. L., Nitric oxide synthase in models of focal ischemia, *Stroke*, 28, 1283, 1997.
44. Park, C. K., Nehls, D. G., Graham, D. I., Teasdale, G. M., and McCulloch, J., The glutamate antagonist MK-801 reduces focal ischemic brain damage in the rat, *Ann. Neurol.*, 24, 543, 1988.
45. Carreau, A., Duval, D., Poignet, H., Scatton, B., Vige, X., and Nowicki, J. P., Neuroprotective efficacy of N omega-nitro-L-arginine after focal cerebral ischemia in the mouse and inhibition of cortical nitric oxide synthase, *Eur. J. Pharmacol.*, 256, 241, 1994.
46. Dalkara, T., Yoshida, T., Irikura, K., and Moskowitz, M. A., Dual role of nitric oxide in focal cerebral ischemia, *Neuropharmacology*, 33, 1447, 1994.
47. Nowicki, J. P., Duval, D., Poignet, H., and Scatton, B., Nitric oxide mediates neuronal death after focal cerebral ischemia in the mouse, *Eur. J. Pharmacol.*, 204, 339, 1991.
48. Dawson, V. L., Kizushi, V. M., Huang, P. L., Snyder, S. H., and Dawson, T. M., Resistance to neurotoxicity in cortical cultures from neuronal nitric oxide synthase defcient mice, *J. Neurosci.*, 16, 2479, 1996.
49. Endres, M., Scott, G. S., Salzman, A. L., Kun, E., Moskowitz, M. A., and Szabó, C., Protective effects of 5-iodo-6-amino-1,2-benzopyrone, an inhibitor of poly(ADP-ribose) synthetase against peroxynitrite-induced glial damage and stroke development, *Eur. J. Pharmacol.*, 351, 377, 1998.
50. Love, S., Barber, R., and Wilcock, G. K., Neuronal accumulation of poly(ADP-ribose) after brain ischaemia, *Neuropathol. Appl. Neurobiol.*, 25, 98, 1999.
51. Tokime, T., Nozaki, K., Sugino, T., Kikuchi, H., Hashimoto, N., and Ueda, K., Enhanced poly(ADP-ribosyl)ation after focal ischemia in rat brain, *J. Cereb. Blood Flow Metab.*, 18, 991, 1998.
52. Takahashi, K. and Greenberg, J. H., The effect of reperfusion on neuroprotection using an inhibitor of poly(ADP-ribose) polymerase, *Neuroreport*, 10, 2017, 1999.
53. Endres, M., Scott, G., Namura, S., Salzman, A. L., Huang, P. L., Moskowitz, M. A., and Szabó, C., Role of peroxynitrite and neuronal nitric oxide synthase in the activation of poly(ADP-ribose) synthetase in a murine model of cerebral ischemia-reperfusion, *Neurosci. Lett.*, 248, 41, 1998.

54. Takahashi, K., Pieper, A. A., Croul, S. E., Zhang, J., Snyder, S. H., and Greenberg, J. H., Post-treatment with an inhibitor of poly(ADP-ribose) polymerase attenuates cerebral damage in focal ischemia, *Brain Res.*, 829, 46, 1999.
55. Sun, A. Y. and Cheng, J. S., Neuroprotective effects of poly(ADP-ribose) polymerase inhibitors in transient focal cerebral ischemia of rats, *Chung Kuo Yao Li Hsueh Pao*, 19, 104, 1998.
56. Milam, K. M. and Cleaver, J. E., Inhibitors of poly(adenosine diphosphate-ribose) synthesis: effect on other metabolic processes, *Science*, 223, 589, 1984.
57. Hunting, D. J. and Gowans, B. J., Inhibition of repair patch ligation by an inhibitor of poly(ADP-ribose) synthesis in normal human fibroblasts damaged with ultraviolet radiation, *Mol. Pharmacol.*, 33, 358, 1988.
58. Hunting, D. J., Gowans, B. J., and Henderson, J. F., Specificity of inhibitors of poly(ADP-ribose) synthesis. Effects on nucleotide metabolism in cultured cells, *Mol. Pharmacol.*, 28, 200, 1985.
59. Hauschildt, S., Scheipers, P., Bessler, W. G., and Mulsch, A., Induction of nitric oxide synthase in L929 cells by tumour-necrosis factor alpha is prevented by inhibitors of poly(ADP-ribose) polymerase, *Biochem. J.*, 288, 255, 1992.
60. Redegeld, F. A., Chatterjee, S., Berger, N. A., and Sitkovsky, M. V., Poly-(ADP-ribose) polymerase partially contributes to target cell death triggered by cytolytic T lymphocytes, *J. Immunol.*, 149, 3509, 1992.
61. Lo, E. H., Bosque-Hamilton, P., and Meng, W., Inhibition of poly(ADP-ribose) polymerase: reduction of ischemic injury and attenuation of N-methyl-D-aspartate-induced neurotransmitter dysregulation, *Stroke*, 29, 830, 1998.
62. Iadecola, C., Bright and dark sides of nitric oxide in ischemic brain injury, *Trends Neurosci.*, 20, 132, 1997.
63. Zingarelli, B., Salzman, A. L., and Szabó, C., Genetic disruption of poly (ADP-ribose) synthetase inhibits the expression of P-selectin and intercellular adhesion molecule-1 in myocardial ischemia/reperfusion injury, *Circ. Res.*, 83, 85, 1998.
64. Cuzzocrea, S., Zingarelli, B., Gilad, E., Hake, P., Salzman, A. L., and Szabó, C., Protective effects of 3-aminobenzamide, an inhibitor of poly(ADP-ribose) synthase in a carrageenan-induced model of local inflammation, *Eur. J. Pharmacol.*, 342, 67, 1998.
65. Szabó, C., Lim, L. H., Cuzzocrea, S., Getting, S. J., Zingarelli, B., Flower, R. J., Salzman, A. L., and Perretti, M., Inhibition of poly(ADP-ribose) synthetase attenuates neutrophil recruitment and exerts antiinflammatory effects, *J. Exp. Med.*, 186, 1041, 1997.
66. Zingarelli, B., Cuzzocrea, S., Zsengeller, Z., Salzman, A. L., and Szabó, C., Protection against myocardial ischemia and reperfusion injury by 3-aminobenzamide, an inhibitor of poly(ADP-ribose) synthetase, *Cardiovasc. Res.*, 36, 205, 1997.
67. Shohami, E., Beit-Yannai, E., Horowitz, M., and Kohen, R., Oxidative stress in closed-head injury: brain antioxidant capacity as an indicator of functional outcome, *J. Cereb. Blood Flow Metab.*, 17, 1007, 1997.
68. Rink, A., Fung, K. M., Trojanowski, J. Q., Lee, V. M., Neugebauer, E., and McIntosh, T. K., Evidence of apoptotic cell death after experimental traumatic brain injury in the rat, *Am. J. Pathol.*, 147, 1575, 1995.
69. Clark, R. S., Kochanek, P. M., Dixon, C. E., Chen, M., Marion, D. W., Heineman, S., DeKosky, S. T., and Graham, S. H., Early neuropathologic effects of mild or moderate hypoxemia after controlled cortical impact injury in rats, *J. Neurotrauma*, 14, 179, 1997.
70. Wada, K., Chatzipanteli, K., Busto, R., and Dietrich, W. D., Role of nitric oxide in traumatic brain injury in the rat, *J. Neurosurg.*, 89, 807, 1998.

71. Wada, K., Chatzipanteli, K., Busto, R., and Dietrich, W. D., Effects of L-NAME and 7-NI on NOS catalytic activity and behavioral outcome after traumatic brain injury in the rat, *J. Neurotrauma*, 16, 203, 1999.
72. McIntosh, T. K., Smith, D. H., Voddi, M., Perri, B. R., and Stutzmann, J. M., Riluzole, a novel neuroprotective agent, attenuates both neurologic motor and cognitive dysfunction following experimental brain injury in the rat, *J. Neurotrauma*, 13, 767, 1996.
73. Palmer, A. M., Marion, D. W., Botscheller, M. L., Swedlow, P. E., Styren, S. D., and DeKosky, S. T., Traumatic brain injury-induced excitotoxicity assessed in a controlled cortical impact model, *J. Neurochem.*, 61, 2015, 1993.
74. Globus, M. Y., Alonso, O., Dietrich, W. D., Busto, R., and Ginsberg, M. D., Glutamate release and free radical production following brain injury: effects of posttraumatic hypothermia, *J. Neurochem.*, 65, 1704, 1995.
75. Wu, W. and Li, L., Inhibition of nitric oxide synthase reduces motoneuron death due to spinal root avulsion, *Neurosci. Lett.*, 153, 121, 1993.
76. Wu, W., Expression of nitric-oxide synthase (NOS) in injured CNS neurons as shown by NADPH diaphorase histochemistry, *Exp. Neurol.*, 120, 153, 1993.
77. Wu, W., Liuzzi, F. J., Schinco, F. P., Depto, A. S., Li, Y., Mong, J. A., Dawson, T. M., and Snyder, S. H., Neuronal nitric oxide synthase is induced in spinal neurons by traumatic injury, *Neuroscience*, 61, 719, 1994.
78. Scott, G. S., Jakeman, L. B., Stokes, B. T., and Szabó, C., Peroxynitrite production and activation of poly(adenosine diphosphate-ribose) synthetase in spinal cord injury, *Ann. Neurol.*, 45, 120, 1999.
79. LaPlaca, M. C., Raghupathi, R., Verma, A., Pieper, A. A., Saatman, K. E., Snyder, S. H., and McIntosh, T. K., Temporal patterns of poly(ADP-ribose) polymerase activation in the cortex following experimental brain injury in the rat, *J. Neurochem.*, 73, 205, 1999.
80. Wallis, R. A., Panizzon, K. L., and Girard, J. M., Traumatic neuroprotection with inhibitors of nitric oxide and ADP-ribosylation, *Brain Res.*, 710, 169, 1996.
81. Whalen, M. J., Clark, R. S. B., Dixon, C. E., Robichaud, P., Marion, D. W., Vagni, V., Graham, S., Virag, L., Hasko, G., Stachlewitz, R., Szabó, C., and Kockhanek, P. M., Reduction of cognitive and motor deficits after traumatic brain injury in mice deficient in poly(ADP-ribose) polymerase, *J. Cereb. Blood Flow Metab.*, 19, 832–842, 1999.
82. Jenner, P., Schapira, A. H., and Marsden, C. D., New insights into the cause of Parkinson's disease, *Neurology*, 42, 2241, 1992.
83. Heikkila, R. E., Nicklas, W. J., Vyas, I., and Duvoisin, R. C., Dopaminergic toxicity of rotenone and the 1-methyl-4-phenylpyridinium ion after their stereotaxic administration to rats: implication for the mechanism of 1-methyl-4-phenyl-1,2,3,6-tetrahydropyridine toxicity, *Neurosci. Lett.*, 62, 389, 1985.
84. Kopin, I. J. and Markey, S. P., MPTP toxicity: implications for research in Parkinson's disease, *Annu. Rev. Neurosci.*, 11, 81, 1988.
85. Langston, J. W., Langston, E. B., and Irwin, I., MPTP-induced parkinsonism in human and non-human primates — clinical and experimental aspects, *Acta Neurol. Scand. Suppl.*, 100, 49, 1984.
86. Kindt, M. V., Heikkila, R. E., and Nicklas, W. J., Mitochondrial and metabolic toxicity of 1-methyl-4-(2′-methylphenyl)-1,2,3,6-tetrahydropyridine, *J. Pharmacol. Exp. Ther.*, 242, 858, 1987.
87. Javitch, J. A. and Snyder, S. H., Uptake of MPP(+) by dopamine neurons explains selectivity of parkinsonism-inducing neurotoxin, MPTP, *Eur. J. Pharmacol.*, 106, 455, 1984.

88. Kitayama, S., Shimada, S., and Uhl, G. R., Parkinsonism-inducing neurotoxin MPP+: uptake and toxicity in nonneuronal COS cells expressing dopamine transporter cDNA, *Ann. Neurol.*, 32, 109, 1992.
89. Przedborski, S., Jackson-Lewis, V., Kostic, V., Carlson, E., Epstein, C. J., and Cadet, J. L., Superoxide dismutase, catalase, and glutathione peroxidase activities in copper/zinc-superoxide dismutase transgenic mice, *J. Neurochem.*, 58, 1760, 1992.
90. Schulz, J. B., Matthews, R. T., Klockgether, T., Dichgans, J., and Beal, M. F., The role of mitochondrial dysfunction and neuronal nitric oxide in animal models of neurodegenerative diseases, *Mol. Cell. Biochem.*, 174, 193, 1997.
91. Ara, J., Przedborski, S., Naini, A. B., Jackson-Lewis, V., Trifiletti, R. R., Horwitz, J., and Ischiropoulos, H., Inactivation of tyrosine hydroxylase by nitration following exposure to peroxynitrite and 1-methyl-4-phenyl-1,2,3,6-tetrahydropyridine (MPTP), *Proc. Natl. Acad. Sci. U.S.A.,* 95, 7659, 1998.
92. Przedborski, S., Jackson-Lewis, V., Yokoyama, R., Shibata, T., Dawson, V. L., and Dawson, T. M., Role of neuronal nitric oxide in 1-methyl-4-phenyl-1,2,3,6- tetrahydropyridine (MPTP)-induced dopaminergic neurotoxicity, *Proc. Natl. Acad. Sci. U.S.A.,* 93, 4565, 1996.
93. Matthews, R. T., Beal, M. F., Fallon, J., Fedorchak, K., Huang, P. L., Fishman, M. C., and Hyman, B. T., MPP+ induced substantia nigra degeneration is attenuated in nNOS knockout mice, *Neurobiol. Dis.*, 4, 114, 1997.
94. Liberatore, G. T., Jackson-Lewis, V., Mandir, A. S., McAuliffe, W. J., Vila, M., Dawson, V. L., Dawson, T. M., and Przedborski, S., Inducible nitric oxide synthase stimulates dopaminergic neurodegeneration in the MPTP model of Parkinson's disease, *Nat. Med.*, 5, 1403, 1999.
95. Mandir, A. S., Przedborski, S., Jackson-Lewis, V., Wang, Z. Q., Simbulan-Rosenthal, C. M., Smulson, M. E., Hoffman, B. E., Guastella, D. B., Dawson, V. L., and Dawson, T. M., Poly(ADP-ribose) polymerase activation mediates 1-methyl-4-phenyl-1, 2,3,6-tetrahydropyridine (MPTP)-induced parkinsonism, *Proc. Natl. Acad. Sci. U.S.A.,* 96, 5774, 1999.
96. Cosi, C., Colpaert, F., Koek, W., Degryse, A., and Marien, M., Poly(ADP-ribose) polymerase inhibitors protect against MPTP-induced depletions of striatal dopamine and cortical noradrenaline in C57B1/6 mice, *Brain Res.*, 729, 264, 1996.
97. Cosi, C. and Marien, M., Decreases in mouse brain NAD+ and ATP induced by 1-methyl-4-phenyl-1,2,3,6-tetrahydropyridine (MPTP): prevention by the poly(ADP-ribose) polymerase inhibitor, benzamide, *Brain Res.*, 809, 58, 1998.
98. Selkoe, D. J., Alzheimer's disease: genotypes, phenotypes, and treatments, *Science*, 275, 630, 1997.
99. Selkoe, D. J., Alzheimer's disease: a central role for amyloid, *J. Neuropathol. Exp. Neurol.*, 53, 438, 1994.
100. Smith, M. A., Hirai, K., Hsiao, K., Pappolla, M. A., Harris, P. L., Siedlak, S. L., Tabaton, M., and Perry, G., Amyloid-beta deposition in Alzheimer transgenic mice is associated with oxidative stress, *J. Neurochem.*, 70, 2212, 1998.
101. Love, S., Barber, R., and Wilcock, G. K., Increased poly(ADP-ribosyl)ation of nuclear proteins in Alzheimer's disease, *Brain*, 122, 247, 1999.

3 Importance of Poly(ADP-Ribose) Polymerase Activation in Myocardial Reperfusion Injury

Basilia Zingarelli

CONTENTS

3.1 Introduction 41
3.2 Poly(ADP-Ribose) Polymerase Activation: A Pathway of Cellular Damage and Dysfunction 42
3.3 Oxidative and Nitrosative Stress and PARP Activation in Myocardial Ischemia and Reperfusion 43
3.4 PARP Activation and Energy Substrate Metabolism during Myocardial Ischemia and Reperfusion: The Role of Mitochondria 46
3.5 PARP Activation and Leukocyte–Endothelial Interactions during Myocardial Ischemia and Reperfusion: The Role of Adhesion Molecules 51
3.6 Conclusion 54
Acknowledgment 56
References 56

3.1 INTRODUCTION

In clinical therapy of acute myocardial infarction, coronary reperfusion has proved to be the only way to limit infarct size by restoring the fractional uptake of oxygen in the heart to maintain the rate of cellular oxidation. However, restoration of flow is accompanied by detrimental manifestations known as "reperfusion injury," thus directly influencing the degree of recovery.[1] Reperfusion injury refers to extremely complex situations that had not occurred during the preceding ischemic period and are mainly recognized as reperfusion arrhythmias, postischemic contractile dysfunction, and lethal cell injury (i.e., necrosis and/or apoptosis).[2] The existence of such damage has clinical relevance, as it would imply the possibility of improving recovery with specific interventions applied only at the time of reperfusion. Experimental research has been directed toward establishing the precise sequence of biochemical events of reperfusion, as such knowledge could lead to rational treatments designed to prevent or delay myocardial cell death. At the present time, there is no simple answer to the question of what determines cell death and the failure to recover cell function after reperfusion.

Recent experimental reports have suggested that the dysfunction is triggered by the endothelial and myocyte generation of a large burst of oxidant molecules and amplified by accumulation of neutrophils into injured tissue.[2] A candidate pathway of oxidant-induced injury involves the nuclear enzyme poly(ADP-ribose) polymerase (PARP), also termed as poly(ADP-ribose) synthetase (PARS) and poly(ADP-ribose) transferase (pPADPRT).[3,4] Once activated, in response to nicks and breaks in the strand DNA, PARP initiates an energy-consuming, inefficient repair cycle by transferring ADP ribose units to nuclear proteins. This process rapidly depletes the intracellular NAD^+ and ATP energetic pools, which slows the rate of glycolysis and mitochondrial respiration leading to cellular dysfunction and death.[5-7] The objective of this chapter is to provide an overview of the experimental evidence implicating PARP as a pathophysiological modulator of myocardial ischemia and reperfusion *in vivo,* to discuss the activation of PARP by reactive metabolites of oxygen and nitrogen, to examine the role of the enzyme in the modulation of leukocyte–endothelial cell interactions, and to discuss how these mechanisms may be involved in the pathophysiology of ischemic heart disease.

3.2 POLY(ADP-RIBOSE) POLYMERASE ACTIVATION: A PATHWAY OF CELLULAR DAMAGE AND DYSFUNCTION

PARP is a chromatin-associated nuclear enzyme, which possesses putative DNA repair function in the nucleus of eukaryotic cells. The enzyme is composed of three functional domains: an N-terminal DNA-binding domain that binds to DNA strand breaks, a central automodification domain containing auto-poly(ADP-ribosyl)ation sites, and a C-terminal catalytic domain. Binding of the N-terminal domain to DNA nicks and breaks activates the C-terminal catalytic domain that, in turn, cleaves NAD^+ into ADP-ribose and nicotinamide. PARP covalently attaches ADP-ribose to various proteins, including the automodification domain of PARP itself, and then extends the initial ADP-ribose group into a nucleic acidlike polymer, poly(ADP)-ribose.[3,7] As reported in other chapters of this book, the process of DNA damage and PARP activation leads to intracellular NAD^+ depletion and alteration in high-energy phosphates, and progresses to a loss of cellular viability postulated as a "suicide phenomenon."[8] Numerous *in vitro* studies in several cellular types have demonstrated that such a suicidal mechanism is responsible for cellular injury in response to oxygen-derived free radicals, nitric oxide (NO) and peroxynitrite.[9-12] In the past several years, much evidence has provided for the involvement of oxygen- and nitrogen-derived reactive species in the pathological events associated with *in vivo* inflammatory processes. It is therefore conceivable to propose PARP as plausible mechanism of injury in diseases characterized by an overwhelming burst of free radicals and oxidants. As widely demonstrated in other chapters of this book, experimental reports have proposed that DNA damage due to reactive species generation leads to PARP activation and cellular energy exhaustion, and underlies the pathogenesis of diabetes, arthritis, and neuroinjury.[10-14] Recently, gene-targeted animals have been developed in which the gene encoding PARP has been disrupted. These deficient mice appear to be resistant to neural damage following vascular stroke,[15] to chemical-induced parkinsonism,[16] and to pancreatic beta-cell destruction and diabetes development.[17] Other studies from the author's laboratory established that activation of PARP also plays a crucial role in *in vivo* vascular and energetic failure in cardiovascular shock and inflammation (the reader is also referred to the other chapters in this book).[5,18-25]

3.3 OXIDATIVE AND NITROSATIVE STRESS AND PARP ACTIVATION IN MYOCARDIAL ISCHEMIA AND REPERFUSION

The concept that myocardial ischemia and reperfusion injury are the result of deleterious actions of reactive oxygen- and nitrogen-derived species has been widely accepted. The production of oxyradicals starts early in the ischemic period in mitochondria with altered redox balance and is further enhanced during reperfusion by a massive activation of xanthine oxidase in endothelial cells, and NADPH oxidase in infiltrated neutrophils.[26] A plethora of nitrogen derivatives, highly reactive and capable of oxidation, are also formed during the early phase of reperfusion and may contribute to tissue injury. A characteristic marker of the occurring nitrosative stress is represented by the chemical alteration of the myocardial protein structure due to nitration of tyrosine residues.[19,27,28] A number of chemical reactions with NO derivatives can yield nitrotyrosine formation. Nitrotyrosine can be formed from the reaction of peroxynitrite, the reaction of nitrite with hypochlorous acid, or the reaction of nitrite with myeloperoxidase and hydrogen peroxide.[29-32] The cytotoxic effects of these reactive species are further increased since an impaired cellular mechanism of the antioxidant network coexists, i.e., such as alteration of superoxide dismutase, glutathione peroxidase, and catalase.[2,26] In this highly reactive milieu, DNA has been shown to be a target for free radical attack, as demonstrated by the occurrence of fragmentation[33] and the formation of 8-hydroxydeoxyguanosine, which increases steadily and progressively as a function of reperfusion time in previously ischemic hearts.[34] Therefore, the phenomenon of reperfusion represents an ideal trigger for PARP activation by DNA damage. The extraordinary protection afforded by PARP inhibition in several experimental models of myocardial ischemia and reperfusion proves this hypothesis substantially (see Table 3.1 for references). We evaluated the role of PARP activation in an *in vivo* acute model of myocardial reperfusion injury in the rat.[19] We demonstrated that peroxynitrite was formed, as evidenced by plasma oxidation of dihydrorhodamine 123 and massive formation of nitrotyrosine in the reperfused tissue. Myocardial ischemia and reperfusion resulted

TABLE 3.1
Protection against Myocardial Ischemia and Reperfusion Injury by Inhibition of PARP *In Vivo*

Animal Species	Mode of PARP Inhibition	Effect of PARP Inhibition	Ref.
Rat	3-aminobenzamide, GPI6150	Reduction of infarct size, reduction of plasma creatine phosphokinase levels, improved myocardial ATP, reduction of neutrophil infiltration, reduction of peroxynitrite formation, reduction of poly(ADP-ribose) formation	19, 35
Rabbit	3-Aminobenzamide, 4-amino-benzamide, 4-amino-1,8-naphtalimide, 1,5-dihydroxy-isoquinoline	Reduction of infarct size, improved contractile function	36
Pig	3-Aminobenzamide	Reduction of infarct size, improved contractile function	37
Mice	Disruption of PARP gene at exon 2	Prevention of myocardial damage, reduction of plasma creatine phosphokinase levels, prevention of ICAM-1 and P-selectin expression, reduction of neutrophil infiltration, reduction of peroxynitrite formation	20

in a marked cellular injury, as measured by an increase of plasma creatine phosphokinase activity and a large infarcted area, and was associated with reduction of the cardiac concentration of ATP. Pharmacological inhibition of PARP activity improved the outcome of myocardial dysfunction, as evidenced by a reduction in plasma levels of creatine phosphokinase and infarct size, and preservation of the ATP pools. Cardiac myeloperoxidase activity and peroxynitrite formation were also reduced by treatment with the PARP inhibitor, thus suggesting a reduction of infiltration of neutrophil granulocytes and the consequent release of reactive species into the reperfused myocardium.[19] In a similar rat model, the activation of PARP has been documented by a massive poly(ADP-ribosyl)ation of reperfused postischemic myocardium only, but not the nonischemic zone. The accumulation of poly(ADP-ribose) correlated with myocardial cell death, whereas PARP inhibition reduced poly(ADP-ribose) formation and was highly effective in decreasing infarct size.[35] Other investigators have confirmed these results in similar experimental models of myocardial ischemia and reperfusion. In rabbit and pig models of myocardial infarction, Thiemermann and colleagues[36,37] have demonstrated that pharmacological inhibitors of PARP, such as nicotinamide and 3-aminobenzamide, given immediately before reperfusion of the ischemic myocardium, dramatically reduced the infarct size. The cardioprotection afforded by the PARP inhibitors was due to a selective inhibition of PARP, since the structurally related but inactive agents, such as 3-aminobenzoic acid and nicotinic acid, did not cause a reduction in infarct size.[36,37]

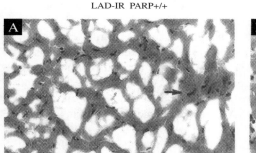
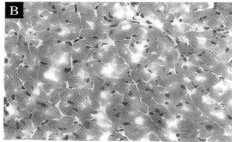

FIGURE 3.1 After 1-h occlusion and 1-h reperfusion of the left anterior descendant coronary artery (LAD-IR), a marked disruption of the myocardial structure characterized by the appearance of extensive necrosis and contraction bands (arrows) was demonstrated in cardiac sections from PARP$^{+/+}$ mice (A). In the PARP$^{-/-}$ mice subjected to LAD-IR, myocardial architecture appeared normal (B). Original magnification ×400. (From Zingarelli, B. et al., *Circ. Res.*, 83, 85, 1998. With permission.)

The development of transgenic mice deprived of a functional gene for PARP has offered a unique opportunity to support the pathophysiological role of the enzyme in myocardial infarction and to unravel some of the cellular mechanisms underlying this disease. Using a murine model of myocardial injury after early reperfusion, we found that absence of a functional PARP gene resulted in a significant prevention of reperfusion injury. Wild-type mice subjected to ligation (1-h) and reperfusion (1-h) of the left anterior descending branch of the coronary artery exhibited high plasma levels of creatine phosphokinase activity and a massive myocardial necrosis associated with neutrophil infiltration as demonstrated by increase in myeloperoxidase activity.[20] When reperfusion after 1-h ischemia was prolonged, wild-type mice also experienced high mortality (only 46% of animals out of 15 survived at 6 h after reperfusion; B. Zingarelli, unpublished observations). In PARP$^{-/-}$ mice, plasma levels of creatine phosphokinase activity were significantly reduced; the histological features of myocardium were typical of a normal architecture or a mild localized necrosis along with a reduction in neutrophil infiltration at 1-h reperfusion (see Figure 3.1 for histology).[20] Survival was markedly improved at prolonged reperfusion: 95% of mice out of 15 survived at 6 h after reperfusion (Z. Yang, B. Zingarelli, and C. Szabó, unpublished observations).

The inhibition of PARP activity and the energy recovery appears also to reverse the hemodynamic dysfunction caused by reperfusion in isolated hearts.[36-40] Thiemermann and colleagues[36,38] have reported that pharmacological inhibitors of PARP significantly reduced the impairment in left ventricular developed pressure and the rise in left ventricular end diastolic pressure in isolated rabbit and rat hearts subjected to no-flow global ischemia and reperfusion. Similar protective effects on myocardial function have been afforded when PARP is genetically disrupted.[39,40] At the end of the reoxygenation, in hearts from wild-type animals there was a significant suppression in the rate of intraventricular pressure development and in the rate of relaxation. In contrast, in the hearts from the PARP knockout animals, no significant suppression of the rate of intraventricular pressure development and relaxation was found when compared with the response to the wild-type hearts (Figure 3.2).[39] Consistent with these findings, it has been reported that poly(ADP-ribose) accumulated in hearts from wild-type animals after *in vitro* global ischemia and reperfusion and was associated with a significant NAD^+ and ATP depletion and contractile dysfunction. Genetic deficiency of PARP preserved the cellular energy pools and improved the recovery of myocardial function.[40] Thus, the extraordinary protection afforded by PARP deletion implies that DNA damage is a crucial event in myocardial dysfunction and death.

In addition to myocardial infarction, PARP is also relevant to other forms of ischemic damage (see Table 3.2 for references). For example, a pathological role for the peroxynitrite–PARP pathway was observed in a rat model of splanchnic ischemia and reperfusion. We found that the production of peroxynitrite was increased at reperfusion, but not during ischemia alone. This increase was associated with increased activity of PARP in intestinal epithelial cells and accumulation of polymorphonuclear cells in the injured tissue. Treatment with the PARP inhibitor 3-aminobenzamide significantly ameliorated the cardiovascular alterations, tissue damage, and neutrophil infiltration associated with splanchnic artery occlusion shock.[24] We also found that 3-aminobenzamide significantly reduced postintubation laryngeal injury in the rat.[41] Furthermore, pharmacological inhibition of PARP dramatically reduces neuronal damage in retinal ischemia elicited by elevating intraocular pressure.[42] Animal and *in vitro* studies also suggest that overactivation of PARP, with resulting depletion of NAD^+, in response to oxidative damage makes a substantial contribution to cell death after brain ischemia.[15,43,44] It is mandatory to emphasize that the oxidant/PARP pathway also has an important pathophysiological relevance in humans, as neuronal accumulation of poly(ADP-ribose), the end product of PARP activity, has been demonstrated after brain ischemia and Alzheimer's disease in humans.[45,46]

3.4 PARP ACTIVATION AND ENERGY SUBSTRATE METABOLISM DURING MYOCARDIAL ISCHEMIA AND REPERFUSION: THE ROLE OF MITOCHONDRIA

The myocardial contraction process is regulated by an efficient conversion of chemical into mechanical energy. At the core of the system are rates of oxidative phosphorylation of adenosine diphosphate (ADP) in the respiratory chain that exactly match rates of adenosine triphosphate (ATP) hydrolysis. Because of the high energy turnover in heart muscle, the cell volume occupied by mitochondria in myocardial cells is greater than in other tissues.[47] Since a linear relation exists between mitochondrial mass and heart rate or myocardial oxygen consumption, it is conceivable that structural and functional damage to mitochondria represent the major deleterious consequences of ischemia.[48] Intact mitochondria maintain a large (up to 180 mV) negative membrane potential across the mitochondrial inner membrane. A decrease in mitochondrial transmembrane potential followed by an intense reactive oxygen

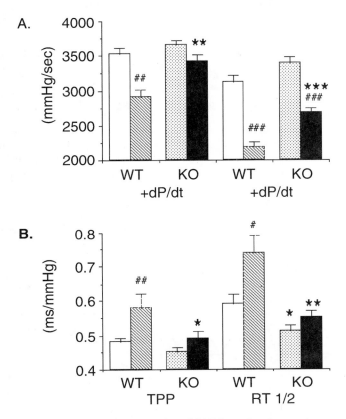

FIGURE 3.2 Effect of the functional integrity of PARS on the changes in myocardial contractility in perfused hearts subjected to hypoxia–reoxygenation *in vitro*. (A) The maximal rates of pressure development (+dP/dt, mmHg/s, left panels) and the maximal rates of relaxation (–dP/dt, mmHg/s, right panels) in wild-type and PARS knockout hearts in control conditions ("before hypoxia") and after 30 min of hypoxia and 30 min of reperfusion ("after hypoxia"). (B) The times to peak pressure (TPP, ms/mmHg, left panels), and the $1/2$ times to relaxation (RT$^{1}/_{2}$, ms/mmHg, right panels) in wild-type and PARS knockout hearts in control conditions ("before hypoxia") and after 30 min of hypoxia and 30 min of reperfusion ("after hypoxia"). Data represent mean ± SEM values for $n = 5$ hearts in each group. *$p < 0.05$, **$p < 0.01$, and ***$p < 0.001$ represent significant differences between the respective wild-type and PARS deficient groups and ##$p < 0.01$ and ###$p < 0.001$ represent significant differences between the respective values before and after hypoxia. (From Grupp, I. L. et al., *J. Mol. Cell. Cardiol.*, 31, 297, 1999.)

TABLE 3.2
Effect of Inhibition of PARP Activity in Tissue Injury Induced by Ischemia and Reperfusion *In Vivo*

Tissue or Organ	Mode of PARP Inhibition	Effect of PARP Inhibition	Ref.
Rat small bowel	3-Aminobenzamide	Amelioration of intestinal damage, reduction of epithelial hypermeability, reduction of neutrophil infiltration, reduction of peroxynitrite formation	24
Rabbit skeletal muscle	3-Aminobenzamide, nicotinamide, 1,5-dihydroxyisoquinoline	Reduction of muscle necrosis	36
Rat retina	3-Aminobenzamide	Reduction of inter-nucleosomal degradation of retinal DNA, reduction of inner retinal tickness	42
Rat brain	3,4-Dihydro-5-[4-(1-piperidinyl)-butoxy]-1(2H)-isoquinolinone	Reduction of infarct size	43
Murine brain	5-Iodo-6-amino-1,2-benzopyrone, disruption of PARP gene at exon 2	Amelioration of neuronal necrosis, reduction of infarct size	44, 15
Rat larynx	3-Aminobenzamide	Amelioration of laryngeal damage, reduction of neutrophil infiltration, reduction of peroxynitrite formation	41

intermediate production and a reduction of mitochondrial mass has been shown to occur during ischemia and reperfusion.[48,49] During ischemia, the acceleration of anaerobic metabolism is not sufficient to compensate for the marked reduction of aerobic metabolism with a consequent marked degradation of the adenine nucleotide pool, thus leaving the mitochondrial carriers in a more fully reduced state. This condition causes an increase of electron leakage from the respiratory chain that leads to the production of superoxide radicals by reacting with molecular oxygen of the inner mitochondrial membrane. Although the restoration of oxygen supply during reperfusion reenergizes the mitochondria, the electron transport is still altered because of the lack of adenine dinucleotides resulting in a larger amount of free radical production in the presence of the ample availability of molecular oxygen.[2,48] Free radical production from the mitochondrial enzymes will be even more pronounced, since mitochondrial superoxide dismutase is inactivated during reperfusion.[26,50] Disruption of mitochondrial function, generation of protons, and subsequent oxidative stress are often followed by an elevated intracellular Na^+ and Ca^{2+} levels, and progressive intracellular acidosis, which will affect the fundamental processes of contraction and excitability. Intracellular Ca^{2+} also activates a number of enzymes (proteases, phospholipase, myofibrillar ATPase), disrupts lysosomal membranes, and further affects mitochondrial oxidative phosphorylation.[51]

In the context of this metabolic derangement, PARP overactivation is expected to mediate an amplification of mitochondrial damage, potentially by accelerating the exhaustion of the cellular energy pool. *In vitro* studies have proved this hypothesis, clearly showing that PARP activation contributes to the mitochondrial alterations in cultured myocytes and other cellular types exposed to oxidants. For example, exposure of thymocytes to hydrogen peroxide, authentic peroxynitrite, and peroxynitrite-generating compounds induces a time- and dose-dependent decrease in mitochondrial transmembrane potential ($\Delta\Psi_m$), which is associated with increase in secondary reactive oxygen intermediate production and loss of cardiolipin, an indicator of mitochondrial membrane damage. Inhibition of PARP by 3-aminobenzamide or 5-iodo-6-amino-1,2-benzopyrone attenuates peroxynitrite-induced $\Delta\Psi_m$ reduction, secondary reactive oxygen intermediate generation, cardiolipin degradation, and intracellular calcium mobilization. Similarly, thymocytes from PARP-deficient animals appear to be protected against the peroxynitrite- and hydrogen peroxide–induced functional and ultrastructural mitochondrial alterations.[52]

In simulated *in vitro* conditions of ischemia and reperfusion, we have also demonstrated that activation of PARP is a major determinant of mitochondrial damage and the consequent myocardial oxidant injury. We have shown that peroxynitrite, hydrogen peroxide, and the NO donor compounds *S*-nitroso-*N*-acetyl-DL-penicillamine (SNAP) and diethyltriamine NONOate all cause a dose-dependent reduction of the mitochondrial respiration of rat cardiac myoblasts. Peroxynitrite and hydrogen peroxide, but not the NO donors, cause activation of cellular PARP activity. The suppression of mitochondrial respiration by peroxynitrite and hydrogen peroxide, but not by the NO donors, is ameliorated by pharmacological inhibition of PARP.[53] In an additional set of experiments we have found that rat myoblasts subjected to hypoxia (1-h) and reoxygenation (1- to 24-h) had a marked increase in PARP activity (at 1 h after reoxygenation) and a reduction in mitochondrial respiration (with maximal inhibition at 6 h), which were reversed by treatment with 3-aminobenzamide. Hypoxia alone did not induce changes in PARP activity or in mitochondrial respiration (Figure 3.3). These results were confirmed using fibroblasts derived from genetically deficient of PARP (PARP-/-) and wild-type mice. In response to hypoxia in wild-type fibroblasts there was an increase in PARP activity, which was further enhanced after reoxygenation and followed by a significant inhibition of the cellular respiration. On the contrary, PARP-/- fibroblasts did not show PARP activation or changes in mitochondrial respiration.[53]

Thiemermann and colleagues[54] have also examined the role of PARP in human cell lines, thus providing useful information relevant to human myocardial infarction. Exposure of human cardiac myoblasts to hydrogen peroxide caused a time- and concentration-dependent reduction in mitochondrial respiration, an increase in cell death, and an increase in PARP activity. The PARP inhibitors, 3-aminobenzamide, 1,5-dehydroxyisoquinoline, or nicotinamide, attenuated the cell injury and death, as well as the increase in PARP activity caused by hydrogen peroxide, whereas the inactive analogues 3-aminobenzoic acid or nicotinic acid were without effect.[54]

Endogenous ADP-ribosylation reactions also appear to be crucial determinants in other forms of oxidant-dependent cardiotoxicity. In this regard, overactivation of PARP, with consequent accelerated NAD^+ catabolism and damaged mitochondrial energy production, has been proposed to mediate the structural and functional changes of the cardiomyopathy induced by zidovudine.[55]

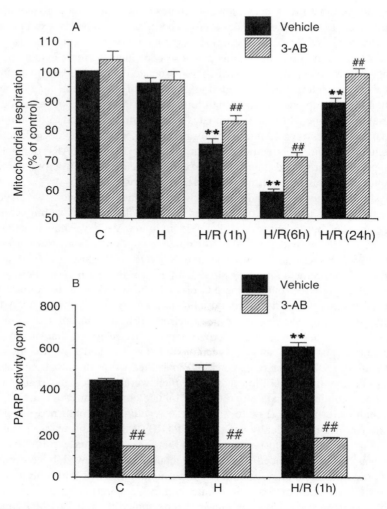

FIGURE 3.3 Effect of hypoxia (H, 1-h) and reoxygenation (H/R) on mitochondrial respiration on PARP activation (A) and mitochondrial respiration (B) in rat myocytes. Solid bars represent values in the presence of vehicle; hatched bars represent data in the presence of the PARP inhibitor 3-aminobenzamide (1 mM). *$p < 0.01$ represents significant changes elicited by hypoxia and reoxygenation when compared with unstimulated controls; †$p < 0.01$ represents protective effects provided by 3-aminobenzamide. (From Gilad, E. et al., *J. Mol. Cell. Cardiol.*, 29, 2585, 1997.)

3.5 PARP ACTIVATION AND LEUKOCYTE–ENDOTHELIAL INTERACTIONS DURING MYOCARDIAL ISCHEMIA AND REPERFUSION: THE ROLE OF ADHESION MOLECULES

It is now well appreciated that infiltration of neutrophils is a crucial event for ischemia and reperfusion injury. In the earliest stages of reperfusion after ischemia, neutrophils moving out of the circulation into inflamed tissue play a physiological role in the destruction of foreign antigens and remodeling of injured tissue. Nevertheless, neutrophils may augment the damage to vascular and parenchymal cellular elements by the release of proteolytic enzymes, free radicals, and proinflammatory mediators.[56] A growing body of experimental data suggests that overactivation of PARP is an important modulator of leukocyte–endothelial cell interactions. A consistent biological effect of the genetic or pharmacological inhibition of PARP during an inflammatory process has been, in fact, represented by the reduction of neutrophil infiltration in the site of injury. This biological event has been reported not only in myocardial ischemia and reperfusion, but also in other experimental models of inflammation such as zymosan- or carrageenan-induced paw edema, pleurisy, arthritis, peritonitis, and colitis.[19-21,23,24] In these studies reduction of neutrophil infiltration is always associated with the reduction of the resultant oxidative and nitrosative stress and with amelioration of tissue damage. The mechanism of regulation of neutrophil trafficking by PARP appears multiple and may involve the maintenance of endothelial integrity and the regulation of the expression of adhesion molecules.

We have found that during experimental myocardial ischemia and reperfusion injury, genetic ablation of PARP inhibits the release of P-selectin from the preformed pools on the endothelial surface and the upregulation of intercellular adhesion molecule-1 (ICAM-1), thus preventing the rolling and the firm adhesion of neutrophils along the endothelium.[20] Therefore, accumulation of neutrophils into the site of injury was markedly reduced in PARP-/- mice, as evidenced by reduction of myeloperoxidase activity (Figure 3.4). *In vitro* experiments confirmed that inhibition of PARP might directly influence the endothelial–neutrophil interaction by affecting endothelial expression of adhesion molecules. Incubation of human endothelial cells with peroxynitrite or immunostimulation with TNF-α induced the expression of P-selectin and upregulation of ICAM-1, respectively. Pharmacological inhibition of PARP by 3-aminobenzamide inhibited the oxidant-dependent expression of P-selectin and the cytokine-mediated upregulation of ICAM-1 (Figure 3.5). The role of PARP activation in adhesion molecule expression was further confirmed with experiments using murine fibroblasts lacking the gene for PARP (PARP-/-) or fibroblasts with a normal genotype (PARP+/+). Stimulation with cytokines induced a significant expression of ICAM-1 in a concentration-dependent manner in PARP+/+, while it elicited a very weak response in PARP-/-. Pharmacological inhibition of PARP by 3-aminobenzamide in normal fibroblasts reduced the cytokine-mediated expression of ICAM-1.[20]

Using an intravital microscopic technique, we have evaluated the kinetics of leukocyte migration into the mesenteric vasculature during an inflammatory pro-

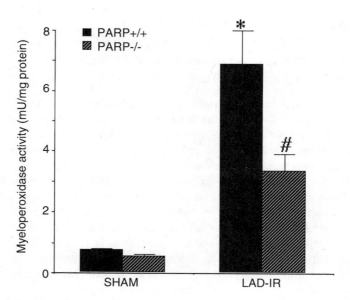

FIGURE 3.4 Neutrophil infiltration after coronary occlusion and reperfusion is reduced in PARP-/- mice. Myeloperoxidase, an enzyme present in neutrophils, was measured as an index of neutrophil infiltration into the injured tissue. Tissue myeloperoxidase activity was significantly enhanced in PARP+/+ mice subjected to myocardial injury (LAD-IR) in comparison with sham-operated PARP+/+ animals (*$p < 0.001$). In contrast, in PARP-/- mice subjected to myocardial injury, the increase of myeloperoxidase was significantly reduced when compared with PARP+/+ animals (#$p < 0.001$). Each data point is the mean ± SEM of six animals for each group. (From Zingarelli, B. et al., *Circ. Res.*, 83, 85, 1998. With permission.)

cess induced by intraperitoneal injection of the phlogogen agent zymosan in mice.[23] Polymorphonuclear cells accumulated in response to zymosan at a high rate of influx. Administration of 3-aminobenzamide reduced cell recruitment in a dose-dependent fashion. Zymosan also induced a marked increase in emigrated leukocytes in mouse mesenteric postcapillary venules at 4 h after intraperitoneal injection. Treatment with 3-aminobenzamide significantly reduced the number of cells emigrated outside the postcapillary venules. These findings were further supported by data obtained in PARP-deficient animals. Relative to wild-type controls, PARP-/- mice had reduced intestinal myeloperoxidase activity in response to intraperitoneal zymosan administration.[23]

The regulation of neutrophil recruitment and expression of adhesion molecules by PARP activation also occurs in cell types other than endothelial cells. For example, inhibition of PARP by nicotinamide and 3-aminobenzamide inhibited cytokine- or mitogen-induced ICAM-1 expression on thyroid cells.[57] Poly(ADP-ribosyl)ation of nuclear proteins in activated granulocytes was associated with an increase of adhesion, which was markedly reduced or even abolished by 3-aminobenzamide treatment.[58] Genetic blockade of PARP also inhibited neutrophil recruitment, oxidant generation, and mucosal injury in murine colitis by reducing upregulation of ICAM-1 in endothelial and intestinal epithelial cells.[21]

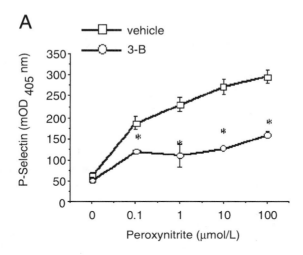
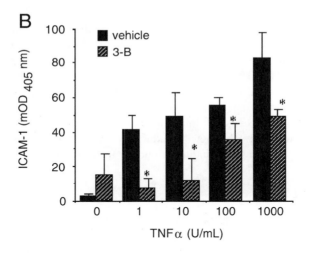

FIGURE 3.5 Pharmacological inhibition of PARP reduces expression of P-selectin (A) and ICAM-1 (B) in human umbilical vein endothelial cells. Expression of P-selectin was induced by 1-h incubation of cells with peroxynitrite (0.1 to 100 μmol/l); expression of ICAM-1 was induced by 4-h incubation with TNF-α (1 to 1000 U/ml). $*p < 0.05$ represents significant effect of 3-aminobenzamide (3-AB, 1 mmol/l) on expression of the adhesion molecules, as determined by a cell surface enzyme immunoassay. Data are expressed as means ± SEM of $n = 6$ to 12 wells of at least 3 separate experiments. mOD_{405} = milliunits of optical density recorded spectrophotometrically at 405 nm. (From Zingarelli, B. et al., *Circ. Res.,* 83, 85, 1998. With permission.)

Potential mechanisms have been postulated to explain the reduction of neutrophil influx after PARP inhibition. PARP-related changes in cellular energetics[3,8,9] and related processes involving calcium sequestration, biosynthetic processes, and maintenance of the normal cell shape and adherence may be involved. Alternatively, or in addition to the energetic changes, poly(ADP-ribosyl)ation may directly play a role in gene expression. In this regard, it has been demonstrated that in *in vitro* human endothelial cells inhibition of PARP activity by 3-aminobenzamide blocked oxidant-induced binding activity of the transcription factor activator protein-1 to the promoter of ICAM-1.[59]

The suppression of neutrophil infiltration and/or the direct inhibition of gene expression may also explain the beneficial effects provided by the absence of functional PARP against the delayed inflammatory response in the late stage of reperfusion. In a murine model of late reperfusion injury (24 h) after a short period of ischemia (30 min), genetic inhibition of PARP exerted cardioprotective effects associated with reduced release of cytokines, such as TNF-α and IL-10, and reduced activity of the inducible NO synthase.[60] These findings are in line with other experimental observations, which propose PARP as a pleiotropic modulator of the gene expression. There is strong evidence, in fact, that poly(ADP-ribosyl)ation is a process of post-translational modification of chromatin proteins in the cell nucleus.[61] It is suggested that it leads to relaxation of chromatin with the consequence that genes become more accessible to the RNA polymerase.[3,62] In this regard, in synovial fibroblasts 3-aminobenzamide, a specific inhibitor of PARS, inhibits the expression of collagenase by IL-1β.[63] Inhibition of PARS activity blocks oxidant-induced transcription of c-fos,[64] inhibits induction of major histocompatibility class II,[65] downregulates DNA-methyltransferase, protein kinase C, and inhibits expression of the *ras* gene in endothelial cells.[66,67] Inhibition of PARP activity also blocks endotoxin-induced expression of mRNA for the inducible NO synthase in immunostimulated macrophages by reducing the binding of nuclear factor κB to its target sequence.[68,69]

3.6 CONCLUSION

The analysis of the current experimental data suggests the existence of a self-amplifying vicious cycle in myocardial ischemia and reperfusion governed by the nuclear enzyme PARP (see representative scheme in Figure 3.6). Early production of oxidants by dysfunctional mitochondria after reperfusion leads to DNA damage and activation of PARP, which in turn causes further derangement of cellular energetic status and induces endothelial injury and activation of adhesion molecules. The loss of the endothelial barrier function is then responsible for the infiltration of neutrophils, which represent a source of more oxidant production. In this positive feedback cycle, reactive species maintain the futile energy-consuming repair cycle by inducing further DNA injury and activation of PARP. The energy dysmetabolism and/or the poly(ADP-ribosyl)ation of important nuclear proteins appear also to affect the post-translation and transcriptional regulatory events of the inflammatory response. The current research has raised the exciting prospect that pharmacological inhibition of PARP may ameliorate the endothelial and myocardial dysfunction by interrupting the vicious cycle at the level of energetic failure, neutrophil infiltration,

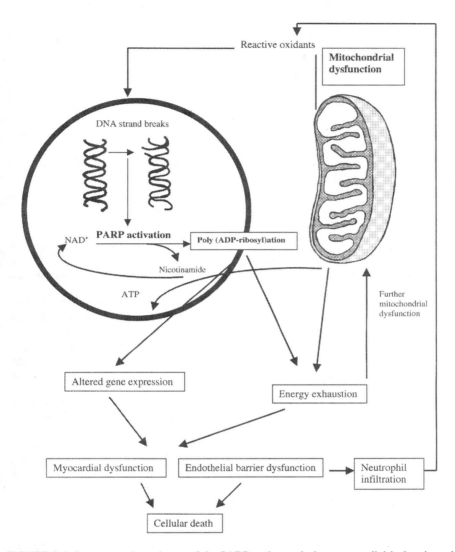

FIGURE 3.6 Representative scheme of the PARP pathway during myocardial ischemia and reperfusion injury. The reduction of oxygen supply during ischemia alters the mitochondrial function, leading to the production of reactive oxidant species. The highly reactive species induce breakage of DNA strands with consequent initiation of poly(ADP-ribosyl)ation by PARP. The activation of PARP rapidly depletes the cellular NAD$^+$ and ATP pools. The cellular energy exhaustion maintains the mitochondria in a reduced state, thereby allowing a further production of reactive oxidants at the reperfusion. Furthermore, depletion of NAD$^+$ and ATP leads to cellular dysfunction, i.e., alteration of the mechanism of relaxation and contraction in the myocytes and alteration of the endothelial adhesiveness. The loss of the endothelial barrier function results in accumulation of neutrophils into the site of injury, which represent an important source of reactive species. The cellular dysfunction is further enhanced by the concomitant poly(ADP-ribosyl)ation of nuclear chromatin, leading to altered gene expression of proinflammatory and adhesion molecules, and culminates in cell death.

and related oxidative damage. It must be noted, however, that the predictability of *in vivo* experimentation in animals and its extrapolation to humans is a function of the genetic kinship. Therefore, additional questions, whether PARP activation has a role in the immunoregulatory and inflammatory process in humans and whether PARP inhibitors may significantly reduce the incidence of fatal and nonfatal myocardial infarction, remain to be addressed.

ACKNOWLEDGMENT

The author thanks Dr. Valina L. Dawson and Dr. Csaba Szabó for their editorial assistance in this manuscript.

REFERENCES

1. Weisfeldt, M. L., Reperfusion and reperfusion injury, *Clin. Res.*, 35, 13, 1987.
2. Ferrari, R., Agnoletti, L., Comini, L., Gaia, G., Bachetti, T., Cargnoni, A., Ceconi, C., Curello, S., and Visioli, O., Oxidative stress during myocardial ischaemia and heart failure, *Eur. Heart J.*, 19, B2, 1998.
3. Ueda, K. and Hayaishi, O., ADP-ribosylation, *Annu. Rev. Biochem.*, 54, 73, 1985.
4. Szabó, C. and Dawson, V. L., Role of poly(ADP-ribose) synthetase in inflammation and ischaemia-reperfusion, *Trends Pharmacol. Sci.*, 19, 287, 1998.
5. Zingarelli, B., O'Connor, M., Wong, H., Salzman, A. L., and Szabó, C., Peroxynitrite-mediated DNA strand breakage activates poly-ADP ribosyl synthetase and causes cellular energy depletion in macrophages stimulated with bacterial lipopolysaccharide, *J. Immunol.*, 156, 350, 1996.
6. Ménissier-de Murcia, J., Niedergang, C., Trucco, C., Ricoul, M., Dutrillaux, B., Mark, M., Oliver, F.J., Masson, M., Dierich, A., LeMeur, M., Walztinger, C., Chambon, P., and de Murcia, G., Requirement of poly(ADP-ribose) polymerase in recovery from DNA damage in mice and in cells, *Proc. Natl. Acad. Sci. U.S.A.*, 94, 7303, 1997.
7. D'Silva, I., Pelletier, J. D., Lagueux, J., D'Amours, D., Chaudhry, M. A., Weinfeld, M., Lees-Miller, S. P., and Poirier, G. G., Relative affinities of poly(ADP-ribose) polymerase and DNA-dependent protein kinase for DNA strand interruptions, *Biochim. Biophys. Acta*, 1430, 119, 1999.
8. Berger, N. A., Oxidant-induced cytotoxicity: a challenge for metabolic modulation, *Am. J. Respir. Cell. Mol. Biol.*, 4, 1, 1991.
9. Cochrane, C. G., Mechanisms of oxidant injury of cells, *Mol. Aspects Med.*, 12, 137, 1991.
10. Zhang, J., Dawson, V. L., Dawson, T. M., and Snyder, S. H., Nitric oxide activation of poly(ADP-ribose) synthetase in neurotoxicity, *Science*, 263, 687, 1994.
11. Radons, J., Heller, B., Burkle, A., Hartmann, B., and Rodriguez, M. L., NO toxicity in islet cells involves polyADP-ribose-polymerase activation and concomitant NAD depletion, *Biochem. Biophys. Res. Commun.*, 199, 1270, 1994.
12. Pieper, A. A., Verma, A., Zhang, J., and Snyder, S. H., Poly(ADP-ribose) polymerase, nitric oxide and cell death, *Trends Pharmacol. Sci.*, 20, 171, 1999.
13. Heller, B., Wang, Z. Q., Wagner, E. F., Radons, J., Burkle, A., Fehsel, K., Burkart, V., and Kolb, H., Inactivation of the poly(ADP-ribose) polymerase gene affects oxygen radical and nitric oxide toxicity in islet cells, *J. Biol. Chem.*, 270, 11176, 1995.

14. Szabó, C., Virag, L., Cuzzocrea, S., Scott, G. S., Hake, P., O'Connor, M. P., Zingarelli, B., Salzman, A., and Kun, E., Protection against peroxynitrite-induced fibroblast injury and arthritis development by inhibition of poly(ADP-ribose) synthase, *Proc. Natl. Acad. Sci. U.S.A.*, 95, 3867, 1998.
15. Eliasson, M. J. L., Sampei, K., Mandir, A. S., Hurn, P. D., Traystman, R. J., Bao, J., Pieper, A., Wang, Z.-Q., Dawson, T. M., Snyder, S. H., and Dawson, V. L., Poly(ADP-ribose) polymerase gene disruption renders mice resistant to cerebral ischemia, *Nat. Med.*, 3, 1090, 1997.
16. Mandir, A. S., Przedborski, S., Jackson-Lewis, V., Wang, Z. Q., Simbulan-Rosenthal, C. M., Smulson, M. E., Hoffman, B. E., Guastella, D. B., Dawson, V. L., and Dawson, T. M., Poly(ADP-ribose) polymerase activation mediates 1-methyl-4-phenyl-1,2,3,6-tetrahydropyridine (MPTP)-induced parkinsonism, *Proc. Natl. Acad. Sci. U.S.A.*, 96, 5774, 1999.
17. Burkart, V., Wang, Z. Q., Radons, J., Heller, B., Herceg, Z., Stingl, L., Wagner, E. F., and Kolb, H., Mice lacking the poly(ADP-ribose) polymerase gene are resistant to pancreatic beta-cell destruction and diabetes development induced by streptozocin, *Nat. Med.*, 5, 314, 1999.
18. Zingarelli, B., Salzman, A. L., and Szabó, C., Protective effects of nicotinamide against nitric oxide-mediated vascular failure in endotoxic shock: potential involvement of polyADP ribosyl synthetase, *Shock*, 5, 258, 1996.
19. Zingarelli, B., Cuzzocrea, S., Zsengeller, Z., Salzman, A. L., and Szabó, C., Inhibition of poly(ADP ribose) synthetase protects against myocardial ischemia and reperfusion injury, *Cardiovasc Res.*, 36, 205, 1997.
20. Zingarelli, B., Salzman, A. L., and Szabó, C., Genetic disruption of poly(ADP-ribose) synthetase inhibits the expression of P-selectin and intercellular adhesion molecule-1 in myocardial ischemia-reperfusion injury, *Circ. Res.*, 83, 85, 1998.
21. Zingarelli, B., Szabó, C., and Salzman, A. L., Blockade of poly(ADP-ribose) synthetase inhibits neutrophil recruitment, oxidant generation, and mucosal injury in murine colitis, *Gastroenterology*, 116, 335, 1999.
22. Szabó, C., Zingarelli, B., and Salzman, A. L., Role of poly-ADP ribosyltransferase activation in the nitric oxide- and peroxynitrite-induced vascular failure, *Circ. Res.*, 78, 1051, 1996.
23. Szabó, C., Lim, L. H. K., Cuzzocrea, S., Getting, S. J., Zingarelli, B., Flower, R. J., Salzman, A. L., and Perretti, M., Inhibition of poly(ADP-ribose) synthetase exerts anti-inflammatory effects and inhibits neutrophil recruitment, *J. Exp. Med.*, 186, 1041, 1997.
24. Cuzzocrea, S., Zingarelli, B., Costantino, G., Szabó, A., Salzman, A. L., Caputi, A. P., and Szabó, C., Beneficial effect of 3-aminobenzamide, an inhibitor of poly (ADP-ribose) synthetase in a rat model of splanchnic artery occlusion and reperfusion, *Br. J. Pharmacol.*, 121, 1065, 1997.
25. Cuzzocrea, S., Caputi, A. P., and Zingarelli, B., Peroxynitrite-mediated DNA strand breakage activates poly(ADP ribose) synthetase and causes cellular energy depletion in carrageenan-induced pleurisy, *Immunology*, 93, 96, 1998.
26. Ferrari, F., Oxygen-free radicals at myocardial level: effects of ischaemia and reperfusion, *Adv. Exp. Med. Biol.*, 366, 99, 1994.
27. Wang, P. and Zweier, J. L., Measurement of nitric oxide and peroxynitrite generation in the postischemic heart, *J. Biol. Chem.*, 271, 29223, 1996.
28. Liu, P., Hock, C. E., Nagele, R., and Wong, P. Y., Formation of nitric oxide, superoxide, and peroxynitrite in myocardial ischemia-reperfusion injury in rats, *Am. J. Physiol.*, 272, H2327, 1997.

29. Beckman, J. S., Oxidative damage and tyrosine nitration from peroxynitrite, *Chem. Res. Toxicol.*, 9, 836, 1996.
30. Eiserich, J. P., Cross, C. E., Jones, A. D., Halliwell, B., and Van der Vliet, A., Formation of nitrating and chlorinating species by reaction of nitrite with hypochlorous acid. A novel mechanism for nitric oxide-mediated protein modification, *J. Biol. Chem.*, 271, 19199, 1996.
31. Eiserich, J. P., Hristova, M., Cross, C. E., Jones, D. A., Freeman, B. A., Halliwell, B., and Van der Vliet, A., Formation of nitric oxide derivatives catalysed by myeloperoxidase in neutrophils, *Nature*, 391, 393, 1998.
32. Halliwell, B., What nitrates tyrosine? Is nitrotyrosine specific as a biomarker of peroxynitrite formation *in vivo*? *FEBS Lett.*, 411, 157, 1997.
33. Maulik, G., Cordis, G. A., and Das, D. K., Oxidative damage to myocardial proteins and DNA during ischemia and reperfusion, *Ann. N.Y. Acad. Sci.*, 30, 793, 1996.
34. Cordis, G. A., Maulik, G., Bagchi, D., Riedel, W., and Das, D. K., Detection of oxidative DNA damage to ischemic reperfused rat hearts by 8-hydroxydeoxyguanosine formation, *J. Mol. Cell. Cardiol.*, 30, 1939, 1998.
35. Walles, T., Pieper, A. A., Li, J-H., Zhang, J., Snyder, S. H., and Zweier, J. L., Demonstration that poly(ADP-ribose) accumulation occurs in the postischemic heart and is associated with myocardial necrosis, *Circulation*, 98, I-260 (Abstr. 1354), 1998.
36. Thiemermann, C., Bowes, J., Myint, F. P., and Vane, J. R., Inhibition of the activity of poly(ADP ribose) synthetase reduces ischemia-reperfusion injury in the heart and skeletal muscle, *Proc. Natl. Acad. Sci. U.S.A.*, 94, 679, 1997.
37. Bowes, J., Ruetten, H., Martorana, P. A., Stockhausen, H., and Thiemermann, C., Reduction of myocardial reperfusion injury by an inhibitor of poly (ADP-ribose) synthetase in the pig, *Eur. J. Pharmacol.*, 359, 143, 1998.
38. Docherty, J. C., Kuzio, B., Silvester, J. A., Bowes, J., and Thiemermann, C., An inhibitor of poly(ADP-ribose) synthetase activity reduces contractile dysfunction and preserves high energy phosphate levels during reperfusion of the ischaemic rat heart, *Br. J. Pharmacol.*, 127, 1518, 1999.
39. Grupp, I. L., Jackson, T. M., Hake, P., Grupp, G., and Szabó, C., Protection against hypoxia-reoxygenation in the absence of poly(ADP-ribose) synthetase in isolated working hearts, *J. Mol. Cell. Cardiol.*, 31, 297, 1999.
40. Walles, T., Wang, P., Pieper, A., Snyder, S. H., and Zweier, J. L., Mice lacking poly(ADP-ribose) polymerase gene show attenuated cellular energy depletion and improved recovery of myocardial function following global ischemia, *Circulation*, 98, I-260 (Abstr. 1355), 1998.
41. Stern, Y., Salzman, A. L., Cotton, R. T., and Zingarelli, B., Protective effect of 3-aminobenzamide, an inhibitor of poly(ADP-ribose) synthetase, against laryngeal injury in rats, *Am. J. Respir. Crit. Care Med.*, 160, 1743, 1999.
42. Lam, T. T., The effect of 3-aminobenzamide, an inhibitor of poly-ADP-ribose polymerase, on ischemia/reperfusion damage in rat retina, *Res. Commun. Mol. Pathol. Pharmacol.*, 95, 241, 1997.
43. Takahashi, K., Pieper, A. A., Croul, S. E., Zhang, J., Snyder, S. H., and Greenberg, J. H., Post-treatment with an inhibitor of poly(ADP-ribose) polymerase attenuates cerebral damage in focal ischemia, *Brain Res.*, 829, 46, 1999.
44. Endres, M., Scott, G. S., Salzman, A. L., Kun, E., Moskowitz, M. A., and Szabó, C., Protective effects of 5-iodo-6-amino-1,2-benzopyrone, an inhibitor of poly(ADP-ribose) synthetase against peroxynitrite-induced glial damage and stroke development, *Eur. J. Pharmacol.*, 351, 377, 1998.

45. Love, S., Barber, R., and Wilcock, G. K., Neuronal accumulation of poly(ADP-ribose) after brain ischaemia, *Neuropathol. Appl. Neurobiol.*, 25, 98, 1999.
46. Love, S., Barber, R., and Wilcock, G. K., Increased poly(ADP-ribosyl)ation of nuclear proteins in Alzheimer's disease, *Brain*, 122, 247, 1999.
47. Page, E. and McCallister, L. P., Quantitative electron microscopic description of heart muscle cells. Application to normal, hypertrophied and thyroxin-stimulated hearts, *Am. J. Cardiol.*, 31, 172, 1973.
48. Ferrari R., The role of mitochondria in ischemic heart disease, *J. Cardiovasc. Pharmacol.*, 28, S1, 1996.
49. Kroemer, G., Zamzamin, N., and Susin, S. A., Mitochondrial control of apoptosis, *Immunol. Today*, 18, 44, 1997.
50. Samaja, M., Motterlini, R., Santoro, F., Dell'Antonio, G., and Corno, A., Oxidative injury in reoxygenated and reperfused hearts, *Free Radical Biol. Med.*, 16, 255, 1994.
51. Taegtmeyer, H., King, L. M., and Jones, B. E., Energy substrate metabolism and myocardial ischemia, and targets for pharmacotherapy, *Am. J. Cardiol.*, 82, 54K, 1998.
52. Virág, L., Salzman, A. L., and Szabó, C., Poly(ADP-ribose) synthetase activation mediates mitochondrial injury during oxidant-induced cell death, *J. Immunol.*, 161, 3753, 1998.
53. Gilad, E., Zingarelli, B., Salzman, A. L., and Szabó, C., Protection by inhibition of poly(ADP-ribose) synthetase against oxidant injury in cardiac myoblasts *in vitro*, *J. Mol. Cell. Cardiol.*, 29, 2585, 1997.
54. Bowes, J., Piper, J., and Thiemermann, C., Inhibitors of the activity of poly(ADP-ribose) synthetase reduce the cell death caused by hydrogen peroxide in human cardiac myoblasts, *Br. J. Pharmacol.*, 124, 1760, 1998.
55. Szabados, E., Fischer, G. M., Toth, K., Csete, B., Nemeti, B., Trombitas, K., Habon, T., Endrei, D., and Sumegi, B., Role of reactive oxygen species and poly-ADP-ribose polymerase in the development of AZT-induced cardiomyopathy in rat, *Free Radical Biol. Med.*, 26, 309, 1999.
56. Frangogiannis, N. G., Youker, K. A., and Entman, M. L., The role of the neutrophil in myocardial ischemia and reperfusion, *EXS*, 76, 263, 1996.
57. Hiromatsu, Y., Sato, M., Tanaka, K., Ishisaka, N., Kamachi, J., and Nonaka, K., Inhibitory effects of nicotinamide on intercellular adhesion molecule-1 expression on culture human thyroid cells, *Immunology*, 80, 330, 1993.
58. Meyer, T., Lengyel, H., Fanick, W., and Hilz, H., 3-Aminobenzamide inhibits cytotoxicity and adhesion of phorbol-ester-stimulated granulocytes to fibroblasts monolayer cultures, *Eur. J. Biochem.*, 197, 127, 1991.
59. Roebuck, K. A., Rahman, A., Lakshminarayanan, V., Janakidevi, K., and Malik, A. B., H_2O_2 and tumor necrosis factor-α activate intercellular adhesion molecule 1 (ICAM-1) gene transcription through distinct *cis*-regulatory elements within the ICAM-1 promoter, *J. Biol. Chem.*, 270, 18966, 1995.
60. Yang, Z., Zingarelli, B., and Szabó, C., Role of poly(ADP-ribose) synthetase in delayed myocardial ischemia-reperfusion injury, *Shock*, 13, 60–65, 2000.
61. Oliver, F. J., Ménissier-de Murcia, J., and de Murcia, G., Poly(ADP-ribose) polymerase in the cellular response to DNA damage, apoptosis, and disease, *Am. J. Hum. Genet.*, 64, 1282, 1999.
62. Althaus, F. R., Hofferer, L., Kleczkoska, H. E., Malanga, M., Naegeli, H., Panzeter, P. L., and Realini, C. A., Histone shuttling by poly ADP-ribosylation, *Mol. Cell. Biochem.*, 138, 53, 1994.

63. Ehrlich, W., Huser, H., and Kroger, H., Inhibition of the induction of collagenase by interleukin-1β in cultured rabbit synovial fibroblasts after treatment with the poly(ADP-ribose)-polymerase inhibitor 3-aminobenzamide, *Rheumatol. Int.*, 15, 171, 1995.
64. Amstad, P. A., Krupitza, G., and Cerutti, P. A., Mechanism of c-fos induction by active oxygen, *Cancer Res.*, 52, 3952, 1992.
65. Qu, Z., Fujimoto, S., and Taniguchi, T., Enhancement of interferon-gamma induced major histocompatibility complex class II gene expression by expressing an antisense RNA of poly(ADP-ribose) synthetase, *J. Biol. Chem.*, 269, 5543, 1994.
66. Bauer, P. I., Kirsten, E., Young, L. J. T., Varadi, G., Csonka, E., Buki, K. G., Mikala, G., Hu, R., Comstock, J. A., Mendeleyev, J., Hakam, A., and Kun, E., Modification of growth-related enzymatic pathways and apparent loss of tumorigenicity of a ras-transformed bovine endothelial cell line by treatment with 5-iodo-6-amino-1,2-benzopyrone, *Int. J. Oncol.*, 8, 239, 1995.
67. Bauer, P. I., Kirsten, E., Varadi, G., Young, L. J., Hakam, A., Comstock, J. A., and Kun, E., Reversion of malignant phenotype by 5-iodo-6-amino-1,2-benzopyrone a non-covalently binding ligand of poly(ADP-ribose) polymerase, *Biochimie*, 77, 374, 1995.
68. Hauschildt, S., Scheipers, P., Bessler, W., Schwarz, K., Ullmer, A., Flad, H. D., and Heine, H., Role of ADP-ribosylation in activated monocytes/macrophages, *Adv. Exp. Med. Biol.*, 419, 249, 1997.
69. Le Page, C., Sanceau, J., Drapier, J. C., and Wietzerbin, J., Inhibitors of ADP-ribosylation impair inducible nitric oxide synthase gene transcription through inhibition of NF-κB activation, *Biochem. Biophys. Res. Commun.*, 243, 451, 1998.

4 Role of Poly(ADP-Ribose) Polymerase Activation in Inflammation

Andrew L. Salzman

CONTENTS

4.1 Introduction 61
4.2 Stimulants of PARP Activation 61
    4.2.1 Role of Nitric Oxide 61
    4.2.2 Role of Superoxide Anion 62
    4.2.3 Role of Peroxynitrite 63
4.3 DNA Single-Strand Breakage and PARP Activation 63
4.4 Regulation of Proinflammatory Signaling Pathways by PARP 64
4.5 Role of PARP Activation in Local Inflammatory Responses 65
4.6 Development of a Working Hypothesis Relating PARP and Inflammation 72
Acknowledgment 73
References 73

4.1 INTRODUCTION

Poly(ADP-ribose) polymerase (PARP) is an abundant nuclear enzyme present throughout the phylogenetic spectrum.[1] Although traditionally regarded as a DNA-repair enzyme, the precise physiological roles of PARP remain undefined.[2] At low basal levels, PARP appears to play diverse roles, participating in DNA repair,[3,4] chromatin relaxation,[5] cell differentiation,[6] DNA replication,[7] transcriptional regulation,[8] control of cell cycle,[9] p53 expression and apoptosis,[10] and transformation.[11] Recent data support the novel concept that under pathological conditions, PARP plays a crucial role in the regulation and generation of the inflammatory response. This chapter summarizes the evidence favoring this new role for PARP and proposes therapeutic opportunities afforded by PARP inhibition.

4.2 STIMULANTS OF PARP ACTIVATION

4.2.1 ROLE OF NITRIC OXIDE

Under pathological conditions of inflammation, free radicals and oxidants are formed by parenchymal cells and infiltrating leukocytes. Two principal free radical species, nitrogen centered and oxygen centered, play a key role as effectors of tissue injury. In addition, oxidants are generated by the conversion of oxygen-centered free radicals to hydrogen peroxide and hydroxyl radical and by the interaction of superoxide anion and nitric oxide (NO) to form peroxynitrite. All of these species are formed during inflammation and play varied roles in the activation of PARP and the mediation of tissue damage.

The nitrogen-centered free radical NO is synthesized from the guanidino group of L-arginine by a family of enzymes termed NO synthases (NOS). Three isoforms have been described and cloned: endothelial cell NOS (ecNOS), brain NOS (bNOS), and an inducible macrophage type NOS (iNOS).[12] Whereas ecNOS (produced by the endothelial cells) is responsible for many physiological effects (e.g., maintaining basal vasodilator tone, inhibiting platelet and neutrophil adhesion and activation), many pathophysiological conditions are associated with the production of *large amounts* of NO, produced by the inducible isoform of NOS (iNOS), with consequent cytotoxic effects.[13,14] iNOS, which generates NO for longer periods of time (several hours to days) and at rates several orders of magnitude greater than the constitutive isoforms,[15] is not present in most tissues to any significant extent under resting conditions, but the enzyme is newly synthesized in response to infection and proinflammatory cytokines.[16-20] Upregulation of NO synthesis has been associated with functional derangements in multiple cell types: iNOS induction produces injury in endothelial cells,[21] induces apoptosis in macrophages,[22] inhibits cellular respiration in vascular smooth muscle cells,[23] and has been associated with apoptosis in necrotizing enterocolitis.[24]

4.2.2 Role of Superoxide Anion

In addition to the nitrogen-centered free radical NO, the oxygen-centered free radical superoxide anion appears to play a major role in the pathogenesis of inflammatory injury. Superoxide anion is generated by xanthine oxidase (XO) and NADPH oxidase from the partial reduction of molecular oxygen. Neutrophils and macrophages are known to produce superoxide free radicals and hydrogen peroxide, which normally are involved in the killing of ingested or invading microbes.[25] Under physiological conditions XO is ubiquitously present in the form of a dehydrogenase (XDH). In mammals, XDH is converted from the NAD-dependent dehydrogenase form to the oxygen-dependent oxidase form, either by reversible sulfhydryl oxidation or irreversible proteolytic modification.[26] XO then no longer uses NAD^+ as an electron acceptor, but transfers electrons onto oxygen, generating superoxide radical, peroxide, and hydroxyl radical as purines are degraded to uric acid.[27,28] Inflammatory activation converts XDH to XO, mainly by oxidizing structurally important sulfhydryls. Inflammation also markedly upregulates the conversion of xanthine dehydrogenase.[29]

The level of XO increases markedly during various types of inflammation (>400-fold in bronchoalveolar fluid in pneumonitis), ischemia–reperfusion injury, and atherosclerosis.[30,31] Plasma levels of XO, due to the spillover of tissue XO into the circulation, may be detected in adult respiratory distress syndrome, ischemia–reperfusion injury, arthritis, sepsis, hemorrhagic shock, and other inflammatory conditions.[26,32,33] Inflammatory-induced histamine release by mast cells and basophils lowers the K_m values for XO substrates, thereby enhancing the activity of XO.[34]

Superoxide radical has been associated with tissue injury in a variety of states in which XDH to XO conversion occurs. Oxygen radicals are involved in the microvascular and parenchymal cell injury associated with organ transplant rejection, circulatory shock, stroke, myocardial infarction, zymosan-induced systemic inflammation, intestinal ischemia, and autoimmune disease.[27,35,36] Extremely elevated levels of XO have been recorded in adult respiratory distress syndrome, influenza, and other inflammatory diseases.[25,30] Superoxide is also readily catalyzed to form the oxidants hydrogen peroxide and hydroxyl radical, which mediate cellular injury to a greater extent than superoxide anion per se. In addition, superoxide readily reacts in a diffusion-limited manner with the nitrogen-centered free radical NO, producing peroxynitrite. Although all oxidants have been implicated in DNA strand breakage, and subsequent activation of PARP, the role of peroxynitrite has received particular attention in various inflammatory processes.

4.2.3 ROLE OF PEROXYNITRITE

In vitro and *in vivo* data demonstrate that NO and superoxide anion may not be the relevant final effector species in a variety of forms of inflammatory injury.[22,23,37-41] Peroxynitrite, formed from their interaction, appears rather to be the proximate cause of cytotoxicity and tissue injury.[41] Peroxynitrite is formed in inflammatory conditions such as inflammatory bowel disease, as well as other inflammatory disorders, as evidenced by the formation of nitrotyrosine, a product of the reaction of peroxynitrite with tyrosine residues.[42] Peroxynitrite has been shown to have a wide variety of toxic actions, including damage to membranes, oxidation of intracellular proteins, and nitrosation of critical tyrosine residues. Both exogenous and endogenously generated peroxynitrite have also been shown to induce DNA strand breaks potently in a variety of cell types, including macrophages,[22,23] vascular smooth muscle cells,[41] pulmonary epithelial cells,[43] and intestinal epithelial cells.[44] In contrast to the potent effects of peroxynitrite on the production of DNA single-strand breaks, its precursor NO has no such effects.[22,23,37] The existence of DNA single-strand breaks has now been confirmed in experimental models of inflammation, and in the clinical setting. In colonic biopsies of patients with ulcerative colitis, for example, there are significantly increased levels of 8-hydroxyguanine, 2-hydroxyadenine, 8-hydroxyadenine, and 2,6-diamino-5-formamido-pyrimidine,[45,46] providing indirect evidence of DNA injury.

4.3 DNA SINGLE-STRAND BREAKAGE AND PARP ACTIVATION

Under basal conditions *in vivo*, PARP activity is relatively quiescent. Under conditions of redox stress, the N-terminal domain of PARP recognizes oxidant-induced DNA single-strand breaks, resulting in a conformational change that activates its C-terminal catalytic domain.[47] Activated PARP cleaves the substrate NAD$^+$ into ADP-ribose and nicotinamide. During this process, PARP covalently attaches ADP-ribose to various proteins, including an automodification domain of PARP itself, and then extends the initial ADP-ribose group into a branched nucleic acidlike homopolymer, poly(ADP-ribose).[48] Simultaneously, poly(ADP-ribose) is degraded by various nuclear enzymes, including poly(ADP-ribose) glycohydrolase

The catalysis of poly(ADP-ribose) results in depletion of NAD, inhibition of glycolysis and mitochondrial respiration, and the ultimate reduction of intracellular high-energy phosphates.[22,38,39,41,49-58] The coincident activities of PARP and poly(ADP-ribose) glycohydrolase may be regarded functionally as an NADase. Loss of intracellular energetic stores has deleterious consequences on a broad range of cellular activities, which may be substantially delayed (hours) from the time of oxidant exposure.

This PARP-dependent loss of intracellular energetics and cellular function has been well demonstrated in studies of multiple cell types. Recently, we investigated the role PARP in intestinal epithelial barrier function, an active process that is highly dependent upon cellular ATP concentration.[59] Exposure of human Caco-2BBe enterocyte cell monolayers to peroxynitrite rapidly induced DNA strand breaks and triggered an energy-consuming pathway catalyzed by PARP.[44] The consequent reduction of cellular stores of ATP and NAD^+ was associated with the development of hyperpermeability of the epithelial monolayer to a fluorescent anionic tracer. Pharmacological inhibition of PARP activity had no effect on the development of peroxynitrite-induced DNA single-strand breaks, but attenuated the decrease in intracellular stores of NAD and ATP and the functional loss of intestinal barrier function. Similar studies, in which peroxynitrite has been generated by the endogenous production of superoxide anion and NO, have implicated PARP activation as a major pathway of oxidant-induced cellular necrosis.

4.4 REGULATION OF PROINFLAMMATORY SIGNALING PATHWAYS BY PARP

Initial studies of the role of PARP in experimental models of reperfusion injury focused on its effects on intracellular energetics and resultant cellular dysfunction. In the last several years, however, a new series of *in vivo* investigations has revealed that inhibition of PARP activation, or its genetic deficiency, has unexpected actions in oxidant-driven diseases that are *not* associated with reperfusion injury. These effects, which are demonstrated in a broad range of inflammatory conditions, appear to be incidental to the effect of PARP activation on intracellular energetics.

PARP activity directly influences inflammation by its effect on the expression, activation, and nuclear translocation of key proinflammatory genes and proteins. The absence of PARP or its pharmacological inhibition has been shown to suppress the activation of MAP kinase,[52] AP-1 complex,[60] and NF-κB.[60] Consequently, PARP inhibition interferes with the expression of proinflammatory genes, such as iNOS[62] and ICAM-1,[60] that are dependent upon these signaling pathways. These observations have been revealed in both *in vitro* systems, examining a broad range of cell types, and in various experimental models of inflammation.

In an experimental model of enterocolitis, PARP knockout mice have nearly total suppression of inflammatory-induced upregulation of ICAM-1 expression. Similarly, PARP inhibition blocks ICAM-1 expression in cultured endothelial cells stimulated *in vitro* by a combination of proinflammatory cytokines.[60] We have also observed that PARP inhibition attenuates IL-1β mediated induction of C3 in cultured intestinal enterocytes. Because C3 is an early component of the complement pathway that generates the potent chemotaxin C5a, PARP inhibition may reduce neutrophil recruitment into inflammatory foci. Experimental evidence indeed demonstrates that PARP inhibition suppresses neutrophil infiltration in experimental models of endotoxicosis, carrageenan-induced pleurisy, and splanchnic ischemia–reperfusion injury.[49,51,53] These findings have also been replicated in models of inflammation utilizing a genetic approach to eliminate PARP activity.[50]

The underlying mechanism by which PARP activation alters gene expression is unknown, but may involve the poly(ADP-ribosyl)ation of transcription factors or the repair of DNA strand breaks which interfere with transcription. PARP may also alter the activation of proinflammatory pathways via its influence on the expression of AP-1, a heterodimer composed of *c-fos* and *jun* factors. High levels of transcriptional activation of human ICAM-1, C3, and *c-fos* require AP-1 binding to 5′ flanking regulatory regions.[60] In cultured cells, PARP inhibition blocks oxidant-induced *c-fos* mRNA expression and AP-1 activation.[60] Since the *c-fos* promoter contains an AP-1 consensus site,[8] *c-fos* activation could trigger a positive-feedback cycle of gene expression. Superoxide anion has also been reported to induce the post-translational poly(ADP-ribosyl)ation of *c-fos*.[8]

In addition to its effect on transcription factor activation, PARP inhibition attenuates IL-1β-mediated iNOS expression in rodent pancreatic beta islet cells and in activated macrophases.[62-65] Similar anti-inflammatory effects have been noted *in vitro* in murine macrophages and *in vivo* in rats. PARP inhibition has been shown, for example, to suppress endotoxin-induced expression of TNF-α, IL-6, iNOS, and cyclooxygenase-2[52] and to elevate the expression of the anti-inflammatory cytokine IL-10.[52] These effects are associated with the near total blockade of endotoxin-mediated induction of MAP kinase activity.[52] Since MAP kinase plays a major role in the pleiotropic transduction of intracellular inflammatory cascades, the anti-inflammatory effects of PARP inhibition may be accounted for at this level of gene regulation. One may also expect that PARP-dependent regulation of NF-κB activation[66,67] has a pleiotropic effect on the expression of proinflammatory genes, given the broad role that NF-κB plays in the transcriptional activation of cytokine and chemokine genes.

4.5 ROLE OF PARP ACTIVATION IN LOCAL INFLAMMATORY RESPONSES

The effect of PARP inhibition, or PARP deficiency, on the expression of proinflammatory genes and signaling pathways *in vitro* led us to explore the potential role of PARP activation on *in vivo* inflammation. These studies have included a broad range of models, ranging from carrageenan-induced paw edema, colitis, arthritis, and endotoxic shock. Pharmacological studies, using diverse inhibitors, have consistently

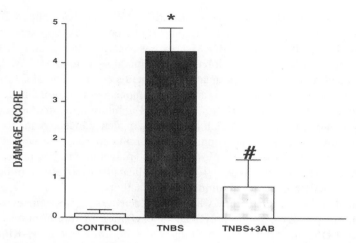

FIGURE 4.1 Effect of 3-AB treatment on the damage score in a rodent model of TNBS-induced colitis. Drug treatment (10 mg/kg twice a day, intravenously) was started 24 h before intracolonic administration of TNBS and continued until the animals were sacrificed. Colonic damage was scored on a 0 (normal) to 10 (severe) scale by four independent observers. Results are mean ± SE of 8 rats for each group. *Significantly different from control rats ($p < 0.05$). #Significantly different from untreated rats subjected to TNBS colitis ($p < 0.05$).

supported data from murine gene deletion models, providing confirmatory support for the role of PARP in the inflammatory process. Our data clearly demonstrate that PARP activation is a critical regulatory step in these experimental models of inflammatory injury.

To examine the role of PARP in a local prototypic inflammatory condition, we pretreated male Wistar rats (250 to 300 g) with the moderately potent PARP inhibitor INH_2BP (5-iodo-6-amino-1,2-benzopyrone) before the injection of carrageenan. This model is characterized by a rapid influx of neutrophils and the formation of edema. Subplantar injection of carrageenan into the rat paw led to a time-dependent increase in paw volume with a maximal response at 3 h. The carrageenan-induced paw edema was significantly reduced by treatment with INH_2BP by 45 ± 3, 48 ± 7, 53 ± 9, and 44 ± 8% at 1, 2, 3, and 4 h, respectively ($p < 0.01$; $n = 6$).

To confirm that PARP plays a relevant role in intestinal inflammatory injury and dysfunction we evaluated the biological effect of *in vivo* pharmacological inhibition of PARP activity in a rat model of trinitrobenzenesulfonic acid (TNBS)-induced colitis, a classic model of hapten-induced autoimmune inflammatory bowel disease. Treatment with the PARP inhibitor 3-aminobenzamide (3-AB) reduced tissue injury (Figure 4.1), as evaluated by the improvement in histology and body weight, reduction in colonic myeloperoxidase activity and prostaglandin levels, and preserved ATP concentration.

The cecum, colon, and rectum showed evidence on gross examination of mucosal congestion, erosion, and hemorrhagic ulcerations 4 days after intracolonic administration of TNBS. The histopathological features included transmural necrosis and edema and a diffuse leukocytic cellular infiltrate in the submucosa. Inhibition of

PARP activation by intraperitoneal administration of 3-AB decreased the extent and severity of the colonic damage. The inflammation of the intestinal tract was also accompanied by a significant loss in body weight in comparison with control rats. PARP inhibition significantly reduced the loss in body weight, which correlated well with the amelioration of colonic injury. Treatment with 3-AB reduced leukocytic infiltration and the elevation of 6-keto-Prostaglandin F $(PGF)_{1\alpha}$ levels, indicative of activation of the inflammatory response by cyclooxygenase. Treatment with 3-AB also reduced the extent of ATP depletion in colonic tissue (Figure 4.2).

Although 3-AB is not a direct scavenger of peroxynitrite (unpublished observations), it reduced the degree of tyrosine nitration in the injured colon in the 3-AB-treated animals. These observations suggest that inhibition of PARP by 3-AB, in an indirect way, influences the quantity of reactive peroxynitrite produced. To exclude a pharmacological effect of 3-AB on oxidant generation, independent of its role as a PARP inhibitor, we utilized a PARP-deficient strain of mice in an experimental model of colitis.

Colitis in PARP[-/-] mice and their littermate wild-type controls was induced by intraluminal administration of TNBS/ethanol, a procedure that induces colonic epithelial injury with ulceration in mice.[64,65] PARP[+/+] mice appeared markedly more sensitive to the injurious effects of TNBS: all developed bloody diarrhea, 50% died within 4 days, and 78% within 7 days after TNBS administration. In contrast, PARP[-/-] mice appeared healthy, with a very mild diarrhea; only 20% died within 7 days. Weight loss was more pronounced in wild-type than in PARP[-/-] mice (Figure 4.3). There were multiple sites of grossly visible colonic mucosal congestion, erosion, and hemorrhagic ulcerations in wild-type animals 4 days after TNBS treatment, whereas the colons of most of the TNBS-treated PARP[-/-] mice were indistinguishable from those of mice treated with 50% ethanol only. At the histological level, TNBS-treated wild-type mice evidenced mucosal erosions, edema, large stretches of denuded epithelia, and a diffuse leukocyte infiltration in the submucosa. The histological features of TNBS-treated PARP[-/-] mice were typical of normal or healing mucosa with an intact epithelium (Figures 4.4 and 4.5).[68]

Colonic injury in PARP[+/+] animals was also associated with increased neutrophil infiltration into the inflamed tissue.[68] The level of myeloperoxidase activity, a marker of neutrophil infiltration, paralleled the increase of tissue malondialdehyde, an indicator of lipid peroxidation (Figure 4.3B). Moreover, a positive staining for nitrotyrosine, a marker of nitrosative injury, was present throughout the inflamed colon in PARP[+/+] animals.[68] TNBS-treated PARP[-/-] mice had substantially less infiltration of neutrophils and formation of malondialdehyde and nitrotyrosine (Figures 4.4 and 4.5).[68]

The diminished neutrophil infiltration in the PARP[+/+] animals implicated PARP in the regulation of proinflammatory signaling pathways.[8,66,67] In support of this concept, TNBS-treated wild-type mice showed immunoreactivity for the neutrophil adhesion molecule ICAM-1 in the endothelium in the submucosal vasculature.[68] In contrast, TNBS-treated PARP[-/-] mice did not reveal upregulated expression of ICAM-1, which was constitutively expressed in the endothelium along the vascular wall.[68] Thus, PARP activation appears to mediate neutrophil recruitment via its effect on the expression of adhesion molecules.

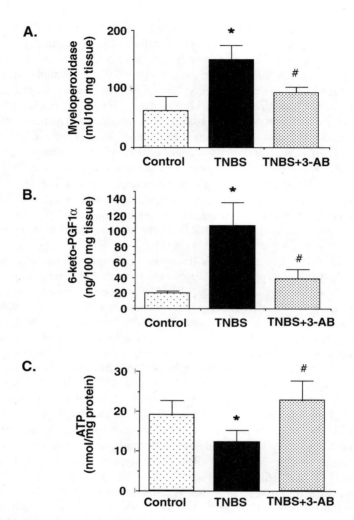

FIGURE 4.2 Effect of 3-AB treatment on myeloperoxidase activity in a rodent model of TNBS-induced colitis. (A) 6-keto-PGF$_{1\alpha}$ and (B) ATP levels (C) 4 days after intracolonic administration of TNBS. Drug treatment (10 mg/kg twice a day, intravenously) was started 24 h before instillation of TNBS and continued until the animals were sacrificed. Results are mean ± SE of 8 rats for each group. *Significantly different from control rats ($p < 0.05$). #Significantly different from untreated rats subjected to TNBS colitis ($p < 0.05$).

The production and role of free radicals and oxidants is also well established in the pathophysiology of autoimmune arthritis.[69-71] Upregulated expression of iNOS and production of toxic quantities of NO have been measured *ex vivo* in chondrocytes obtained from experimental animals and clinical biopsies.[72-76] In addition, plasma levels of nitrite/nitrate (the breakdown products of NO) and plasma and synovial

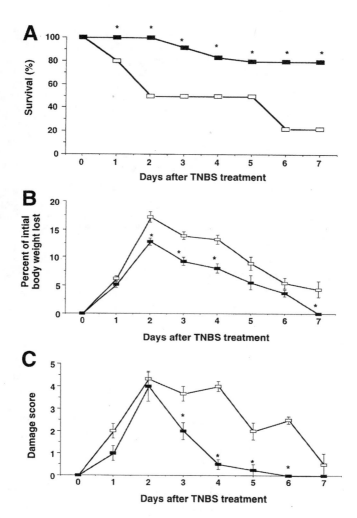

FIGURE 4.3 Severity of TNBS-induced colitis is reduced in PARP$^{-/-}$ mice. (A) Survival is significantly improved in TNBS-treated PARP$^{-/-}$ mice in comparison to the high mortality rate of the wild-type group. (B) Weight loss (expressed as percent of initial body weight lost) is significantly reduced after TNBS in PARP$^{-/-}$ mice. (C) Colon damage was scored on a 0 (normal) to 8 (severe) scale by two independent observers and was markedly reduced in PARP$^{-/-}$ mice. Each data point is the mean ± SEM of 4 to 20 animals for each group. *Significantly different from TNBS-treated wild-type mice ($p < 0.05$). Open squares, PARP$^{+/+}$; closed squares, PARP$^{-/-}$. (From Zingarelli, B. et al., *Gastroenterology*, 116, (2), 335, 1999. With permission.)

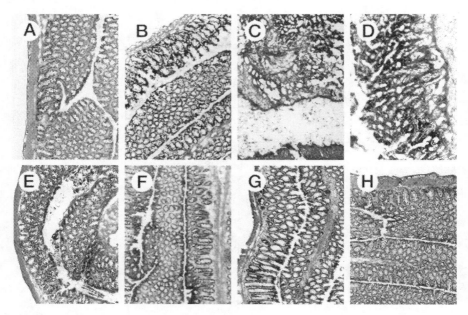

FIGURE 4.4 Time course of changes in colonic architecture after TNBS administration in PARP$^{+/+}$ mice. (A) Representative colonic sections from nontreated mice showed normal architecture (day 0). At days (B) 1 and (C) 2 after TNBS administration, a marked disruption of the structure occurred starting from the epithelium toward the submucosa. At days (D) 3 and (E) 4, the extensive mucosal necrosis was associated with a massive infiltration of inflammatory cells. At days (F) 5, (G) 6, (H) 7, edema and inflammatory cells were still present in the submucosa, whereas a healing process started in the epithelium. Numbers indicate days after TNBS treatment. A similar pattern was seen in five to six different tissue sections in each experimental group (original magnification, 100×). (From Zingarelli, B. et al., *Gastroenterology*, 116, (2), 335, 1999. With permission.)

levels of nitrotyrosine are increased in clinical disease.[77] Experimental studies in a murine model of collagen-induced arthritis have also demonstrated increased nitrotyrosine formation.[78] Given the existence of redox stress in experimental and clinical arthritis, we hypothesized that PARP activation could represent a final common effector mechanism of injury. Similar to the findings in enterocolitis, there are now direct experimental data demonstrating that PARP activation plays a role in the pathophysiology of joint inflammation. Blockade of PARP activity with the weak inhibitor nicotinamide, or with nicotinic acid amide, reduced the onset of the disease in a murine model of arthritis.[79-81] In experimental systems, PARP inhibitors have been shown to prevent both the incidence of joint inflammation as well as the progression of established collagen induced arthritis.[78,79] Furthermore, treatment with the PARP inhibitor 5-iodo-6-amino-1,2-benzopyrone produced substantial protection in a murine model of collagen-induced arthritis and reduced the incidence and severity of joint disease.[78] Vehicle-treated arthritic animals, in contrast, revealed signs of severe suppurative arthritis, with massive infiltration of neutrophils, macrophages, and lymphocytes. PARP inhibition markedly reduced the extent of neu-

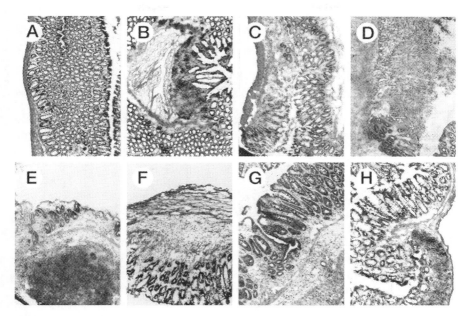

FIGURE 4.5 Time course of changes in colonic architecture after TNBS administration in PARP−/− mice. (A) Representative colonic sections from nontreated mice showed normal architecture (day 0). At days (B) 1, (C) 2, and (D) 3 after TNBS administration, a disruption of the structure occurred at the epithelium. At days (E) 4, (F) 5, (G) 6, and (H) 7, the mucosal and submucosal architecture appeared normal and/or typical of a healing process. Numbers indicate days after TNBS treatment. A similar pattern was seen in five to six different tissue sections in each experimental group (original magnification, 100×). (From Zingarelli, B. et al., *Gastroenterology,* 116, (2), 335, 1999. With permission.)

trophil infiltration into the larger joints, and decreased necrosis and hyperplasia of the synovium.[78] Taken together, these data provide strong support for the concept that PARP activation mediates the leukocytic trafficking and end-organ damage in models of inflammatory joint disease.

In addition to the important role of PARP activation in mediating tissue injury in inflammatory diseases, PARP-independent processes (neutrophil activation, complement activation, fibrinolysis, platelet-activating factor production, free radical independent cytotoxic effects of cytokines, etc.) undoubtedly contribute to the pathophysiology of inflammatory diseases. The experimental data in rodent models strongly suggest, however, that PARP inhibition represents a broadly effective therapeutic approach to the management of inflammation. On theoretical grounds as well, inhibition of PARP is likely to have a therapeutic advantage of targeting final common effectors of tissue injury, rather than proximal initiating triggers. Approaches (e.g., glucocorticoids) that block early steps in the inflammatory process are often incompletely effective because their mode of action requires a near total blockade of the signaling cascade. Parallel pathways of injury that are missed may be capable of substituting for the blocked pathways. Inhibition of the PARP activity, in contrast, interferes with both multiple early signaling pathways and a final common effector pathway of injury, and may therefore have a more comprehensive and effective anti-inflammatory action.

4.6 DEVELOPMENT OF A WORKING HYPOTHESIS RELATING PARP AND INFLAMMATION

We propose the following working hypothesis to place PARP activation in the context of inflammatory injury: Proinflammatory cytokines stimulate free radical formation by stimulating XO activity and *de novo* iNOS expression and by recruiting activated neutrophils that express NADPH oxidase. As a consequence, the oxidants peroxynitrite, hydrogen peroxide, and hydroxyl radical are formed from the interaction of superoxide and NO and by iron-catalyzed oxidation of superoxide. Oxidant stress induces AP-1 formation and generates DNA single-strand breaks. DNA strand breaks then activate PARP, which in turn potentiates NF-κB activation and AP-1 expression, resulting in greater expression of the AP-1 and NF-κB dependent genes, such as iNOS, ICAM-1, TNF-α, and C3. Generation of C5a (derived from C3), in combination with increased endothelial expression of ICAM-1, recruits more activated leukocytes to inflammatory foci, producing greater oxidant stress. The cycle is thus renewed as the increase in oxidant stress triggers more DNA strand breakage. The proposed cycle of inflammatory activation will be augmented in systems where PARP-dependent MAP kinase activation and NF-κB translocation contribute importantly to free radical and oxidant formation and granulocyte recruitment. According to this proposed model, which still requires validation but is supported by multiple lines of evidence, PARP occupies a critical position in a positive-feedback loop of inflammatory injury.

NAD depletion induced by PARP activation is likely to accelerate this positive-feedback cycle by preventing the energy-dependent reduction of oxidized glutathione, the chief intracellular antioxidant and most abundant thiol in eukaryotic cells.[82] NAD is the precursor for NADP, a cofactor that plays a critical role in bioreductive synthetic pathways and the maintenance of reduced glutathione pools.[83] Depletion of reduced glutathione, as a consequence of intracellular energetic failure or overwhelming oxidant exposure,[84] leaves further oxidant stress unopposed, resulting in greater DNA strand breakage.

Based on the above experimental *in vitro* and *in vivo* data, PARP appears to occupy a critical position in a self-amplifying cycle of oxidant-driven damage. It is appropriate then to ask whether PARP inhibition is a candidate for clinical treatment of inflammation. A variety of PARP inhibitors are in preclinical development, many with potency that greatly exceeds the prototypic agents used in experimental proof-of-concept studies of inflammation. A debate is ongoing as to the appropriate type of inflammation that may be best suited to PARP inhibitory therapy. PARP inhibitors may be particularly useful in the treatment of acute inflammatory disorders, such as circulatory shock, where issues of potential toxicity related to chronic administration are unimportant. Although the exact physiological role of PARP is still unclear, and remains a matter of dispute, it is logical to suppose it plays an important role since it is one of the most abundant proteins in the nucleus. PARP has been implicated in many physiological housekeeping functions, such as gene repair, transcription, and cell cycling. Until such time as its true physiological functions are more precisely defined, there should exist considerable caution in the long-term administration of PARP inhibitors to humans. Chronic *in vitro* administration of high doses of the PARP inhibitor nicotinamide, for example, has been shown in experimental models to reduce β-islet cell function.[85] Because nicotinamide is a weak inhibitor, and has other activities beyond PARP inhibition, these data do not directly imply that chronic inhibition of the enzyme is problematic. PARP inhibition has also been associated with an increase in sister chromatid exchange,[86] a concerning finding that raises the risk of malignant transformation. PARP activation clearly leads to cell death, and some have argued that its physiological role is to eliminate genetically damaged cells, thereby reducing oncogenic potential.[87] Indeed, PARP inhibition has been shown to facilitate the rapid ligation of DNA excision-repair patches[88] and to suppress malignant transformation in cells with DNA damage induced by irradiation and chemical carcinogens.[89] Whether chronic PARP inhibition predisposes to cancer is open to question, and we are not aware of any data showing that the PARP-deficient murine strain has a greater risk of malignancy. Perhaps this conclusion would be different in the setting where PARP deficiency or inhibition was associated with a long period of chronic inflammation, redox stress, and associated DNA single-strand breakage. These questions must be properly addressed prior to the long-term use of a PARP inhibitor to treat chronic inflammatory conditions, such as diabetes, arthritis, and enterocolitis, among others.

As with all anti-inflammatory therapies, the use of PARP inhibitors could interfere with the appropriate recruitment of the immune response to eradicate bacterial pathogens. Studies to evaluate the effect of PARP inhibition, or its deficiency, on the course of live bacterial infection are currently under way. Given the broad effects of PARP inhibition on proinflammatory cytokine and chemokine expression, it will be critical, prior to clinical introduction of a PARP inhibitor, to gauge accurately the potential for PARP inhibition to increase infectious risk.

ACKNOWLEDGMENT

This work was supported by a grant from the National Institutes of Health (RO10-M57407).

REFERENCES

1. Lautier, D., Lageux, J., Thibodeau, J., Ménard, L., and Poirier, G.G., Molecular and biochemical features of poly (ADP-ribose) metabolism. *Mol. Cell. Biochem.,* 122:171–193, 1993.
2. Wang, Z.Q., Auer, B., Sting, L., Berghammer, H., Haidacher, D., Schweiger, M., and Wagner, E.F., Mice lacking ADPRT and poly (ADP-ribosylation) develop normally but are susceptible to skin disease. *Genes Dev.,* 9:510–520, 1995.
3. Durkacz, B.W., Omidiji, O., Gray, D.A., and Shall, S., (ADP-ribose)n participates in DNA excision repair. *Nature,* 283:593–596, 1980.

4. Satoh, M.S. and Lindahl, T., Role of poly(ADP-ribose) formation in DNA repair. *Nature*, 356:356–358, 1992.
5. Poirier, G.G., de Murcia, G., Jongstra-Bilen, J., Niedergang, C., and Mandel, P., Poly(ADP-ribosyl)ation of polynuclesomes causes relaxation of chromatin structure. *Proc. Natl. Acad. Sci. U.S.A.*, 79:3423–3427, 1982.
6. Ohashi, Y., Ueda, K., Hayaisha, O., Ikai, K., and Niwa, O., Induction of murine teratocarcinmoa cell differentiation by suppression of poly(ADP-ribose) synthesis. *Proc. Natl. Acad. Sci. U.S.A.*, 81:7132–7136, 1984.
7. Simbulan-Rosenthal, C.M., Rosenthal, D.S., Hilz, H., Hickey, R., Malkas, L., Applegren, N., Wu, Y., Bers, G., and Smulson, M.E., The expression of poly(ADP-ribose) polymerase during differentiation-linked DNA replication reveals that it is a component of the multiprotein DNA replication complex. *Biochemistry*, 35:11622–11633, 1996.
8. Amstad, P.A., Krupitza, G., and Cerutti, P.A., Mechanism of c-fos induction by active oxygen. *Cancer Res.*, 52:3952–3960, 1992.
9. Berger, N.A., Kaichi, A.S., Steward, P.G., Klevecz, R.R., Forrest, G.L., and Gross, S.D., Synthesis of poly(adenosine diphosphate ribose) in synchronized Chinese hamster cells. *Exp. Cell Res.*, 117:127–135, 1978.
10. Whitacre, C.M., Hashimoto, H., Tsia, M.L., Chatterjee, S., Berger, S.J., and Berger, N.A., Involvement of NAD-poly(ADP-ribose) metabolism in p53 regulation and its consequences. *Cancer Res.*, 55:3697–3701, 1995.
11. Kun, E., Kirsten, E., Milo, G.E., Kurian, P., and Kumari, H.L., Cell cycle-dependent intervention by benzamide of carcinogen-induced neoplastic trasnformation and *in vitro* poly(ADP-ribosyl)ation of nuclear proteins in human fibroblasts. *Proc. Natl. Acad. Sci. U.S.A.*, 80:7219–7223, 1983.
12. Koizumi, T., Gupta, R., Banerjee, M., and Newman, J.H., Changes in pulmonary vascular tone during exercise. *J. Clin. Invest.*, 94 (6):2275–2282, 1994.
13. Nathan, C., Nitric oxide as a secretory product of mammalian cells. *FASEB J.*, 6:3051–3064, 1992.
14. Ischiropoulos, H., al-Mehdi, A.B., and Fisher, A.B., Reactive species in ischemic rat lung injury: contribution of peroxynitrite. *Am. J. Phys.*, 269:L158–L164, 1995.
15. Nicolson, A.G., Haites, N.E., McKay, N.G., Wilson, M.W., MacLeod, A.M., and Benjamin, N., Induction of nitric oxide synthase in human mesangial cells. *Biochem. Biophys. Res. Commun.*, 193:1269–1274, 1993.
16. Tepperman, B.L., Brown, J.F., and Whittle, B.J., Nitric oxide synthase induction and intestinal epithelial cell viability in rats. *Am. J. Phys.*, 265:G214–G218, 1993.
17. Boughton-Smith, N.K., Evans, S.M., Laszlo, F., Whittle, B.J.R., and Moncada, S., The induction of nitric oxide synthase and intestinal vascular permeability by endotoxin in the rat. *Br. J. Pharmacol.*, 110:1189–1195, 1993.
18. Wilson, K.T., Ciancio, M.J., and Chang, E.B., Inducible nitric oxide synthase mRNA expression is increased in intestinal mucosa of endotoxemic rats and is inhibited by dexamethasone. *Gastroenterology*, 106:A793 (Abstr.), 1994.
19. Petros, A., Bennett, D., and Vallance, P., Effect of nitric oxide synthase inhibitors on hypotension in patients with septic shock. *Lancet*, 338:1557–1558, 1991.
20. Suschek, C., Rothe, H., Fehsel, K., Enczmann, J., and Kolb-Bachofen, V., Induction of a macrophage-like nitric oxide synthase in cultured rat aortic endothelial cells. *J. Immunol.*, 151:3283–3291, 1993.
21. Estrada, C., Gomez, C., Martin, C., Moncada, S., and Gonzalez, C., Nitric oxide mediates tumor necrosis factor-*a* toxicity in endothelial cells. *Biochem. Biophys. Res. Commun.*, 186:475–482, 1992.

22. Szabó, C., Zingarelli, B., O'Connor, M., and Salzman, A.L., DNA strand breakage, activation of poly-ADP ribosyl synthetase, and cellular energy depletion are involved in the cytotoxicity in macrophages and smooth muscle cells exposed to peroxynitrite. *Proc. Natl. Acad. Sci. U.S.A.*, 93:1753–1758, 1996.
23. Szabó, C. and Salzman, A.L., Endogenous peroxynitrite is involved in the inhibition of cellular respiration in immuno-stimulated J774.2 macrophages. *Biochem. Biophys. Res. Commun.*, 209:739–743, 1995.
24. Ford, H.R., Watkins, S., Reblock, K.K., Teramana, C., and Rowe, M.I., The role of inflammatory cytokines and nitric oxide in the pathogenesis of necrotizing enterocolitis. *J. Ped. Surg.*, 32:275–282, 1997.
25. Oda, T., Akaike, T., Hamamoto, T., Suzuki, F., Hirano, T., and Maeda, H., Oxygen radicals in influenza-induced pathogenesis and treatment with pyran polymer-conjugated SOD. *Science,* 244:974–976, 1989.
26. Tan, S., Yokoyama, Y., Dickens, E., Cash, T.G., Freeman, B.A., and Parks, D.A., Xanthine oxidase activity in the circulation of rats following hemorrhagic shock. *Free Radical Biol. Med.,* 15:407–414, 1993.
27. McCord, J.M., Oxygen-derived free radicals in postischemic tissue injury. *N. Engl. J. Med.,* 312:159–163, 1985.
28. Miesel, R., Zuber, M., Sanocka, D., Graetz, R., and Kroeger, H., Effects of allopurinol on *in vivo* suppression of arthritis in mice and *ex vivo* modulation of phagocytic production of oxygen radicals in whole human blood. *Inflammation,* 18:597–612, 1994.
29. Engerson, T.D., McKelvey, T.G., Rhyne, D.B., Boggio, E.B., Snyder, S.J., and Jones, H.P., The conversion of xanthine dehydrogenase to oxidase in ischaemic rat tissue. *J. Clin. Invest.,* 79:1564–1570, 1987.
30. Akaike, T., Ando, M., Tatsuya, O., Doi, T., Ijiri, S., Araki, S., and Maeda, H., Dependence on O_2 generation by xanthine oxidase of pathogenesis of influenza virus infection in mice. *J. Clin. Invest.,* 85:739–745, 1990.
31. Mohacsi, A., Kozlovszky, B., Kiss, I., Seres, I., and Fulop, T., Jr., Neutrophils obtained from obliterative atherosclerotic patients exhibit enhanced resting respiratory burst and increased degranulation in response to various stimuli. *Biochim. Biophys. Acta,* 1316:210–216, 1996.
32. Grum, C.M., Ragsdale, R.A., Ketai, L.H., and Simon, R.H., Plasma hypoxanthine and exercise. *Am. Rev. Respir. Dis.,* 136:98–101, 1987.
33. Friedl, H.P., Smith, D.J., Till, G.O., Thomson, P.D., Louis, D.S., and Ward, P.A., Ischemia-reperfusion in humans. Appearance of xanthine oxidase activity. *Am. J. Pathol.,* 136:491–495, 1990.
34. Friedl, H.P., Till, G.O., Trentz, O., and Ward, P.A., Role of histamine, complement and xanthine oxidase in thermal injury of the skin. *Am. J. Pathol.,* 135:203–217, 1989.
35. Parks, D.A., Bulkley, G.B., and Granger, D.N., Role of oxygen free radicals in shock, ischemia, and organ preservation. *Surgery,* 94:428–432, 1983.
36. Demling, R., LaLonde, C., Youn, Y.K., Daryani, R., Campbell, C., and Knox, J., Lung oxidant changes after zymosan peritonitis: relationship between physiologic and biochemical changes. *Am. Rev. Respir. Dis.,* 146:1272–1278, 1992.
37. Szabó, C., Zingarelli, B., O'Connor, M., and Salzman, A.L., Peroxynitrite, but not nitric oxide or superoxide, causes DNA strand breakage, activates poly-ADP-ribosyl synthetase, and depletes cellular energy stores in J774 macrophages and rat aortic smooth muscle cells *in vitro*. *Endothelium,* 3:S46 (Abstr.), 1995.

38. Szabó, C., Zingarelli, B., and Salzman, A.L., Peroxynitrite-mediated activation of poly-ADP ribosyl synthetase contributes to the vascular failure in shock, in *The Pathophysiology of Shock: Proceedings of the Third International Shock Meeting*, Elsevier Scientific Publishers, New York, 1995.
39. Zingarelli, B., O'Connor, M., Wong, H., Salzman, A.L., and Szabó, C., Peroxynitrite-mediated DNA strand breakage activates poly-ADP ribosyl synthetase and causes cellular energy depletion in macrophages stimulated with bacterial lipopolysaccharide. *J. Immunol.*, 156:350–358, 1996.
40. Szabó, C., Salzman, A.L., and Ischiropoulos, H., Peroxynitrite-mediated oxidation of dihydrorhodamine 123 occurs in early stages of endotoxic and hemorrhagtic shock and ischemia-reperfusion injury. *FEBS Lett.*, 372:229–232, 1995.
41. Szabó, C., Zingarelli, B., and Salzman, A.L., Role of poly-ADP ribosyltransferase activation in the vascular contractile and energetic failure elicited by exogenous and endogenous nitric oxide and peroxynitite. *Circ. Res.*, 78:1051–1063, 1996.
42. Singer, I.I., Kawka, D.W., Scott, S., Weidner, J.R., Mumford, R.A., Riehl, T.E., and Stenson, W.F., Expression of inducible nitric oxide synthase and nitrotyrosine in colonic epithelium in inflammatory bowel disease. *Gastroenterology*, 111:871–885, 1996.
43. Szabó, C., Saunders, C., O'Connor, M., and Salzman, A.L., Peroxynitrite causes energy depletion and increases permeability via activation of poly-ADP ribosyl synthetase in pulmonary epithelial cells. *Am. J. Respir. Mol. Biol.* 60:105–109, 1996.
44. Kennedy, M.S., Denenberg, A., Szabó, C., and Salzman, A.L., Poly(ADP-ribose) synthetase (PARP) mediates increased permeability induced by peroxynitrite in Caco-2BBe cells. *Gastroenterology*, 114:510–518, 1998.
45. Cochrane, C.G., Damage to DNA by reactive oxygen and nitrogen species: role in inflammatory diosease and progression to cancer. *Biochem. J.*, 313:17–29, 1996.
46. Lih-Brody, L., Powell, S.R., Collier, K.P., Reddy, G.M., Cerchia, R., Kahn, E., Weissman, G.S., Katz, S., Floyd, R.A., McKinley, M.J. et al., Increased oxidative stress and decreased antioxidant defenses in mucosa of inflammatory bowel disease. *Dig. Dis. Sci.*, 41:2078–2086, 1996.
47. Zhang, J., Dawson, V.L., Dawson, T.M., and Snyder, S.H., Nitric oxide activation of poly(ADP-ribose) synthetase in neurotoxicity. *Science*, 263:687–689, 1994.
48. Ueda, K. and Hayaishi, O., ADP-ribosylation. *Annu. Rev. Biochem.*, 54:73–100, 1985.
49. Cuzzocrea, S., Zingarelli, B., Constantino, G., Szabó, A., Salzman, A.L., Caputi, A.P., and Szabó, C., Beneficial effects of 3-aminobenzamide, an inhibitor of poly(ADP-ribose) synthetase in a rat model of splanchnic artery occlusion and reperfusion. *Br. J. Pharmacol.*, 121:1065–1074, 1997.
50. Szabó, C., Lim, L.H., Cuzzocrea, S., Getting, S.J., Zingarelli, B., Flower, R.J., Salzman, A.L., and Perretti, M., Inhibition of poly(ADP-ribose) synthetase attenuates neutrophil recruitment and exerts antiinflammatory effects. *J. Exp. Med.*, 186:1041–1049, 1997.
51. Cuzzocrea, S., Zingarelli, B., Gilad, E., Hake, P., Salzman, A.L., and Szabó, C., Protective effects of 3-aminobenzamide, an inhibitor of poly(ADP-ribose) synthase in a carrageenan-induced model of local inflammation. *Eur. J. Pharmacol.*, 342:67–76, 1998.
52. Szabó, C., Wong, H., Bauer, P.I., Kirsten, E., O'Connor, M., Zingarelli, B., Mendeleyev, J., Hasko, G., Sylvester, E., Salzman, A.L. et al., Regulation of components of the inflammatory response by 5-iodo-6-amino-1,2-benzopyrone, an inhibitor of poly(ADP-ribose) synthetase and pleiotropic modifier of cellular signal pathways. *Int. J. Oncol.*, 10:1093–1104, 1997.

53. Zingarelli, B., Salzman, A.L., and Szabó, C., Protective effects of nicotinamide against nitric oxide mediated vascular failure in endotoxic shock: potential involvement of poly ADP ribosyl synthetase. *Shock,* 5:258–264, 1996.
54. Szabó, C., Cuzzocrea, S., Zingarelli, B., O'Connor, M., and Salzman, A.L., Endothelial dysfunction in endotoxic shock: importance of the activation of poly(ADP ribose) synthetase (PARP) by peroxynitrite. *J. Clin. Invest.,* 100:723–735, 1997.
55. Zingarelli, B., Cuzzocrea, S., Zsengeller, Z., Salzman, A.L., and Szabó, C., Beneficial effect of inhibition of poly-ADP ribose synthetase activity in myocardial ischemia-reperfusion injury. *Cardiovasc. Res.,* 36:205–215, 1997.
56. Gilad, E., Zingarelli, B., Salzman, A.L., and Szabó, C., Protection by inhibition of poly(ADP-ribose) synthetase against oxidant injury in cardiac myoblasts *in vitro*. *J. Mol. Cell. Cardiol.,* 29:2585-2597, 1997.
57. Zingarelli, B., Ischiropoulos, H., Salzman, A.L., and Szabó, C., Amelioration by mercaptoethylguanidine of the vascular and energetic failure in haemorrhagic shock in the anesthetised rat. *Eur. J. Pharmacol.,* 338:55–65, 1997.
58. Cuzzocrea, S., Zingarelli, B., O'Connor, M., Salzman, A.L., Caputi, A.P., and Szabó, C., Role of peroxynitrite and activation of poly(ADP-ribose) synthetase in the vascular failure induced by zymosan-activated plasma. *Br. J. Pharmacol.,* 122:493–503, 1997.
59. Unno, N., Menconi, M.J., Salzman, AS.L., Smith, M., Hagan, S., Ezzel, R.M., and Fink, M.P., Hypermeability and ATP depletion induced by chronic hypoxia or glycolytic inhibition in Caco-2BBe monolayers. *Am. J. Phys.,* 270:G1010–G1021, 1996.
60. Roebuck, K.A., Rahman, A., Lakshminarayanan, V., Janakidevi, K., and Malik, A.B., H_2O_2 and tumor necrosis factor-alpha activate intercellular adhesion molecule 1 (ICAM-1) gene transcription through distinct cis-regulatory elements within the ICAM-1 promoter. *J. Biol. Chem.,* 270:18996–18974, 1995.
61. Szabó, C., Zingarelli, B., Cuzzocrea, S., and Salzman, A.L., Poly (ADP-ribose) synthetase modulates expression of P-selectin and ICAM-1 in myocardial ischemia-reperfusion injury. *Jpn. J. Pharmacol.,* 75 (Suppl.):101P (Abstr.), 1997.
62. Akabane, A., Kato, I., Takasawa, S., Unno, M., Yonekura, H., Yoshimoto, T., and Okamaoto, H., Nicotinamide inhibits IRF-1 mRNA induction and prevents IL-1 beta-induced nitric oxide synthase expression in pancreatic beta cells. *Biochem. Biophys. Res. Commun.,* 215:524–530, 1995.
63. LeClaire, R.D., Kell, W.M., Sadik, R.A., Downs, M.B., and Parker, G.W., Regulation by SEB-induced NO production in endothelial cells. *Infect. Immun.,* 63:539–546, 1995.
64. Fujimura, M., Tominaga, T., and Yoshimoto, T., Nicotinamide inhibits iNOS mRNA in primary rat glial cells. *Neurosci. Lett.,* 228:107–110, 1997.
65. Hauschildt, S., Scheipers, P., Bessier, W., Schwarz, K., Ullmer, A., Flad, H.D., and Heine, H., Role of ADP-ribosylation in activated monocytes/macrophages. *Adv. Exp. Med. Biol.,* 419:249–252, 1997.
66. Le Page, C., Sanceau, J., Drapier, J.C., and Wietzerbin, J., Inhibitors of ADP-ribosylation impair inducible nitric oxide synthase gene transcription through inhibition of NF kappa B activation. *Biochem. Biophys. Res. Commun.,* 243:451–457, 1998.
67. Oliver, F.J., Ménissier-de Murcia, J., Nacci, C., Decker, P., Andriantsitohaina, R., Muller, S., De la Rubia, G., Stoclet, J.C., and de Murcia, G., Resistance to endotoxic shock as a consequence of defective NF-κB activation in poly(ADP-ribose) polymerase-1 deficient mice. *EMBO J.,* 18:4446–4454, 1999.
68. Zingarelli, B., Szabó, C., and Salzman, A.L., Blockade of poly(ADP-ribose) synthetase inhibits neutrophil recruitment, oxidant generation and mucosal injury in colitis. *Gastroenterology,* 116:335–345, 1999.

69. Oyanagui, Y., Nitric oxide and superoxide radical are involved in both initiation and development of adjuvant arthritis in rats. *Life Sci.,* 54:PL285–PL259, 1994.
70. Santos, L. and Tipping, P.G., Attenuation of adjuvant arthritis in rats by treatment with oxygen radical scavengers. *Immunol. Cell. Biol.,* 72:406–414, 1994.
71. Kaur, H. and Halliwell, B., Evidence for nitric oxide-mediated oxidative damage in chronic inflammation. Nitrotyrosine in serum and synovial fluid from rheumatoid patients. *FEBS Lett.,* 350:9–12, 1994.
72. Hauselmann, H.J., Opplinger, L., Michel, B.A., Stefanovic-Racic, M., and Evans, C.H., Nitric oxide and proteoglycan synthesis by human articular chondrocytes in alginate culture. *FEBS Lett.,* 352:361–364, 1994.
73. Sakurai, H., Kohsaka, H., Liu, M.F., Higashiyama, H., Hirata, Y., Kanno, K., Saito, I., and Miyaska, N., Nitric oxide production and inducible nitric oxide synthase expression in inflammatory arthritides. *J. Clin. Invest.,* 96:2357–2363, 1995.
74. Murrell, G.A., Jang, D., and Williams, R.J., Nitric oxide activates metalloprotease enzymes in articular cartilage. *Biochem. Biophys. Res. Commun.,* 206:15–21, 1995.
75. Grabowski, P.S., Macpherson, H., and Ralston, S.H., Nitric oxide production in cells derived from the human joint. *Br. J. Rheumatol.,* 35:207–212, 1996.
76. Hayashi, T., Abe, E., Yamate, T., Taguchi, Y., and Jasin, H.E., Nitric oxide production by superficial and deep articular chondrocytes. *Arthritis Rheumatol.,* 40:261–269, 1997.
77. Farrell, A.J., Blake, D.R., Palmer, R.M., and Moncada, S., Increased concentrations of nitrite in synovial fluid and serum samples suggest increased nitric oxide synthesis in rheumatic diseases. *Annu. Rheumatol. Dis.,* 51:1219–1222, 1992.
78. Szabó, C., Virág, L., Cuzzocrea, S., Scott, G.J., Hake, P., O'Connor, M., Zingarelli, B., Salzman, A.L., and Kun, E., Protection against peroxynitrite-induced fibroblast injury and arthritis development by inhibition of poly(ADP-ribose) synthetase. *Proc. Natl. Acad. Sci. U.S.A.,* 95:3867–3872, 1998.
79. Miesel, R., Kurpisz, M., and Kroger, H., Modulation of inflammatory arthritis by inhibition of poly(ADP ribose) polymerase. *Inflammation,* 19:379–387, 1996.
80. Ehrlich, W., Huser, H., and Kroger, H., Inhibition of the induction of collagenase by interleukin-1 beta in cultured rabbit synovial fibroblasts after treatment with the poly(ADP-ribose)-polymerase inhibitor 3 aminobenzamide. *Rheumatol. Int.,* 15:171–172, 1997.
81. Kroger, K.D., Miesel, R., Dietrich, A., Ohde, M., Rajnavolgyi, E., and Ockenfels, H., Synergistic effects of thalidomide and poly(ADP-ribose) polymerase inhibition on type II collagen-induced arthritis in mice. *Inflammation,* 20:203–215, 1998.
82. Martensson, J., Jain, A., and Meister, A., Gluthatione is required for intestinal function. *Proc. Natl. Acad. Sci. U.S.A.,* 87:1715–1719, 1990.
83. Berger, N.A., Oxidant-induced cytotoxicity: a challenge for metabolic modulation. *Am. J. Respir. Cell. Mol. Biol.,* 4:1–3, 1991.
84. Schoenberg, M.H. and Beger, H.G., Oxygen radicals in intestinal ischemia and reperfusion. *Chem. Biol. Inter.,* 76:141–161, 1990.
85. Reddy, S., Salari-Lak, N., and Sandler, S., Long-term effects of nicotinamide-induced inhibition of poly(adenosine diphosphate-ribose) polymerase activity in rat pancreatic islets exposed to interleukin-1 beta. *Endocrinology,* 136:1907–1912, 1995.
86. Oikawa, A., Tohda, H., Kanai, M., Miwa, M., and Sugimura, T., Inhibitors of poly(adenosine diphosphate ribose) polymerase induce sister chromatid exchanges. *Biochem. Biophys. Res. Commun.,* 97:1311–1316, 1980.

87. Nagele, A., Poly(ADP-ribosyl)ation as a fail-safe, transcription-independent, suicide mechanism in acutely DNA-damaged cells: a hypothesis. *Radiat. Environ. Biophys.,* 34:251–254, 1995.
88. Cleaver, J.E. and Park, S.D., Enhanced ligation of repair sites under conditions of inhibition of poly(ADP-ribose) synthesis by 3-aminobenzamide. *Mutat. Res.,* 173:287–290, 1986.
89. Borek, C., Morgan, W.F., Ong, A., and Cleaver, J.E., Inhibition of malignant transformation in vitro by inhibitors of poly(ADP-ribose) synthesis. *Proc. Natl. Acad. Sci. U.S.A.,* 81:243–247, 1984.

5 Poly(ADP-Ribose) Polymerase Activation and the Pathogenesis of Circulatory Shock

Csaba Szabó

CONTENTS

5.1 Endotoxic Shock 81
    5.1.1 Vascular Contractile Failure in Endotoxic Shock 82
    5.1.2 Endothelial Dysfunction in Endotoxic Shock 84
    5.1.3 Cellular Energetic Failure, Organ Injury and Dysfunction, and Mortality in Endotoxic Shock 86
5.2 Hemorrhagic Shock and Trauma 87
5.3 Splanchnic Occlusion Shock 88
5.4 Zymosan-Induced Nonseptic Shock and Multiple Organ Failure Models 89
5.5 Regulation of Gene Expression by PARP: Implications for the Pathogenesis of Circulatory Shock 92
5.6 Conclusions and Implications 94
Acknowledgment 97
References 97

5.1 ENDOTOXIC SHOCK

Various forms of circulatory shock are associated with a reduced responsiveness of arteries and veins to exogenous or endogenous vasoconstrictor agents (vascular hyporeactivity), myocardial dysfunction, and disrupted intracellular energetic processes, culminating in multiple organ failure and death. Some of these alterations previously have been suggested to be related to nitric oxide (NO) overproduction, due to the activation of the endothelial isoform of NO synthase (ecNOS) in the early stage and expression of a distinct inducible isoform of NOS (iNOS) in the late stage.[1-3]

In circulatory shock, proinflammatory cytokines invoke the stimulation of oxygen-centered free radical production. Therefore, it is not surprising that various forms of circulatory shock are associated with the production of peroxynitrite, a reactive product formed by the rapid reaction of superoxide and NO. The production of peroxynitrite (evidenced as increased nitrotyrosine immunoreactivity or increased oxidation of the fluorescent probe dihydrorhodamine 123 to rhodamine 123) recently has been demonstrated in endotoxin shock and in hemorrhagic shock.[4-6]

As overviewed in other sections of this book (e.g., see Chapter 7 by Virág), DNA single-strand breakage (in response to reactive oxidant and free radical species) is the obligatory trigger of activation of poly(ADP-ribose) polymerase (PARP). When activated, PARP catalyzes the cleavage of NAD^+ into ADP-ribose and nicotinamide. PARP covalently attaches ADP-ribose to various nuclear proteins, such as histones and PARP itself. Activation of PARP can rapidly deplete NAD^+, slowing the rate of glycolysis, electron transport, and ATP formation, resulting in cell dysfunction and cell death. Pharmacological inhibition of PARP activity has been shown to protect against cell damage in response to exogenously or endogenously produced peroxynitrite. The above-described metabolic derangements generally lead to cell death via the necrotic pathway.[7,8] Endogenous triggers of this pathway in shock include hydrogen peroxide and peroxynitrite, potent inducers of DNA single-strand breakage (Table 5.1). This chapter summarizes the evidence supporting the view that PARP-related alterations play a significant role in the pathogenesis of circulatory shock of various types.

5.1.1 Vascular Contractile Failure in Endotoxic Shock

The vascular contractile failure associated with circulatory shock is closely related to overproduction of NO within the blood vessels. Expression of iNOS within the vascular smooth muscle cells has been implicated in the pathogenesis of vascular hyporeactivity during various forms of shock.[2,9] In these investigations, the evidence for the role of NO in the development of vascular hyporeactivity was based on experiments where pharmacological inhibitors of NOS restored the contractility of the blood vessels. However, recent studies have demonstrated that a superoxide dismutase mimetic, similar to an NOS inhibitor, offers a significant protection against the suppression of the vascular contractility of the thoracic aorta in a rat model of endotoxic shock.[10] This is an effect that is in marked contrast with the previously described NO–superoxide interactions in normal blood vessels, where superoxide dismutase is known to prolong the half-life of NO and thereby enhance NO-mediated relaxant responses.[11,12] The most obvious explanation for the protective effect of both NOS inhibitors and superoxide dismutase mimetics against the endotoxin-induced vascular hyporeactivity is related to peroxynitrite generation. Indeed, peroxynitrite exposure can cause vascular hyporeactivity in vascular rings, an effect that is associated with PARP activation in the blood vessels[13] and can be prevented by pharmacological inhibition of PARP.[13,14]

Is PARP involved in the peroxynitrite-induced vascular hyporeactivity in endotoxic shock? In studies in anesthetized rats, inhibition of PARP with 3-aminobenzamide and nicotinamide reduced the suppression of the vascular contractility of the thoracic aorta in *ex vivo* experiments.[13,15] These findings are similar to the *in vitro* results with authentic peroxynitrite, which also causes a vascular hyporeactivity in

TABLE 5.1
Potential Triggers of PARP Activation in Circulatory Shock

Reactive Species	Selected Sources	Diffusible?	Relationship to PARP Activation	Overproduced in Shock?
Superoxide anion	Mitochondrial respiratory chain, activated neutrophils, macrophages, xanthine oxidase, lipid peroxidation, catechol auto-oxidation, redox cycling	Yes	Does not induce DNA single-strand breakage of PARP activation; although not a particularly potent oxidant, it can be cytotoxic via a number of PARP-independent mechanisms	Yes
Hydrogen peroxide	Xanthine oxidase, mitochondrial respiratory chain, catechol auto-oxidation, redox cycling	Yes	Potent trigger of DNA single-strand breakage and PARP activation via the generation of hydroxyl radical in the vicinity of DNA; potent oxidizing agent; can also be cytotoxic via a number of PARP-independent mechanisms	Yes
Hydroxyl radical	From hydrogen peroxide, via the classic Fenton reaction; monoamine oxidase; in response to ionizing radiation; catechol auto-oxidation; redox cycling	No	Potent trigger of DNA single-strand breakage and PARP activation; can also be cytotoxic via a number of PARP-independent mechanisms; the decomposition of peroxynitrite yields a hydroxyl-radical-like intermediate, which, however, does not appear to be "free" hydroxyl radical	Yes
Hypochlorous acid	Activated macrophages	Yes	Does not induce DNA single-strand breakage of PARP activation; can be cytotoxic via a number of PARP-independent mechanisms	Yes
Nitric oxide	NOS, acidification of nitrite, nonenzymatic pathways	Yes	Does not induce DNA single-strand breakage of PARP activation, unless it combines with superoxide to produce peroxynitrite; although not a particularly potent oxidant, can be cytotoxic via a number of PARP-indepedent mechanisms; it can also exert marked antioxidant and cytoprotective actions	Yes
Peroxynitrite	Reaction of NO and superoxide; NOS can produce peroxynitrite under conditions of low cellular L-arginine	Yes	Potent trigger of DNA single-strand breakage and PARP activation; can also be cytotoxic via a number of PARP-independent mechanisms	Yes
Nitroxonium anion	NO	Yes	Trigger of DNA single- and double-strand breakage; may trigger PARP activation; can also be cytotoxic via a number of PARP-independent mechanisms	Probably

thoracic aortic rings, and which is also reduced by pharmacological inhibition of PARP.

At present, it is not clear to what extent the "pure" NO-mediated (guanylyl cyclase–related) vasorelaxant mechanisms vs. the peroxynitrite-mediated (and, in part, PARP-related) vasorelaxant mechanisms contribute to the vascular hyporeactivity in various forms of shock, and in various phases of shock. In also remains to be investigated whether there are interactions between the cGMP-related and the peroxynitrite-related pathways.

5.1.2 Endothelial Dysfunction in Endotoxic Shock

Peroxynitrite production has been suggested to contribute to endothelial injury in ischemia–reperfusion, circulatory shock, and atherosclerosis.[10,16-18] In fact, Villa and co-workers[19] directly demonstrated that infusion of authentic peroxynitrite into isolated perfused hearts results in an impairment of the endothelium-dependent relaxations. In light of these previous studies, the question has arisen whether the endothelial dysfunction associated with endotoxic shock is also related to peroxynitrite formation, and, if so, whether the PARP pathway plays a role in the process.

Ample evidence supports a role for peroxynitrite formation in the pathogenesis of the endothelial dysfunction in endotoxic shock. For instance, we have recently observed that a manganese–mesoporphyrin cell-permeable superoxide dismutase analogue and peroxynitrite scavenger protect against the development of endothelial dysfunction in endotoxic shock.[10] Similarly, selective inhibition of iNOS by canavanine[20] or by guanidinoethyldisulfide[21] provided protection against the endothelial dysfunction and the endothelial morphological damage in vessels obtained from rats subjected to endotoxic shock. One possible rationalization of the above-mentioned results is that iNOS-derived NO combines with superoxide (the latter derived from activated neutrophils and other sources) to form peroxynitrite, which, in turn, causes endothelial injury.

Current data demonstrating protective effects of 3-aminobenzamide against the development of endothelial dysfunction in vascular rings obtained from rats with endotoxic shock[22] suggest that DNA strand breakage and PARP activation occur in endothelial cells during shock and that the subsequent energetic failure reduces the ability of the cells to generate NO in response to acetylcholine-induced activation of the muscarinic receptors on the endothelial membrane (Figure 5.1). Similar findings were reported in pulmonary vessels challenged with endotoxin.[23] Indeed, several lines of *in vitro* data demonstrate DNA injury, PARP activation, and consequent cytotoxicity in endothelial cells exposed to hydroxyl radical generators,[24-26] or in response to peroxynitrite.[22] The relative contribution of peroxynitrite vs. hydroxyl radical in the PARP activation and endothelial injury in shock remains to be further investigated, as both species are known to be produced in shock, and the available scavengers (such as superoxide dismutase analogues) would be expected to reduce the production of both peroxynitrite and hydroxyl radical. The data with iNOS inhibitors demonstrating protection against the endothelial injury in endotoxic shock[20,22] would favor the contribution of an NO-derived species, such as peroxynitrite. Considering the existence of synergistic cytotoxic interactions of hydrogen

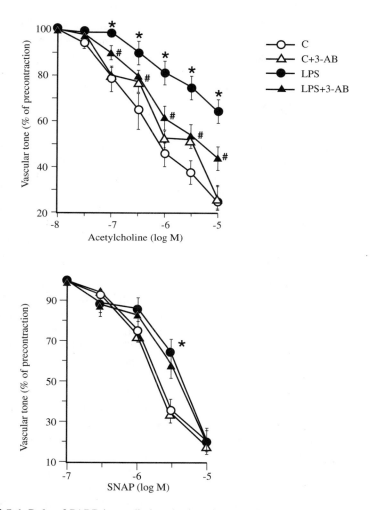

FIGURE 5.1 Role of PARP in mediating the impairment of endothelium-dependent relaxations in endotoxic shock. Endothelium-dependent relaxations to acetylcholine and endothelium-independent relaxations to the NO donor compound S-N-acetyl-penicillamine (SNAP) in thoracic aortic rings from control animals (C, open circles), in rings from control animals treated with the PARP inhibitor 3-aminobenzamide (C + 3-AB, open triangles), in rings from rats treated with bacterial lipopolysaccharide (LPS, closed circles), and in rings treated with LPS and 3-aminobenzamide (LPS + 3-AB, closed triangles). There was a significant impairment of the endothelium-dependent relaxations in endotoxemia (*$p < 0.05$) and a significant prevention of this impairment by the PARP inhibitor 3-aminobenzamide (#$p < 0.05$). Data represent means ± SEM of n = 6 to 8 vascular rings. (From Szabó, C. et al., *J. Clin. Invest.*, 100, 723–735, 1997. With permission.)

peroxide and peroxynitrite,[22] it is also conceivable that a similar synergism may exist *in vivo*, and both peroxynitrite and hydroxyl radical contribute to the activation of PARP and the subsequent endothelial injury *in vivo*.

5.1.3 Cellular Energetic Failure, Organ Injury and Dysfunction, and Mortality in Endotoxic Shock

Substantial evidence now suggests that NO (or a related species, such as peroxynitrite) plays a role in the cellular energetic changes and the related organ dysfunction associated with endotoxic shock. This conclusion is based chiefly on the results of pharmacological studies in which inhibition of NO synthesis, especially by agents that are selective toward iNOS, reduce cellular injury and improve organ function in shock.[27] It is noteworthy that in the same experimental models of rodent endotoxic shock, the cell-permeable superoxide dismutase analogue MnIII tetrakis (4-benzoic acid) porphyrin[28] also reduced the endotoxin-induced depression of mitochondrial respiration in peritoneal macrophages *ex vivo*,[10] thereby suggesting that peroxynitrite, rather than NO per se, plays a role in these alterations.

Peroxynitrite-induced activation of the PARP pathway also has been implicated in the pathophysiology of the cellular energetic failure associated with endotoxic shock by demonstration of increased DNA strand breakage, decreased intracellular NAD^+ and ATP levels, and mitochondrial respiration in peritoneal macrophages obtained from rats subjected to endotoxic shock.[15,29] This cellular energetic failure was reduced by pretreatment of the animals with the PARP inhibitors 3-aminobenzamide or nicotinamide.[15,29]

Probably a similar cellular energetic failure and cellular dysfunction is responsible for the PARP-dependent intestinal epithelial hyperpermeability in endotoxic shock.[30] In fact, *in vitro* studies demonstrate that intestinal and pulmonary epithelial cells challenged with peroxynitrite or hydrogen peroxide develop "leaky" characteristics, an effect that can be inhibited by pharmacological inhibition of PARP.[31,32]

In contrast to these encouraging results in peritoneal macrophages, it appears that the PARP pathway only plays a limited role in the hepatic dysfunction associated with endotoxic shock. In an endotoxic shock model in the rat, inhibition of PARP with 3-aminobenzamide and nicotinamide did not affect the alterations in most parameters of liver injury, whereas inhibition of PARP with 1,5-dihydroxyisoquinoline resulted in a marginal protective effect.[33] These observations are perhaps not surprising when considering the fact that in *in vitro* studies, the oxidant-induced injury in cultured hepatocytes is not prevented by pharmacological inhibition of PARP.[34] The exact reason inhibition of PARP does not affect the course of the oxidant injury in hepatocytes remains to be further investigated. Although hepatic injury does not appear to be largely PARP mediated in endotoxic shock, pharmacological inhibition of PARP, either with 3-aminobenzamide[13] or with the novel potent PARP inhibitor 5-iodo-6-amino-1,2,-benzopyrone,[35] improves survival rate in mice challenged with high-dose endotoxin (Figure 5.2). Similarly, PARP-deficient mice show a massive survival benefit when challenged with lethal doses of endotoxin, as compared with wild-type mice.[36] Based on these observations, one may suggest that, in response to pharmacological inhibition of PARP, the improved hemodynamic

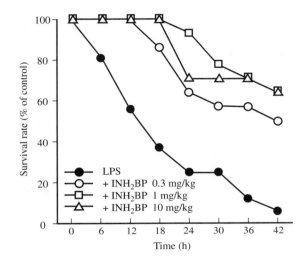

FIGURE 5.2 Role of PARP in endotoxin-induced mortality changes. INH$_2$BP improves survival in mice subjected to endotoxic shock: effect of INH$_2$BP pretreatment (0.3 to 10 mg/kg) on endotoxin-induced (120 mg/kg i.p.) mortality in mice; n = 7 to 8 animals in each group. (From Szabó, C. et al., *Int. J. Oncol.*, 10, 1093–1101, 1997. With permission.)

status due to improved vascular function, and possibly the improved cellular energetic status in some organs, does result in an overall survival benefit in this condition.

5.2 HEMORRHAGIC SHOCK AND TRAUMA

Recent investigations tested the effects of pharmacological inhibitors of PARP in rodent and porcine models of hemorrhagic shock. In a rat model of severe hemorrhagic shock without resuscitation, rats treated with the PARP inhibitor 5-iodo-6-amino-1,2-benzopyrone exhibited a significant improvement in survival rate: the drug increased the 50% survival time from 30 min to approximately 70 min. Moreover, treated animals tended to maintain higher mean arterial blood pressure values. Similar to the protective effect of 5-iodo-6-amino-1,2-benzopyrone, the NOS inhibitor N^G-methyl-L-arginine and a cell-permeable superoxide dismutase mimetic also improved survival times and improved blood pressure during hemorrhagic shock.[37] Similarly, in a porcine model of hemorrhagic shock (resuscitated with Ringer's lactate), inhibition of PARP with 3-aminobenzamide prevented the deterioration of mean arterial pressure and cardiac output during the resuscitation phase, thereby prolonging survival. These observations suggest that the cardiovascular decompensation in hemorrhagic shock is, at least in part, mediated by activation of PARP.[38] Because the PARP inhibitor neither reduced afterload nor increased heart rate, the basis for the increased cardiac index presumably represents an increase in stroke volume secondary to an augmentation of venous return, or an inotropic effect. The improved myocardial performance in the animals treated with the PARP inhibitor was also associated with a slight effect on the left atrial pressure during resuscitation.

Therefore, vasoconstrictor and venous capacitance-decreasing effects of the PARP inhibitor cannot be excluded.

In a more recent study in the rat, the effect of PARP inhibition has been tested in a combined trauma/hemorrhage/resuscitation model. Inhibition of PARP with 3-aminobenzamide prevented the development of vascular hyporeactivity *ex vivo*.[39]

With respect to trauma, another recent study has demonstrated the role of PARP in a head trauma model in the mouse. PARP-deficient mice had a faster recovery and better neurological performance than wild-type animals subjected to severe head trauma induced by a blunt device.[40] Similarly, the activation of PARP has been demonstrated in spinal cord trauma.[41] Although an *in vitro* study points toward the possibility that PARP activation plays a central role in the process,[42] the effect of PARP inhibition or PARP deficiency in this latter model remains to be tested *in vivo*.

5.3 SPLANCHNIC OCCLUSION SHOCK

A recent set of studies implicated the role of the peroxynitrite–PARP pathway in the pathophysiological changes associated with splanchnic occlusion shock.[43] In a rat model of splanchnic occlusion shock, which was induced in rats by clamping both the superior mesenteric artery and the celiac trunk for 45 min, followed by release of the clamp (reperfusion), there was a marked increase in the oxidation of dihydrorhodamine 123 to rhodamine (a marker of peroxynitrite-induced oxidative processes) in the plasma of the splanchnic occlusion (SAO)-shocked rats after reperfusion, but not during ischemia alone. Immunohistochemical examination demonstrated a marked increase in the immunoreactivity to nitrotyrosine, a specific "footprint" of peroxynitrite, in the necrotic ileum in shocked rats. In addition, in *ex vivo* studies in aortic rings from shocked rats, there was a reduction in the contractions to noradrenaline and impaired responsiveness to a relaxant effect to acetylcholine (vascular hyporeactivity and endothelial dysfunction, respectively). Splanchnic artery ischemia and reperfusion also resulted in a marked increase in epithelial permeability. 3-Aminobenzamide treatment significantly reduced ischemia/reperfusion injury in the bowel as evaluated by histological examination and significantly improved mean arterial blood pressure, improved contractile responsiveness to norepinephrine, enhanced the endothelium-dependent relaxations, and reduced the reperfusion-induced increase in epithelial permeability.[43] 3-Aminobenzamide also prevented the infiltration of neutrophils into the reperfused intestine, as evidenced by reduced myeloperoxidase activity; improved the histological status of the reperfused tissues; reduced the production of peroxynitrite during reperfusion; and improved survival.[43] These results demonstrate that the PARP inhibitor 3-aminobenzamide exerts multiple protective effects in splanchnic artery occlusion/reperfusion shock, and suggest that peroxynitrite and/or hydroxyl radical produced during the reperfusion phase trigger DNA strand breakage, PARP activation, and subsequent cellular dysfunction. The vascular endothelium is likely to represent an important cellular site of protection by 3-aminobenzamide in SAO shock. The reduced neutrophil infiltration and the reduced nitrotyrosine staining after inhibition of PARP occurs, despite the fact that 3-aminobenzamide is not a direct scavenger of peroxynitrite and does not inhibit the synthesis or action of its precursors. Its effect, therefore, may be related to the interruption of positive-feedback cycles; i.e., inhibition of PARP, in an indirect way, influences the amounts of reactive peroxynitrite produced. Although the mechanism of this action clearly requires further work, the following scenario should be considered. Because 3-aminobenzamide reduced neutrophil influx into the reperfused myocardium, it is expected that a subsequent reduced production of neutrophil-derived oxidants (including superoxide and peroxynitrite) occurs. It is well established that myocardial ischemia and reperfusion are associated with neutrophil accumulation with a subsequent burst of oxygen free radical production, activation of inflammation, excessive calcium entry, and, ultimately, cell death. Activation and accumulation of polymorphonuclear cells is one of the initial events of tissue injury, which triggers the release of oxygen free radicals, arachidonic acid metabolites, and lysosomal proteases, with subsequent more-pronounced infiltration of neutrophils into the reperfused tissues, excessive calcium entry, and, ultimately, cell death. Peroxynitrite and hydroxyl radical are known to exert cytotoxic effects to the vascular endothelium, and the mechanism of this injury, at least in part, is mediated by PARP activation (see above). Since endothelium-derived NO is a potent inhibitor of both neutrophil aggregation and adherence, an improvement of the endothelial function by 3-aminobenzamide would reduce the infiltration of neutrophils during reperfusion, thus resulting in reduced peroxynitrite formation and protection against the tissue injury. In other words, the following positive-feedback cycle may be present in splanchnic occlusion shock: early hydroxyl radical and peroxynitrite production → PARP-related endothelial injury → neutrophil infiltration → more hydroxyl and peroxynitrite production. Inhibition of PARP would intercept this cycle at the level of endothelial injury. This model would explain the reduction of nitrotyrosine immunoreactivity and dihydrorhodamine oxidation during reperfusion in the 3-aminobenzamide-treated rats: reduced neutrophil infiltration leads to reduced peroxynitrite generation. Another possibility to explain the reduced neutrophil infiltration after PARP inhibition may be related to inhibition of adhesion receptor molecule expression,[45] possibly as a result of inhibition of activation of signal transcription pathways governed by activation of nuclear factor κB.[36]

5.4 ZYMOSAN-INDUCED NONSEPTIC SHOCK AND MULTIPLE ORGAN FAILURE MODELS

Zymosan is a wall component of the yeast *Saccharomyces cerevisiae*. Injection of zymosan into experimental animals is known to produce an intense inflammatory response. Intraperitoneal injection of zymosan represents a convenient model to induce a nonseptic form of shock. This model, similar to most other forms of shock, is associated with organ failure, vascular dysfunction, proinflammatory mediator production, neutrophil recruitment, and mortality. In 1997, we demonstrated in zymosan-induced models of shock and inflammation that inhibition of PARP (by a pharmacological approach or by the use of genetically engineered animals) reduces neutrophil recruitment and accumulation into inflammatory tissue sites[44] (Figure 5.3). Extravasated neutrophils become activated once inside the inflammatory sites to secrete a variety of substances such as growth factors, chemokines and cytokines,

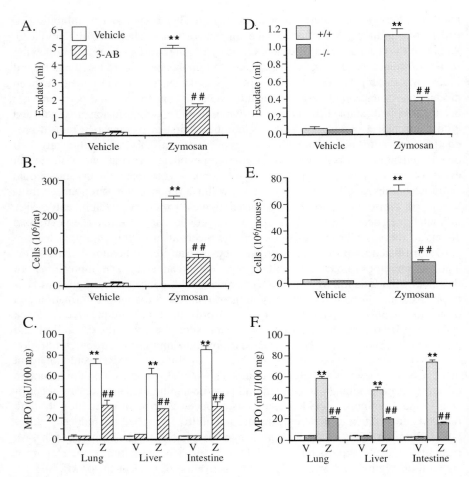

FIGURE 5.3 Effects of PARP inhibition on the course of the zymosan-induced multiple organ failure in rats (panels A to C) and mice (panels D to F). Data represent values at 18 h after vehicle (V) or zymosan (Z) administration (500 mg/kg i.p. for 18 h). (A and D) Exudate volumes; (B and E) neutrophil counts in the exudate; (C and F) myeloperoxidase (MPO) activities in lung, liver, and small intestine. In rats, PARP was inhibited by treatment with 3-aminobenzamide (10 mg/kg i.v. 10 min prior to zymosan administration, and repeated at every 6 h). In mice, responses in PARP$^{+/+}$ wild-type controls and PARP$^{-/-}$ animals were compared. Data are mean ± SEM of 5 to 6 animals in each group. **$p < 0.01$ represents significant increase in exudate volume, cell number, or MPO activity in the 3-aminobenzamide treated group. ##$p < 0.01$ represents significant reduction the various parameters in the group in which PARP was inhibited. (From Szabó, C. et al., *J. Exp. Med.*, 186, 1041–1049, 1997. With permission.)

complement components, proteases, NO, reactive oxygen metabolites, and peroxynitrite, which are important mediators of tissue injury. We hypothesized that prevention of neutrophil-dependent inflammatory pathways is likely to contribute to the reduced fluid extravasation and improved histological status after inhibition of PARP, and we proposed that a reduced neutrophil recruitment represents an important additional mechanism for the anti-inflammatory effects provided by inhibition of PARP. What, then, is the mechanism of the protection against neutrophil recruitment provided by inhibition of PARP? Data obtained in a study using isolated mesenteries inflamed with zymosan provided evidence that the effects of PARP inhibition are, at least in part, due to interference with neutrophil postadhesion phenomena.[44] The strongest indication for this conclusion derived from our experiments utilizing intravital microscopy, which allowed the characterization of temporally related processes, such as rolling, adhesion, and emigration. In this model, challenge with zymosan produced a significant degree of cell adhesion and emigration in the mouse mesenteric microcirculation. Treatment of mice with 3-aminobenzamide did not modify zymosan-induced cell adhesion to any extent, but the drug suppressed the degree of cell emigration. This clearly indicated that 3-aminobenzamide was affecting postadhesion phenomena such that only the number of emigrated cells was altered when analysis was done at a fixed time point. To investigate this phenomenon further, an appropriate protocol was set up to monitor the adherent leukocytes in "real time." Inflammation in the mouse mesentery was induced by zymosan, and postcapillary venules with a congruous number of adherent cells were selected, such that their fate could be monitored following intravenous challenge with 3-aminobenzamide or vehicle. Under these conditions, the PARP inhibitor produced a marked phenomenon of detachment (Figure 5.4).[44] The molecular mechanism of this action remains to be characterized. In light of the recent observations demonstrating that PARP regulates the activation of the transcription factor nuclear factor κB[36] (see also below), we find it conceivable that interference with the signal-transduction pathway of a yet unidentified surface adhesion molecule may mediate the observed effect of PARP inhibition.

Zymosan is also a convenient way to induce vascular failure *in vitro* or *in vivo*. In an *in vitro* study, incubation of rat aortic smooth muscle cells with zymosan-activated plasma induced the production of nitrite, the breakdown product of NO, due to the expression of iNOS over 6 to 24 h. In addition, zymosan-activated plasma triggered the production of peroxynitrite in these cells, as measured by the oxidation of the fluorescent dye dihydrorhodamine 123 and by nitrotyrosine Western blotting.[46] Incubation of the smooth muscle cells with zymosan-activated plasma induced DNA single-strand breakage and PARP activation. These effects were reduced by inhibition of NOS with N^G-methyl-L-arginine, and by glutathione (a scavenger of peroxynitrite). As expected, 3-aminobenzamide inhibited the zymosan-activated plasma-induced activation of PARP.[46] Incubation of thoracic aortae with zymosan-activated plasma *in vitro* caused a reduction of the contractions of the blood vessels to noradrenaline (vascular hyporeactivity) and elicited a reduced responsiveness to the endothelium-dependent vasodilator acetylcholine (endothelial dysfunction). Preincubation of the thoracic aortae with 3-aminobenzamide prevented the development of vascular hyporeactivity in response to zymosan-activated plasma. Moreover, glutathione and 3-aminobenzamide treatment protected against the zymosan-activated plasma-induced development of endothelial dysfunction.[46] Similarly, in *ex vivo* experiments, thoracic aorta rings of zymosan-treated rats showed a reduced contraction to noradrenaline and reduced responsiveness to the relaxant effect to acetylcholine (vascular

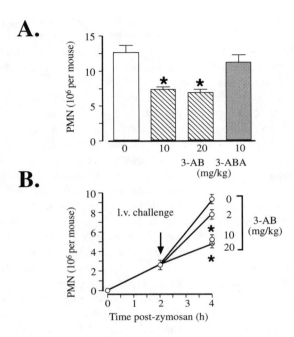

FIGURE 5.4 The PARP inhibitor 3-aminobenzamide inhibits neutrophil accumulation in murine peritoneal cavities in response to zymosan challenge. (A) Mice received various intravenous doses of 3-aminobenzamide (3-AB, 10 or 20 mg/kg) or its inactive structural analogue 3-aminobenzoic acid (3-ABA, 10 mg/kg) immediately before the i.p. administration of zymosan (12.5 mg/kg). Peritoneal cavities were washed 4 h later and the number of PMN in the lavage fluids quantified. Data are mean ± SEM of 6 to 12 mice per group. *$p < 0.05$ vs. control group (treated with vehicle). (B) Mice were treated with zymosan (1 mg i.p.) at time 0. Various doses of 3-aminobenzamide were given intravenously 2 h later and the number of PMN accumulated into peritoneal cavities quantified 4 h post-zymosan. Some mice were killed 2 h post-zymosan. Data are mean ± SEM of 8 to 14 mice per group. * $p < 0.05$ vs. control group (dose 0). (From Szabó, C. et al., *J. Exp. Med.*, 186, 1041–1049, 1997. With permission.)

hyporeactivity and endothelial dysfunction, respectively).[47] Treatment of zymosan-shocked rats with 3-aminobenzamide or nicotinamide, inhibitors of PARP, reduced peroxynitrite production and ameliorated cardiovascular dysfunction.[47]

5.5 REGULATION OF GENE EXPRESSION BY PARP: IMPLICATIONS FOR THE PATHOGENESIS OF CIRCULATORY SHOCK

It appears that PARP plays an important role in the regulation of gene expression and cell differentiation.[48,49] Under basal conditions, PARP is closely associated to DNA (especially at regions of cruciform DNA, bent DNA, and in A-T rich regions)[50] and regulates histone shuttling and nucleosomal unfolding.[51] PARP appears to be more frequently associated with transcriptionally active regions of chromatin.[52,53] In fact, a recent report proposes that PARP acts as a functional component of the positive cofactor 1 activity, its function being the enhancement activator-dependent transcription processes.[54]

Using pharmacological inhibitors of PARP, it has been demonstrated that the activity of PARP is required for the expression of the major histocompatibility complex class II gene,[55-57] ras, c-myc,[48,58] DNA methyltransferase gene,[58] and protein kinase C.[58] Moreover, in several independent lines of investigations, it has been demonstrated that pharmacological inhibition of PARP — with nicotinamide, 3-aminobenzamide, and 5-iodo-6-amino-1,2-benzopyrone (INH_2BP) — suppresses the expression of mRNA of iNOS.[13,29,59,60] In studies using INH_2BP, inhibition of iNOS expression in RAW macrophages was indicated by the inhibition of nitrite production, iNOS mRNA expression, and iNOS protein expression.[35] The regulation appeared to occur in the early stage of iNOS induction, since the PARP inhibitor used gradually lost its effectiveness when applied at increasing times *after* the stimulus for iNOS induction. Interestingly, in transfection studies using INH_2BP in murine RAW macrophages, it was found that this PARP inhibitor suppressed the transcription of iNOS, when cells transfected with the full-length (–1592 bp) promoter construct with INH_2BP. However, similar cotreatment of cells transfected with the –367 bp deletional construct did not significantly reduce the lipopolysaccharide (LPS)-mediated increase in luciferase activity.[35] The regulation by PARP of iNOS induction also occurred in whole animals challenged with LPS: pretreatment, but not post-treatment, of the animals with the inhibitor suppressed the LPS-induced increase in plasma nitrite/nitrate concentrations and reduced the LPS-induced increase in iNOS expression in the lung.[35]

Thus, from these experimental data it appears that PARP, via a not-yet-characterized mechanism, appears to regulate the process of iNOS expression. Inhibition of this process may represent an additional mode of beneficial action of PARP inhibition in various forms of inflammation. However, interpretation of the above data was hampered by pharmacological problems. For instance, in the studies quoted above, extremely high concentrations (10 to 30 mM) of the PARP inhibitors 3-aminobenzamide and nicotinamide were required to see suppression of iNOS induction, and the high concentrations of these agents may have had additional pharmacological actions, such as inhibition of total protein and RNA synthesis, and/or free radical scavenging actions.[13,24,59,60] On the other hand, the more potent PARP inhibitor INH_2BP effectively suppressed the expression of iNOS at lower concentrations (100 to 300 μM). However, in the case of INH_2BP several additional modes of action had to be considered, as this agent is a known inducer of alkaline phosphatases, with a secondary, pleiotropic modulation of cellular responses[35,58]: these effects may well include inhibition of MAP kinase activity, which, in its own right, may suppress the process of iNOS induction. Experiments in cells or animals with ablation of the PARP gene were required to address the question definitely whether inhibition of PARP per se suppresses the process of iNOS induction. In the first such study, using wild-type and PARP-deficient fibroblasts stimulated with bacterial endotoxin and interferon γ, we have demonstrated that the expression of iNOS gene is markedly suppressed in the absence of PARP.[61] Similar results were recently observed *in vivo*, in mice challenged with bacterial lipopolysaccharide, where circulating nitrite/nitrate levels and tissue iNOS protein expression were suppressed in the absence of PARP gene.[36] It is noteworthy that in *in vivo* experiments, PARP inhibitors have beneficial effects in concentrations that do not affect the expression of iNOS,[13,22,29] or even in conditions where iNOS is not even expressed, such as the early phases of myocardial and splanchnic reperfusion.[43,62,63] Therefore, it appears that inhibition of iNOS expression is not obligatory for the protective effect of PARP inhibitors in inflammation and reperfusion injury. Nevertheless, it is possible that, during more chronic administration, pharmacological inhibitors of PARP may suppress the expression of iNOS, and thereby reduce the generation of NO and peroxynitrite.

In addition to the iNOS gene, studies using wild-type and PARP-deficient mice established that PARP is regulating the expression of a whole host of genes. For example, in mice challenged with endotoxin, the absence of PARP gene suppressed the production of tumor necrosis factor α, and interferon γ (but not of interleukin 6).[36] The central mechanism whereby these effects may be governed on the level of signal transduction may be related to suppression of activation of the transcription factor nuclear factor κB, a key element in the signal transduction pathway of many proinflammatory cytokines[36] (Figure 5.5). The suppression of the expression of nuclear factor κB in the absence of functional PARP was somewhat predicted by prior *in vitro* studies demonstrating the same phenomenon in cultured macrophages.[45]

5.6 CONCLUSIONS AND IMPLICATIONS

Current strategies aimed at limiting NO-mediated cell/organ injury in shock include agents that inhibit the induction of iNOS; NOS enzyme inhibitors, preferably with selectivity for iNOS; agents that scavenge or inactivate NO; and agents that limit substrate or cofactor availability for iNOS. Less attention has been directed to strategies that interfere with intracellular cytotoxic pathways initiated by NO or its toxic derivatives. Direct and indirect experimental evidence reviewed in this chapter supports the view that peroxynitrite-induced DNA strand breakage and PARP activation contribute greatly to the pathophysiology of shock. Based on the data reviewed in this chapter, we conclude that PARP is a central mediator of circulatory shock, with multiple roles in mediating early and delayed pathophysiological pathways of shock (Figure 5.6). Pharmacological inactivation of PARP represents a novel, therapeutically viable strategy to limit cellular injury and improve the outcome of circulatory shock, as well as a variety of other pathophysiological conditions associated with peroxynitrite production. The viability of this potential therapeutic strategy is strengthened by recent observations demonstrating that the absence of PARP does not compromise DNA repair.[64] PARP inhibition is unlikely to interfere with the important antimicrobial effects of NO, since invading bacteria do not contain PARP. Because inhibition of iNOS has immunosuppressive effects and may facilitate the reoccurrence of latent infections in the treated organism, especially during a chronic treatment regime,[65,66] inhibition of PARP may have fewer side effects in this respect when compared with inhibition of iNOS. Additional theoretical advantages of PARP inhibitors as agents for the resuscitation in circulatory shock include (1) no obvious hemodynamic effects in control animals, (2)

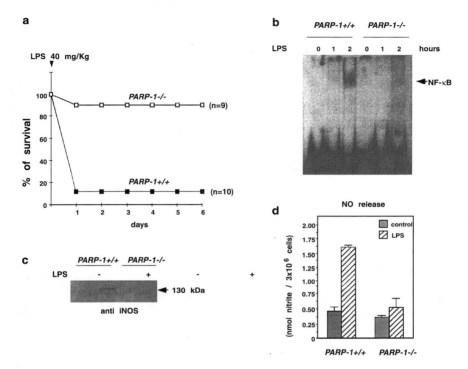

FIGURE 5.5 PARP-deficient mice are resistant to endotoxin-induced mortality, show reduced activation of nuclear factor κB, and express reduced amounts of iNOS. (a) Survival of wild-type and PARP-deficient mice after i.p. injection of bacterial LPS (40 mg/kg i.p.). The number of mice is given in the plot. (b) Nuclear factor κB activation in peritoneal macrophages from LPS-challenged mice. (c) Expression of iNOS at 18 h after endotoxin treatment. (d) Nitrite release in primary cultured murine macrophages from wild-type and PARP-deficient mice. (From Oliver, F.J. et al., *EMBO J.*, 18, 4446–4454, 1995. With permission.)

no side effects in control or shocked animals, and (3) the fact that the PARP inhibitor targets a delayed process of cell death permits the treatment to be effective even with a delayed start of administration.

It is noteworthy that the vast majority of the evidence implicating the role of PARP in peroxynitrite-induced toxicity and in other forms of cellular oxidant injury was obtained using pharmacological inhibitors, such as 3-aminobenzamide. This agent is a prototypical PARP inhibitor that has been used in a large number of investigations to inhibit the catalytic activity of PARP when DNA single-strand breakage was triggered by oxyradicals or by peroxynitrite. In these studies, 3-aminobenzamide provided cytoprotective effects but did not interfere with the development of DNA strand breakage, supporting the view that the agent, at 1 to 3 mM, does not scavenge oxyradicals or peroxynitrite.[29,67] There is also ample prior evidence that 3-aminobenzamide, in low millimolar concentrations, does not directly inhibit oxyradical-triggered oxidative processes.[25,68-71] Moreover, 3-aminobenzamide is not an inhibitor of NO synthase and does not scavenge NO.[29,67] Thus, it appears that 3-

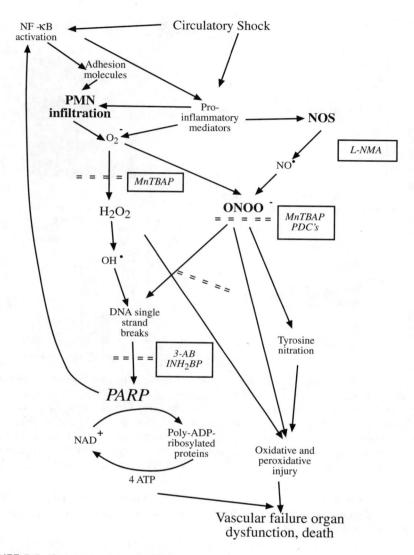

FIGURE 5.6 (See caption on page 97.)

aminobenzamide, indeed, exerts its protective actions by inhibition of the catalytic activity of PARP, rather than by interfering with proximal processes (such as the direct actions of peroxynitrite or oxyradicals). Nevertheless, it has been repeatedly put forward that 3-aminobenzamide may act as a hydroxyl radical scavenger,[72,73] thereby contributing to cytoprotection. Clearly, follow-up studies with cells and animals with ablated PARP gene are needed to confirm the conclusions of the studies with pharmacological inhibitors of PARP. In the first such study, Heller and co-workers[74] have, indeed, observed that islets of PARP$^{-/-}$ mice are resistant to NO and oxidant-related injury, when compared with the response in islets of wild-type mice.

FIGURE 5.6 Proposed scheme of PARP-dependent and PARP-independent cytotoxic pathways involving NO·, hydroxyl radical (OH·), and peroxynitrite (ONOO⁻) in shock. Endotoxin or other stimuli associated with various forms of circulatory shock trigger the release of proinflammatory mediators, which, in turn, induce the expression iNOS. NO, in turn, combines with superoxide to yield peroxynitrite. Hydroxyl radical (produced from superoxide via the iron-catalyzed Haber–Weiss reaction) and peroxynitrite or peroxynitrous acid induce the development of DNA single-strand breakage, with consequent activation of PARP. Depletion of the cellular NAD⁺ leads to inhibition of cellular ATP-generating pathways, leading to cellular dysfunction. NO alone does not induce DNA single-strand breakage, but may combine with superoxide (produced from the mitochondrial chain or from other cellular sources) to yield peroxynitrite. Under conditions of low cellular L-arginine NOS may produce both superoxide *and* NO, which then can combine to form peroxynitrite. PARP activation, in a not-yet-characterized fashion, promotes the activation of nuclear factor κB, and the expression of proinflammatory mediators, adhesion molecules, and iNOS. There are PARP-independent, parallel pathways of cellular metabolic inhibition, and these pathways can be activated by NO, hydroxyl radical, superoxide, and peroxynitrite (alone or in combination or synergy). For instance, peroxynitrite can induce cell injury via protein tyrosine nitration. PARP activation promotes adhesion receptor expression, postadhesion phenomena, and neutrophil (PMN) recruitment, thereby triggering positive-feedback cycles. The various interrelated pathways of shock can be interrupted by various pharmacological agents; some of them, such as the NOS inhibitor N^G-methyl-L-arginine or L-NMA, the superoxide dismutase mimetic and peroxynitrite scavenger porphyrinic compound MnTBAP, the peroxynitrite decomposition catalyst compounds or PDCs, and selected PARP inhibitors 3-aminobenzamide (3-AB) and INH₂BP, are presented on the figure. See text for further details.

Similarly, we observed that lung fibroblasts from PARP⁻/⁻ mice are protected from peroxynitrite-induced cell injury when compared with the fibroblasts of the corresponding wild-type animals.[61] Finally, PARP knockout mice were found to be remarkably resistant to endotoxin-induced mortality.[36] It remains to be dissected, to what extent inhibition of proinflammatory gene expression vs. inhibition of terminal cytotoxic pathways is responsible for the protection against shock seen in the absence of functional PARP.

ACKNOWLEDGMENT

This work was supported by grants from the National Institutes of Health (R29GM54773 and R01GM60915).

REFERENCES

1. Szabó, C. and Thiemermann, C., Role of nitric oxide in hemorrhagic, traumatic, and anaphylactic shock and thermal injury. *Shock,* 2, 145–155, 1994.
2. Szabó, C., Alterations in nitric oxide production in various forms of circulatory shock. *New Horizons,* 3, 2–32, 1995.
3. Kilbourn, R.G., Szabó, C., and Traber, D., Beneficial versus detrimental effects of nitric oxide synthase inhibitors in circulatory shock: lessons learned from experimental and clinical studies. *Shock,* 7, 235–246, 1997.

4. Szabó, C., Salzman, A.L., and Ischiropoulos, H., Endotoxin triggers the expression of an inducible isoform of NO synthase and the formation of peroxynitrite in the rat aorta *in vivo*. *FEBS Lett.*, 363, 235–238, 1995.
5. Szabó, C., Salzman, A.L., and Ischiropoulos, H., Peroxynitrite-mediated oxidation of dihydrorhodamine 123 occurs in early stages of endotoxic and hemorrhagic shock and ischemia–reperfusion injury. *FEBS Lett.*, 372, 229–232, 1995.
6. Wizemann, T., Gardner, C., Laskin, J., Quinones, S., Durham, S., Goller, N., Ohnishi, T., and Laskin D., Production of nitric oxide and peroxynitrite in the lung during acute endotoxemia. *J. Leukoc. Biol.*, 56, 759–768, 1994.
7. Szabó, C. and Dawson, V.L., Role of poly(ADP-ribose) synthetase activation in inflammation and reperfusion injury. *Trends Pharmacol. Sci.*, 19, 287–298, 1998.
8. Szabó, C., Role of poly (ADP-ribose) synthetase in inflammation. *Eur. J. Pharmacol.*, 350, 1–19, 1998.
9. Rees, D.D., Role of nitric oxide in the vascular dysfunction of septic shock. *Biochem. Soc. Trans.*, 23, 1025–1029, 1995.
10. Zingarelli, B., Day, B.J., Crapo, J., Salzman, A.L., and Szabó, C., The potential involvement of peroxynitrite in the pathogenesis of endotoxic shock. *Br. J. Pharmacol.*, 120, 259–267, 1997.
11. Gryglewski, R.J., Palmer, R.M., and Moncada, S., Superoxide anion is involved in the breakdown of endothelium-derived vascular relaxing factor. *Nature*, 320, 454–456, 1996.
12. Rubanyi, G.M. and Vanhoutte, P.M., Superoxide anions and hyperoxia inactivate endothelium-derived relaxing factor. *Am. J. Physiol.*, 250, H822–H827, 1996.
13. Szabó, C., Zingarelli, B., and Salzman, A.L., Role of poly-ADP ribosyltransferase activation in the nitric oxide- and peroxynitrite-induced vascular failure. *Circ. Res.*, 78, 1051–1063, 1996.
14. Chabot, F., Mitchell, J.A,, Quinlan, G.J., and Evans, T.W., Characterization of the vasodilator properties of peroxynitrite on rat pulmonary artery: role of poly (adenosine 5′-diphosphoribose) synthase. *Br. J. Pharmacol.*, 121, 485–490, 1997.
15. Zingarelli, B., Salzman, A.L, and Szabó, C., Protective effects of nicotinamide against nitric oxide-mediated vascular failure in endotoxic shock: potential involvement of polyADP ribosyl synthetase. *Shock*, 5, 258–264, 1996.
16. Crow, J.P. and Beckman, J.S., The role of peroxynitrite in nitric oxide-mediated toxicity. *Curr. Top. Microbiol. Immunol.*, 196, 57–73, 1995.
17. White, C.R., Brock, T.A., Chang, L.Y., Crapo, J., Briscoe, P., Ku, D., Bradley, W.A., Gianturco, S.H., Gore, J., and Freeman, B.A., Superoxide and peroxynitrite in atherosclerosis. *Proc. Natl. Acad. Sci. U.S.A.*, 9, 1044–1048, 1994.
18. White, C.R., Darley-Usmar, V., Berrington, W.R., McAdams, M., Gore, J.Z., Thompson, J.A., Parks, D.A., Tarpey, M.M., and Freeman, B.A., Circulating plasma xanthine oxidase contributes to vascular dysfunction in hypercholesterolemic rabbits. *Proc. Natl. Acad. Sci. U.S.A.*, 93, 8745–8749, 1996.
19. Villa, L.M., Salas, E., Darley-Usmar, M., Radomski, M.E.W., and Moncada, S., Peroxynitrite induces both vasodilatation and impaired vascular relaxation in the isolated perfused rat heart. *Proc. Natl. Acad. Sci. U.S.A.*, 91, 12383–12387, 1994.
20. Fatehi-Hassanabad, Z., Burns, H., Aughey, E.A., Paul, A., Plevin, R., Parratt, J.R., and Furman, B.L., Effects of L-canavanine, an inhibitor of inducible nitric oxide synthase, on endotoxin mediated shock in rats. *Shock*, 6, 194–200, 1996.
21. Szabó, C., Bryk, R., Zingarelli, B., Southan, G.J., Gahman, T.C., Bhat, V., Salzman, A.L., and Wolff, D.J., Pharmacological characterization of guanidinoethyldisulphide (GED), a novel inhibitor of nitric oxide synthase with selectivity towards the inducible isoform. *Br. J. Pharmacol.*, 118, 1659–1668, 1996.

22. Szabó, C., Cuzzocrea, S., Zingarelli, B., O'Connor, M., and Salzman, A.L., Endothelial dysfunction in endotoxic shock: importance of the activation of poly(ADP ribose) synthetase (PARS) by peroxynitrite. *J. Clin. Invest.*, 100, 723–735, 1997.
23. Pulido, E.J., Shames, B.D., Selzman, C.H., Barton, H.A., Banerjee, A., Bensard, D.D., and McIntyre, R.C., Inhibition of poly(ADP-ribose) synthetase attenuates endotoxin-induced dysfunction of pulmonary vasorelaxation. *Am. J. Physiol.*, 277, L769–776, 1999.
24. Spragg, R.G., DNA strand break formation following exposure of bovine pulmonary artery and aortic endothelial cells to reactive oxygen products. *Am. J. Respir. Cell. Mol. Biol.*, 4, 4–10, 1991.
25. Thies, R.L. and Autor, A.P., Reactive oxygen injury to cultured pulmonary artery endothelial cells: mediation by polyADP-ribose polymerase activation causing NAD depletion and altered energy balance. *Arch. Biochem. Biophys.*, 286, 353–363, 1991.
26. Junod, A.F., Jornot, L., and Petersen, H., Differential effects of hyperoxia and hydrogen peroxide on DNA damage, polyadenosine diphosphate-ribose polymerase activity, and nicotinamide adenine dinucleotide and adenosine triphosphate contents in cultured endothelial cells and fibroblasts. *J. Cell. Physiol.*, 140, 177–185, 1991.
27. Southan, G.J. and Szabó, C., Selective pharmacological inhibition of distinct nitric oxide synthase isoforms. *Biochem. Pharmacol.*, 51, 383–394, 1995.
28. Szabó, C., Day, B.J., and Salzman, A.L., Evaluation of the relative contribution of nitric oxide and peroxynitrite to the suppression of mitochondrial respiration in immunostimulated macrophages, using a novel mesoporphyrin superoxide dismutase analog and peroxynitrite scavenger. *FEBS Lett.*, 381, 82–86, 1996.
29. Zingarelli, B., O'Connor, M., Wong, H., Salzman, A.L., and Szabó, C., Peroxynitrite-mediated DNA strand breakage activates poly-ADP ribosyl synthetase and causes cellular energy depletion in macrophages stimulated with bacterial lipopolysaccharide. *J. Immunol.*, 156, 350–358, 1996.
30. Szabó, A., Salzman, A.L., and Szabó, C., Inhibition of poly (ADP-ribose) synthetase ameliorates endotoxin-induced intestinal and pulmonary leakage. *Life Sci.*, 63, 2133–2139, 1998.
31. Szabó, C., Saunders, C., O'Connor, M., and Salzman, A.L., Peroxynitrite causes energy depletion and increases permeability via activation of poly-ADP ribosyl synthetase in pulmonary epithelial cells. *Am. J. Respir. Mol. Cell. Biol.*, 16, 105–109, 1997.
32. Kennedy, M., Szabó, C., and Salzman, A.L., Activation of poly(ADP-ribose) synthetase (PARS) mediates cytotoxicity induced by peroxynitrite in human intestinal epithelial cells. *Gastroenterology*, 114, 510–518, 1998.
33. Wray, G.M., Hinds, C.J., and Thiemermann, C., Effects of inhibitors of poly(ADP-ribose) synthetase activity on hypotension and multiple organ dysfunction caused by endotoxin. *Shock*, 10, 13–9, 1998.
34. Yamamoto, K., Tsukidate, K., and Farber, J.L., Differing effects of the inhibition of polyADP-ribose polymerase on the course of oxidative cell injury in hepatocytes and fibroblasts. *Biochem. Pharmacol.*, 46, 483–491, 1993.
35. Szabó, C., Wong, H.R., Bauer, P.I., Kirsten, E., O'Connor, M., Zingarelli, B., Mendeleyev, J., Hasko, G., Vizi, E.S., Salzman, A.L., and Kun, E., Regulation of components of the inflammatory response by 5-iodo-6-amino-1,2-benzopyrone, and inhibitor of poly(ADP-ribose) synthetase and pleiotropic modifier of cellular signal pathways. *Int. J. Oncol.*, 10, 1093–1101, 1997.
36. Oliver, F.J., Ménissier-de Murcia, J., Nacci, C., Decker, P., Andriantsitohaina, R., Muller, S., De la Rubia, G., Stoclet, J.C., and de Murcia, G., Resistance to endotoxic shock as a consequence of defective NF-κB activation in poly (ADP-ribose) polymerase-1 deficient mice. *EMBO J.*, 18, 4446–4454, 1999.

37. Szabó, C., Potential role of the peroxynitrite-poly(ADP-ribose) synthetase pathway in a rat model of severe hemorrhagic shock. *Shock,* 9, 341–344, 1998.
38. Szabó, A., Hake, P., Salzman, A.L., and Szabó, C., Inhibition of poly(ADP-ribose) synthetase exerts protective effects in a porcine model of hemorrhagic shock. *Shock,* 10, 347–353, 1998.
39. St. John, J., Barbee, R.W., Sonin, N., Clemens, M.G., and Watts, J.A., Inhibition of poly(ADP-ribose) synthetase improves vascular contractile responses following trauma-hemorrhage and resuscitation. *Shock,* 12, 188–195, 1999.
40. Whalen, M.J., Clark, R.S.B., Dixon, C.E., Robichaud, P., Marion, D.W., Vagni, V., Graham, S., Virág, L., Haskó, G., Stachlewitz, R., Szabó, C., and Kochanek, P., Reduction in deficits of memory and motor function after traumatic brain injury in mice deficient in poly(ADP-ribose) polymerase. *J. Cereb. Blood Flow Metab.,* 19, 832–842, 1999.
41. Scott, G.S., Jakeman, L., Stokes, B., and Szabó, C., Production of peroxynitrite and activation of poly(ADP-ribose) synthetase in spinal cord trauma. *Ann. Neurol.,* 45, 120–124, 1999.
42. Wallis, R.A., Panizzon, K.L., and Girard, J.M., Traumatic neuroprotection with inhibitors of nitric oxide and ADP-ribosylation. *Brain Res.,* 710, 169–177, 1996.
43. Cuzzocrea, S., Zingarelli, B., Costantino, G., Szabó, A., Salzman , A.L., Caputi, A.P., and Szabó, C., Beneficial effects of 3-aminobenzamide, an inhibitor of poly(ADP-ribose) synthetase in a rat model of splanchnic artery occlusion and reperfusion. *Br. J. Pharmacol.,* 121, 1065–1074, 1997.
44. Szabó, C., Lim, L.H., Cuzzocrea, S., Getting, S.J., Zingarelli, B., Flower, R.J., Salzman, A.L., and Perretti, M., Inhibition of poly(ADP-ribose) synthetase exerts anti-inflammatory effects and inhibits neutrophil recruitment. *J. Exp. Med.,* 186, 1041–1049, 1997.
45. Zingarelli, B., Salzman. A.L., and Szabó, C., Genetic disruption of poly(ADP ribose) synthetase inhibits the expression of P-selectin and intercellular adhesion molecule-1 in myocardial ischemia–reperfusion injury. *Circ. Res.,* 83, 85–94, 1998.
46. Cuzzocrea, S., Zingarelli, B., O'Connor, M., Salzman, A.L., and Szabó, C., Role of peroxynitrite and activation of poly(ADP-ribose) synthetase in the vascular failure induced by zymosan-activated plasma. *Br. J. Pharmacol.,* 122, 493–503, 1997.
47. Cuzzocrea, S., Zingarelli, B., and Caputi, A.P., Role of peroxynitrite and poly(ADP-ribosyl) synthetase activation in cardiovascular derangement induced by zymosan in the rat. *Life Sci.,* 63, 923–933, 1998.
48. Nagao, M., Nakayasu, M., Aonuma, S., Shima, H., and Sugimura, T., Loss of amplified genes by poly(ADP-ribose) polymerase inhibitors. *Environ. Health Perspect.,* 93, 169–174, 1991.
49. Smulson, M.E., Kang,V.H., Ntambi, J.M., Rosenthal, D.S., Ding, R., and Simbulan, C.M., Requirement for the expression of poly(ADP-ribose) polymerase during the early stages of differentiation of 3T3-L1 preadipocytes, as studied by antisense RNA induction. *J. Biol. Chem.,* 270, 119–127, 1995.
50. Sastry, S.S., Buki, K., and Kun, E., Binding of adenosine diphosphoribosyltransferase to the termini and internal regions of linear DNA. *Biochemistry,* 28, 5680–5687, 1989.
51. Althaus, F.R., Hofferer, L., Kleczkowska, H.E., Malanga, M., Naegeli, H., Panzeter, P.L., and Realini, C.A., Histone shuttling by poly ADP-ribosylation. *Mol. Cell. Biochem.* 138, 53–9, 1994.
52. Hough, C.J. and Smulson, M.E., Association of poly(adenosine diphosphate ribosylated) nucleosomes with transcriptionally active and inactive regions of chromatin. *Biochemistry,* 23, 5016–5023, 1994.

53. de Murcia, G., Huletsky, A., and Poirier, G.G., Modulation of chromatin structure by polyADP ribosylation. *Biochem. Cell Biol.,* 66, 626–635, 1998.
54. Meisterernst, M., Stelzer, G, and Roeder, R.G., Poly(ADP-ribose) polymerase enhances activator-dependent transcription *in vitro. Proc. Natl. Acad. Sci. U.S.A.,* 94, 2261–2265, 1997.
55. Hiromatsu, Y., Sato, M., Yamada, K, and Nonaka, K., Nicotinamide and 3-aminobenzamide inhibit recombinant human interferon-γ-induced HLA-DR antigen expression, but not HLA-A, B, C antigen expression, on cultured human thyroid cells. *Clin. Endocrinol.,* 36, 91–95, 1992.
56. Taniguchi, T., Nemoto, Y., Ota, K., Imai, T., and Tobari, J., Participation of poly(ADP-ribose) synthetase in the process of norepinephrine-induced inhibition of major histocompatibility complex class II antigen expression in human astrocytoma cells. *Biochem. Biophys. Res. Commun.,* 193, 886–889, 1993.
57. Qu, Z., Fujimoto, S., and Taniguchi, T., Enhancement of interferon-γ induced major histocompatibility complex class II gene expression by expressing an antisense RNA of poly(ADP-ribose) synthetase. *J. Biol. Chem.,* 269, 5543–5547, 1994.
58. Bauer, P.I., Kirsten, E., Young, L.J.T., Varadi, G., Csonka, E., Buki, K.G., Mikala, G., Hu, R., Comstock, J.A., Mendeleyev, J., Hakam, A., and Kun, E., Modification of growth related enzymatic pathways and apparent loss of tumorigenicity of a *ras*-transformed bovine endothelial cell line by treatment with 5-iodo-6-amino-1,2-benzopyrone. *Int. J. Oncol.,* 8, 239–252, 1996.
59. Hauschildt, S., Scheipers, P., Bessler, W.G., and Mulsch, A., Induction of nitric oxide synthase in L929 cells by tumour-necrosis factor α is prevented by inhibitors of poly(ADP-ribose) polymerase. *Biochem. J.,* 288, 255–260, 1992.
60. Pellat-Seceunyk, K., Wietzerbin, J., and Drapier, J.C., Nicotinamide inhibits nitric oxide synthase mRNA induction in activated macrophages. *Biochem. J.,* 297, 53–58, 1994.
61. Szabó, C., Virág, L., Cuzzocrea, S., Scott, G.J., Hake, P., O'Connor, M.P, Zingarelli, B., Salzman, A.L., and Kun, E., Protection against peroxynitrite-induced fibroblast injury and arthritis development by inhibition of poly(ADP-ribose) synthetase. *Proc. Natl. Acad. Sci. U.S.A.,* 95, 3867–3872, 1998.
62. Zingarelli, B., Cuzzocrea, S., Zsengeller, Z., Salzman, A.L., and Szabó, C., Beneficial effect of inhibition of poly-ADP ribose synthetase activity in myocardial ischemia-reperfusion injury. *Cardiovasc. Res.,* 36, 205–215, 1997.
63. Thiemermann, C., Bowes, J., Myint, F.P., and Vane, J.R., Inhibition of the activity of poly(ADP ribose) synthetase reduces ischemia-reperfusion injury in the heart and skeletal muscle. *Proc. Natl. Acad. Sci. U.S.A.,* 94, 679–683, 1997.
64. Wang, Z.Q., Auer, B., Stingl, L., Berghammer, H., Haidacher, D., Schweiger, M., and Wagner, E.F., Mice lacking ADPRT and poly(ADP-ribosyl)ation develop normally but are susceptible to skin disease. *Genes Dev.,* 9, 510–520, 1995.
65. Stenger, S., Donhauser, N., Thuring, H., Rollingh, M., and Bogdan, C., Reactivation of latent leishmaniasis by inhibition of inducible nitric oxide synthase. *J. Exp. Med.,* 183, 1501–1514, 1996.
66. Augusto, O., Linares, E., and Giorgio, S., Possible roles of nitric oxide and peroxynitrite in murine leishmaniasis. *Braz. J. Med. Biol. Res.,* 29, 853–862, 1996.
67. Zhang, J., Dawson, V.L., Dawson, T.M., and Snyder, S.H., Nitric oxide activation of poly(ADP-ribose) synthetase in neurotoxicity. *Science,* 263, 687–689, 1994.
68. Berger, S.J., Sudar, D.C., and Berger, N.A., Metabolic consequences of DNA damage: DNA damage induces alterations in glucose metabolism by activation of poly(ADP-ribose) polymerase. *Biochem. Biophys. Res. Commun.,* 134, 227–232, 1996.

69. Schraufstatter, I., Hinshaw, D., Hyslop, P., Spragg, R., and Cochrane, C., Oxidant injury of cells. DNA strand breaks activate polyadenosine diphosphate-ribose polymerase and lead to depletion of nicotinamide adenine dinucleotide. *J. Clin. Invest.*, 77, 1312–1330, 1986.
70. Schraufstatter, I.U., Hyslop, P.A., Jackson, J.H., and Cochrane, C.G., Oxidant-induced DNA damage of target cells. *J. Clin. Invest.*, 82, 1040–1050, 1988.
71. Berger, N.A., Oxidant-induced cytotoxicity: a challenge for metabolic modulation. *Am. J. Respir. Cell. Mol. Biol.*, 4, 1–3, 1991.
72. Wilson, G.L., Patton, N.J., McCord, J.M., Mullins, D.W., and Mossman, B.T., Mechanisms of streptozotocin- and alloxan-induced damage in rat B cells. *Diabetologia*, 27, 587–591, 1984.
73. Farber, L.J., Kyle, M.E., and Coleman, J.B., Biology of disease. Mechanisms of cell injury by activated oxygen species. *Lab. Invest.*, 62, 670–679, 1990.
74. Heller, B., Wang, Z.Q., Wagner, E.F., Radons, J., Burkle, A., Fehsel, K., Burkart, V., and Kolb, H., Inactivation of the poly(ADP-ribose) polymerase gene affects oxygen radical and nitric oxide toxicity in islet cells. *J. Biol. Chem.*, 270, 11176–11180, 1995.

6 Role of Poly(ADP-Ribose) Polymerase in the Pathogenesis of Pancreatic Islet Cell Death and Type 1 Diabetes

Volker Burkart

CONTENTS

6.1 Introduction 104
6.2 Type 1 Diabetes 104
    6.2.1 Definition 104
    6.2.2 Pathogenesis 104
6.3 Beta Cell–Damaging Cellular Effector Mechanisms 106
    6.3.1 Cellular Sources of Beta Cell-Damaging Inflammatory Mediators 106
    6.3.2 Radical-Induced Primary Damage in the Beta Cell 107
    6.3.3 Consequence of Primary DNA Damage in the Beta Cell 107
6.4 Inhibition of PARP Activity for Beta Cell Protection 108
    6.4.1 Pharmacological PARP Inhibition 108
        6.4.1.1 Models to Study the Effects of PARP Inhibition 108
        6.4.1.2 Effect of PARP Inhibition on Diabetes Development 109
        6.4.1.3 Effect of PARP Inhibition on Beta Cell Susceptibility 110
    6.4.2 PARP-Deficient Animals 112
        6.4.2.1 Animal Models with PARP Deficiency 112
        6.4.2.2 Effect of PARP Deficiency on Diabetes Development 113
        6.4.2.3 Effect of PARP Deficiency on Beta Cell Susceptibility 113
6.5 Potential Involvement of PARP in Islet Cell Death Pathways 115
    6.5.1 Interaction of PARP Activity with Mitochondrial Function 115
    6.5.2 Potential Role of PARP in Apoptosis or Necrosis of Beta Cells 116

    6.5.3   PARP-Independent Pathways of Beta Cell Death 116
6.6   Clinical Trials to Preserve Beta Cell Function by PARP Inhibition 118
    6.6.1   PARP Inhibition after Onset of Type 1 Diabetes............................. 118
    6.6.2   PARP Inhibition for the Prevention of Type 1 Diabetes 118
6.7   Summary and Outlook.. 119
Acknowledgment 120
References... 120

6.1 INTRODUCTION

In the last two decades considerable progress has been achieved in the elucidation of the immunological processes involved in the pathogenesis of type 1 (insulin-dependent) diabetes. The application of newly available methodologies allowed novel experimental approaches leading to the characterization of critical steps in the destruction of insulin-producing pancreatic beta cells. Studies on inflammatory islet cell destruction identified nitric oxide (NO) as a major beta cell toxic mediator and described DNA strand breaks as critical primary damage triggering beta cell death. Recent studies provided unequivocal evidence for the hypothesis that the activation of the nuclear enzyme poly(ADP-ribose) polymerase (PARP) with subsequent intracellular NAD^+ depletion represent the determining metabolic steps in the chain of events causing progressive inflammatory beta cell death and diabetes in the mouse model studied.

6.2 TYPE 1 DIABETES

6.2.1 Definition

Human type 1 (insulin-dependent) diabetes mellitus (IDDM) is the most common chronic disease in children and young adults. The manifestation of type 1 diabetes, which generally occurs before the age of 20 years, is characterized by the sudden occurrence of severe hyperglycemia and/or ketacidosis accompanied by polyuria, polydipsia, and weight loss.[1] These symptoms result from a marked hypoinsulinemia due to absolute, persistent deficiency of insulin secretion after an irreversible loss of insulin-producing beta cells from the pancreatic islets of Langerhans.

During the last decades evidence has accumulated that chronic immunological processes are responsible for the progredient destruction of the beta cells, finally leading to a near complete loss of the insulin secretory capacity and the manifestation of diabetes in the affected individual.[2-5]

6.2.2 Pathogenesis

Human studies that examine the pathogenetic processes involved in beta cell destruction revealed signs of humoral and cellular immune reactivity directed against beta cells long before diabetes manifestation (Figure 6.1). In the serum of a large proportion of "prediabetic" individuals (auto-)antibodies can be detected with specificity

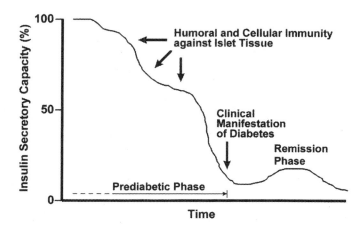

FIGURE 6.1 Hypothetical time course of the events involved in the reduction of the pancreatic beta cell mass and the development of type 1 diabetes.

to beta cell antigens such as insulin,[6,7] proinsulin,[8] protein tyrosine phosphatase 2,[9,10] or glutamic acid decarboxylase.[11-13] A few human studies available now report the presence of inflammatory cells, like macrophages and lymphocytes, within islets of pancreas biopsy specimens taken from individuals at risk to develop type 1 diabetes or after recent onset of the disease.[14-16] These findings strongly support the view that immune reactions directed against autologous beta cells are responsible for the gradual loss of the insulin secretory capacity preceding the clinical manifestation of diabetes. After disease manifestation, followed by an occasional remission period with reduced insulin requirements, the patients become strictly insulin dependent for the rest of their lives (see Figure 6.1).

Based on these findings, extensive studies were performed to evaluate the beta cell–damaging potential of the humoral and cellular components of the islet-directed immune attack during the pathogenesis of type 1 diabetes.[17,18] These investigations progressed rapidly once rodent models with spontaneous development of diabetes similar to human type 1 diabetes became available.[19,20] Initially, the most conclusive observations came from experiments in rodents rendered diabetic by the application of the beta cell–specific toxin streptozocin[19,21,22] (Figure 6.2). Later, the findings of these studies could be extended and refined by the use of BioBreeding (BB) rats[20] and nonobese diabetic (NOD) mice,[19,23] which spontaneously develop diabetes.

Diabetes development in these rodent models was found to share basic features with the pathogenesis of the human disease: before clinical manifestation of the disease, signs of islet-directed immune reactivity are observed, such as the appearance of autoantibodies against beta cell components[24] and the progressive infiltration of the islets with immune cells (insulitis).[19,20] Investigations in these animal models confirmed previous observations in humans, strongly indicating that the destruction of beta cells results mainly from cellular immune reactions rather than humoral immune phenomena.

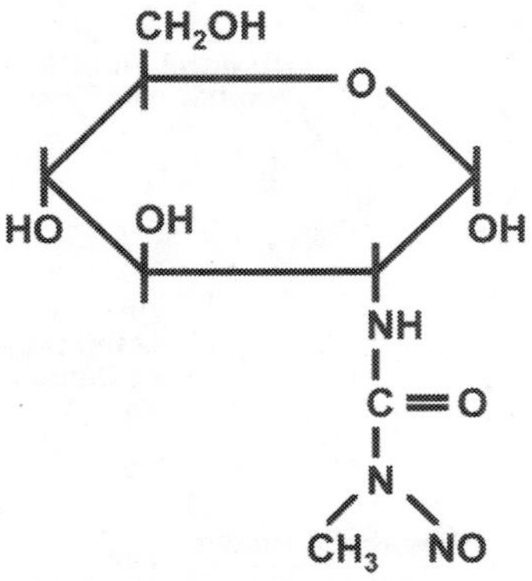

FIGURE 6.2 Structural formula of streptozocin (2-deoxy-2-(3-methyl-3-nitrosoureido)-D-gluco-pyranose).

6.3 BETA CELL–DAMAGING CELLULAR EFFECTOR MECHANISMS

Meanwhile, several *in vitro* and *in vivo* studies have provided strong evidence for islet-directed cellular immune reactions as important mediators of beta cell destruction. The infiltration of the islets with immune cells prior to diabetes manifestation appears to be the most prominent morphological sign of the cell-mediated immune attack against islets. During this inflammatory process, macrophages, as early islet-infiltrating cells,[25] play a major role as beta cell–damaging effector cells[26] and also as regulators of the initiation and progression of the T-lymphocyte-mediated immune response against islet cells.[27-29]

6.3.1 Cellular Sources of Beta Cell–Damaging Inflammatory Mediators

During the initial phase of insulitis, which is dominated by the presence of macrophages surrounded by destroyed beta cells,[30] a variety of inflammatory mediators are released,[31-33] which exert islet cell–damaging activities ranging from functional disturbances[34] to the induction of cell death.[35] Meanwhile, reactive radicals, especially NO,[36,37] were identified as major beta cell toxic mediators.[38] High concentrations of NO can be formed by the inducible isoform of NO synthase (iNOS) in activated islet-infiltrating macrophages[39] and in islet capillary endothelial cells exposed to inflammatory cytokines, such as interleukin-1β (IL-1β), tumor necrosis factor α (TNF-α), or interferon γ (IFNγ).[40-42] Another potential source of toxic radicals during islet inflammation is the beta cell itself, which is able to produce large amounts of NO after induction of its intracellular iNOS by IL-1β alone[43-46] or in combination with other inflammatory cytokines.[47] Besides NO release, the formation of reactive oxygen intermediates (ROI) may also play a role in beta cell destruction. Interestingly, macrophages from diabetes-prone BB rats were found to produce considerably increased amounts of ROI compared with macrophages from normal Wistar rats or diabetes-resistant BB rats.[48]

6.3.2 Radical-Induced Primary Damage in the Beta Cell

In the attempt to identify the initial events of inflammatory beta cell death, isolated rodent and human pancreatic islets, islet cells, or cells of beta cell lines were exposed to inflammatory mediators such as reactive radicals and cytokines, which are assumed to be present in the inflamed islet tissue.

When cultivated cells were exposed to chemically generated radicals to mimick macrophage- or endothelial cell–derived NO or ROI, severe DNA damage could be detected within 1 h. DNA strand breaks, detected by *in situ* nick translation, clearly preceded islet cell death for several hours. Exposure to chemical NO donors or to ROI, enzymatically generated by xanthine oxidase, resulted in cell lysis only 5 to 6 h after the initiation of radical formation.[49] In these studies, islet cell lysis was determined by the incapability of the cell to exclude trypan blue, reflecting the loss of membrane integrity as a definite sign of irreversible lethal cell damage.[50] In addition to DNA damage, delayed appearance of mitochondrial dysfunction was observed after exposure to enzymatically generated ROI.[51] Moreover, NO and ROI can react together to form the highly toxic peroxynitrite, which is able to induce rapid DNA damage in cultivated human and rat islet cells.[52]

As mentioned above, exposure of islet cells to inflammatory cytokines induces the formation of NO by iNOS within the beta cell.[45] Intracellularly generated NO can cause a variety of metabolic dysfunctions in beta cells, e.g., the reduction of the insulin secretory capacity[34,53] and the impairment of mitochondrial DNA[54] and enzyme systems.[44] However, similar to the findings after exposure to extracellularly generated NO, the most prominent damage observed after cytokine-induced generation of endogenous NO appears to be the rapid formation of nuclear DNA strand breaks.[55] Isolated rat islets exhibited extensive DNA damage already after 15 min of IL-1β exposure, as could be demonstrated by the electrophoretic separation of DNA fragments in the "comet assay."[56] In beta cell lines, the cytokine combination IL-1β, TNF-α, and IFNγ induced DNA fragmentation after 3 to 6 h, preceding cell lysis for several hours.[35,57]

These observations clearly identify radical-induced DNA fragmentation as a prominent primary damage in beta cells exposed to a variety of inflammatory mediators. In addition, several studies provide evidence for mitochondrial dysfunction and damage as potentially critical events that precede and probably contribute to cell death.

6.3.3 Consequence of Primary DNA Damage in the Beta Cell

A hypothesis of the possible consequences of primary DNA damage for the survival of beta cells was originally advanced by Okamoto and co-workers (Uchigata et al.[58]) based on their experimental work with rodent pancreatic islets and islet cells. When they exposed isolated islets and islet cells to the beta cell toxin streptozocin, they observed a rapid appearance of DNA fragments and a decrease of intracellular NAD$^+$ contents, strongly indicating a role for the NAD$^+$-consuming enzyme PARP (EC 2.4.2.30), which is activated by DNA strand breaks.[59] Consequently, Okamoto and his group investigated the involvement of PARP in this process by applying pharmaceutical inhibitors of the enzyme. In fact, they could effectively protect islet cells from streptozocin-induced damage by suppressing PARP activity. Although PARP inhibition by nicotinamide, picolinamide, or benzamide derivatives did not affect streptozocin-induced formation of primary DNA damage, the NAD$^+$ contents and the proinsulin synthesis of the islet cells were well preserved.[58,60] They could further extend their observations to the involvement of PARP in the induction of beta cell damage *in vivo*. The administration of diabetogenic doses of streptozocin to rats caused extensive DNA strand breaks and markedly decreased intracellular NAD$^+$ levels in pancreatic islets within 20 min after injection.[61] Pretreatment of rats with the PARP inhibitors nicotinamide and 3-aminobenzamide was found to protect the islets from the streptozocin-induced decrease of proinsulin biosynthesis.[62] From these observations Okamoto and his co-workers proposed the following sequence of initial events leading to the inhibition of proinsulin synthesis by beta cell–damaging agents with alkylating or radical-forming properties: islet DNA strand breaks → stimulation of nuclear poly(ADP-ribose) synthetase → depletion of intracellular NAD → inhibition of proinsulin synthesis.[58,61]

6.4 INHIBITION OF PARP ACTIVITY FOR BETA CELL PROTECTION

The hypothesis by Uchigata et al.,[58] outlined above, was confirmed and refined by numerous *in vivo* and *in vitro* studies aiming to elucidate the role of PARP in the first steps of beta cell damage and in the development of diabetes.

6.4.1 PHARMACOLOGICAL PARP INHIBITION

6.4.1.1 Models to Study the Effects of PARP Inhibition

Initial investigations were performed in animal models of streptozocin-induced diabetes in rodents, which had been established[63-65] and are still used extensively to study the mechanisms involved in the destruction of pancreatic beta cells.[19,66] Streptozocin, an antibiotic produced by *Streptomyces achromogenes*, contains a glucose moiety (see Figure 6.2) and therefore is taken up rather selectively by the beta cell via the glucose transporter GLUT-2.[67,68] When the substance decomposes within the beta cell it causes rapid and considerable damage to nuclear DNA by alkylation.[59] More recently it could be demonstrated that streptozocin also may cause cell damage via the intracellular formation of NO.[69-71]

The injection of a single dose of 160 to 200 mg streptozocin/kg body weight (BW) to mice or 50 to 80 mg streptozocin/kg BW to rats causes acute beta cell toxicity and results in persisting hyperglycemia about 2 days after application.[21] Therefore, the administration of a single high dose of streptozocin represents a useful experimental approach to target a toxic attack, involving primary nuclear

DNA damage, to the pancreatic beta cells *in vivo*. The application of multiple subdiabetogenic doses of streptozocin in mice (5 × 35 to 40 mg streptozocin/kg BW on 5 consecutive days) elicits an immune response against autologous pancreatic islet structures, resulting in insulitis, beta cell destruction, and the development of hyperglycemia.[64]

Both models of streptozocin-induced diabetes and animal models with spontaneous diabetes development (NOD mouse, BB rat) were used in studies that aimed at the protection of beta cells and prevention of diabetes development by pharmacological inhibition of PARP. In view of a future application in human (pre)diabetic individuals, the studies mainly focused on the evaluation of the preventive or therapeutic potential of well-tolerated pharmacological PARP inhibitors, especially nicotinamide, a B vitamin, which is supposed to cause no severe side effects even during long-term application.[72] The *in vivo* studies were accompanied by *in vitro* experiments to investigate the role of PARP in inflammatory islet cell death on the level of beta cells and their subcellular components.

6.4.1.2 Effect of PARP Inhibition on Diabetes Development

The intraperitoneal application of nicotinamide at a dose of 500 mg/kg BW preserved beta cell function and protected from diabetes development in mice treated with streptozocin at a single high dose of 200 mg/kg BW. The animals were protected from hyperglycemia,[73,74] their islets showed preserved NAD^+ levels,[75] and their beta cells were largely protected from destruction[76] (Table 6.1). In several studies in the rat, hyperglycemia was induced by a single intravenous or intraperitoneal injection of 50 to 65 mg streptozocin/kg BW. The administration of nicotinamide in a concentration range from 75 to 1000 mg/kg BW resulted in a dose-dependent reduction of hyperglycemia[73,77] and a preservation of the islet NAD^+ contents.[78] Nicotinamide treatment preserved islet morphology and insulin levels in plasma and pancreas.[79-81] In streptozocin-treated rats (65 mg/kg BW) nicotinamide and 3-aminobenzamide showed comparable, dose-dependent effects on the reduction of hyperglycemia and on the preservation of insulin in plasma and pancreas.[79] Nicotinamide was also found to protect against diabetes induced by the intraperitoneal injection of multiple low doses of streptozocin (5 × 40 mg/kg BW injected on 5 consecutive days). Animals treated with nicotinamide (200 or 500 mg/kg BW) showed a reduced hyperglycemia,[65,82] whereas only weak effects on the development of insulitis were noted.[83]

The potential of nicotinamide to protect against diabetes was further evaluated in animal models with spontaneous diabetes development (Table 6.2). In the NOD mouse, the development of spontaneous and cyclophosphamide-accelerated diabetes could be successfully prevented by nicotinamide treatment (500 mg/kg BW). In addition, nicotinamide was able to reduce the severity of insulitis.[84-87] In contrast, in several studies in the BB rat, oral or intraperitoneal application of nicotinamide showed no consistent effect on the development of hyperglycemia and diabetes.[88-92] From the partially contradictory observations in the animal models with spontaneous diabetes development it may be concluded that the disease processes in NOD mice and BB rats comprise distinct stages, which exhibit different sensitivity toward modulation by nicotinamide.[72]

TABLE 6.1
Effect of Pharmacological PARP Inhibitors on Beta Cell Function and Diabetes Development in Streptozocin-Induced Diabetes

SZ Dosage (mg/kg BW)	PARP Inhibitor (mg/kg BW)	Species	Effects	Ref.
200, iv	NA, 500, ip	Mouse	Prevention of hyperglycemia	73
200, iv	NA, 500, ip	Mouse	Protection from beta cell destruction	76
	NA, 500, ip	Mouse	Preservation of islet NAD$^+$ contents	75
200, ip	NA, 500, ip	Mouse	Prevention of hyperglycemia	74
50, iv	NA, 500, ip	Rat	Prevention of hyperglycemia	73
50, ip	NA, 250–1000, ip	Rat	Prevention of hyperglycemia	77
65, iv	NA, 250, ip	Rat	Preservation of islet NAD$^+$ contents	78
	NA, 3-AB, 75–300, ip	Rat	Dose-dependent reduction of hyperglycemia, preservation of plasma and pancreas insulin levels	79
50, iv	NA, 350, ip	Rat	Prevention of hyperglycemia, partial preservation of plasma and pancreas insulin levels	80
55, ip	NA, 1000, im, o	Rat	Prevention of hyperglycemia, preservation of islet morphology	81
5 × 40, ip	NA, 500 ip	Mouse	Decrease of hyperglycemia	65
	NA, 500, sc	Mouse	Reduction of hyperglycemia, weak effect on insulitis	83
	NA, 200, ip	Mouse	Reduction of hyperglycemia	82

Abbreviations: 3-AB, 3-aminobenzamide; BW, bodyweight; NA, nicotinamide; im, intramuscular; ip, intraperitoneal; iv, intravenous; sc, subcutaneous; o, oral; SZ, streptozocin.

6.4.1.3 Effect of PARP Inhibition on Beta Cell Susceptibility

The findings from the *in vivo* studies were supported by *in vitro* experiments to elucidate at a molecular and cellular level the mechanisms involved in the protection from diabetes development observed after the application of PARP inhibitors in animals. These studies analyzed the effects of pharmacological PARP inhibitors on cultivated isolated islets, islet cells, or beta cell lines exposed to chemical beta cell toxins and to inflammatory mediators.

As mentioned before, the application of streptozocin results in the formation of DNA strand breaks resulting in a dose-dependent increase of PARP activity in beta cells.[93] The administration of nicotinamide to inhibit PARP activation was unable to prevent the induction of primary DNA damage,[58,60] but significantly inhibited the synthesis of poly(ADP-ribose) in a dose-dependent manner[94] and preserved intracellular NAD$^+$ levels.[95] Furthermore, nicotinamide considerably protected islet cells from the streptozocin-induced impairment of metabolic activity,[96] as demonstrated by the preserved insulin secretory capacity of pancreatic islet cells exposed to the beta cell toxin in the presence of the PARP inhibitory drug.[94] Interestingly, there is also evidence for an increased number of DNA strand breaks in streptozocin-treated

TABLE 6.2
Effect of NA Administration on Beta Cell Function and Diabetes Development in Animal Models with Spontaneous Diabetes

NA Dosage (mg/kg BW)	Animal Model	Effects	Ref.
500, sc	NOD mouse	Prevention of hyperglycemia, reduction of insulitis, partial therapeutic effects	84
	NOD mouse, Cy-treated	Prevention of diabetes development, reduction of insulitis	85
500, o	NOD mouse	Prevention of diabetes development, reduction of insulitis	86
	NOD mouse	Delay of diabetes onset, decrease of diabetes incidence and intensity of insulitis	87
670, ip	BB rat	No effect on diabetes incidence	88
500, ip	BB rat	Slight, nonsignificant decrease of IDDM	89
500, o	BB rat	Slight, nonsignificant decrease of IDDM	90
	BB rat	Prevention of hyperglycemia, reduction of diabetes incidence	91
500, ip	BB rat	Slight, nonsignificant decrease of IDDM	92

Abbreviations: BB, BioBreeding rat; BW, body weight; Cy, cyclophosphamide; IDDM, insulin-dependent diabetes mellitus; ip, intraperitoneal; NA, nicotinamide; NOD, nonobese diabetic mouse; o, oral; sc, subcutaneous.

islet cells that are exposed to nicotinamide,[97] a finding that could be explained by the reduced DNA repair capacity of the cell after PARP inhibition.

More recently, the protective potential of pharmacological PARP inhibition was tested in rat islet cells exposed to chemically generated ROI or NO, which have been identified as important beta cell toxic inflammatory mediators.[49,98] Although nicotinamide failed to prevent the formation of primary DNA damage in radical-exposed cells, the presence of the drug effectively inhibited the activation of PARP as assessed by the complete lack of poly(ADP-ribose) formation and by the preservation of intracellular NAD$^+$ concentrations. Finally, nicotinamide and 3-aminobenzamide improved the survival of rat islet cells exposed to chemically generated NO[98,99] or to ROI enzymatically generated by xanthin oxidase.[49,51]

The presence of nicotinamide in the cultures of islets or islet cells was also found to improve the survival and functional activity of beta cells exposed to inflammatory cytokines (IL-1β, TNF-α, IFNγ) which mainly act via the induction of NO formation. IL-1β-induced NO production in isolated rat pancreatic islets results in the appearance of an increased number of DNA strand breaks, elevated nuclear PARP activity, decreased intracellular NAD$^+$ contents, and functional impairment of the beta cell.[100] Cultivation in the presence of nicotinamide partially improved IL-1β-mediated reduction of basal[100] and glucose-induced insulin secretion.[101] The preservation of the functional activity in IL-1β-exposed beta cells by nicotinamide may be explained by the observation that the drug is able to inhibit

IL-1β-induced NO production in cultivated rat islets and cells of the insulinoma line RINm5F, probably due to suppression of protein biosynthesis.[102,103] Furthermore, nicotinamide was able to reduce the deleterious effects of a mixture of the inflammatory cytokines IL-1β, TNF-α, and IFNγ on human islets. In this study, nicotinamide counteracted the cytokine-induced NO formation and the decrease of insulin content and of glucose-induced insulin secretion.[104]

The strong islet cell protective potential of nicotinamide could even be demonstrated in a more complex model of inflammatory islet cell damage, in which isolated rat islet cells were cocultivated with syngeneic activated macrophages. The presence of nicotinamide in the coculture dose-dependently protected the islet cells from macrophage-induced lysis,[105-107] which is mainly mediated via the release of NO.[36] Interestingly, nicotinamide was found to inhibit NOS mRNA induction in activated macrophages,[108] a side effect of the drug, which may also contribute to the beta cell protective effect observed when nicotinamide is present during coculture.

However, sustained inhibition of PARP activity by long-term exposure to nicotinamide may also cause adverse effects on rodent islets or islet cells under *in vitro* conditions. Cultivation of mouse islets with nicotinamide for 1 week partially inhibited glucose-stimulated insulin release and decreased the glucose oxidation rate and the ATP contents.[109] Moreover, nicotinamide failed to prevent the IL-1β-induced impairment of insulin release in rat islets after 6 days of culture.[110]

Nevertheless, the results from the *in vivo* studies in animal models and from the *in vitro* experiments with isolated islets and islet cells strongly indicate that the protection from diabetes development observed after the application of PARP inhibitors in rodents is mediated by the protection of the pancreatic beta cells against damage by inflammatory mediators. However, the studies with pharmacological PARP inhibitors do not allow a final conclusion on the importance of PARP in beta cell damage and diabetes development, since, meanwhile, numerous side effects of the drugs were described which may contribute to their protective potential. Although the PARP inhibitor nicotinamide does not directly interact with NO, the substance acts as an effective scavenger of beta cell–damaging ROI.[94] Furthermore, nicotinamide was found to decrease NO production in human pancreatic islets.[104] *In vitro*, nicotinamide is usually added at doses of 5 to 25 m*M* to islets or islet cells. Clearly, these are concentrations at which a host of pharmacological effects are to be considered. For example, nicotinamide, at higher doses, is able to inhibit PARP-unrelated enzymatic activities[72] and to interfere with transcriptional[108] and translational processes,[111,112] as was shown for the expression of the iNOS gene.

6.4.2 PARP-Deficient Animals

6.4.2.1 Animal Models with PARP Deficiency

To avoid undesired pharmacological effects of PARP inhibitors, a new strategy was applied by specific genetic inactivation of PARP.

In one approach the targeting vector contained a *Hind*III fragment of the PARP gene in which a *Kpn*I fragment was replaced by a neo cassette containing a TGA stop codon. The fragment was electroporated into D3 embryonic stem (ES) cells and fused in frame to the second exon of the PARP gene. Stably transfected cell clones

PARP in the Pathogenesis of Pancreatic Islet Cell Death and Type 1 Diabetes

with homologous recombination events were selected and used for the generation of mice with homozygous PARP deficiency ($PARP^{-/-}$) on a mixed genetic background of 129 SV and C57BL/6.[113] In another approach, the targeting vector was constructed to contain an *AccI–XhoI* fragment spanning the first PARP exon. The PARP fragment was interrupted by insertion of a neo cassette with a flanking diphtheria toxin A fragment cassette. The construct was electroporated into J1 ES cells, and neomycin-resistant clones were selected for the generation of $PARP^{-/-}$ mice on an ICR background.[114,115] In the two approaches for PARP gene disruption, the successful targeting of the alleles was confirmed by Southern blot analysis of genomic DNA from the mice.

Studies with both knockout strains provided unequivocal evidence for PARP activation and subsequent NAD^+ depletion as the dominant metabolic events in beta cell destruction and diabetes development induced by streptozocin.

6.4.2.2 Effect of PARP Deficiency on Diabetes Development

The intraperitoneal injection of a single high dose of streptozocin (160 to 200 mg/kg BW) induced severe hyperglycemia in wild-type mice with normal PARP expression (Table 6.3). A day after streptozocin administration, the blood glucose levels transiently dropped to less than 50 mg/dl, indicating uncontrolled insulin release from beta cells damaged by acute streptozocin toxicity.[116] After 2 days the mice developed hyperglycemia (blood glucose > 200 mg/dl) which persisted for 60 days.[115] Interestingly, $PARP^{-/-}$ mice were completely protected from streptozocin-induced hyperglycemia. They did not show any significant deviations from the physiological blood glucose level until 21 days after streptozocin treatment.[115-117] In PARP heterozygous ($PARP^{+/-}$) mice, diabetes development was delayed. Within the first week after streptozocin administration these mice remained normoglycemic.[116,117] Thereafter, their blood glucose concentrations rose to the levels of the diabetic wild-type mice.[117]

Immunohistochemical analysis of the pancreas proved that the resistance to streptozocin-induced diabetes in PARP-deficient mice is based on the preservation of islet structure, number of insulin-containing beta cells, and total pancreatic insulin contents.[115-117] Streptozocin-induced primary DNA damage was detectable in the islets of pancreatic sections from streptozocin-treated, wild-type $PARP^{+/-}$ and $PARP^{-/-}$ mice. However, PARP enzymatic activity showed a marked increase only in islets of $PARP^{+/+}$ mice. In $PARP^{+/-}$ islets, PARP activity was substantially reduced and in $PARP^{-/-}$ islets it was completely abolished.[117]

6.4.2.3 Effect of PARP Deficiency on Beta Cell Susceptibility

To investigate whether the destruction of beta cells is the consequence of PARP activation, isolated islet cells were exposed to streptozocin *in vitro* for 18 h. As expected, streptozocin induced a dose-dependent increase of lysis in the wild-type islet cells and a significant decrease in the NAD^+ contents to 32% of the initial level. In contrast, $PARP^{-/-}$ and $PARP^{+/-}$ islet cells were largely resistant to streptozocin-induced lysis and their NAD^+ contents were well preserved.[116]

The observations on streptozocin-induced diabetes and beta cell death in $PARP^{-/-}$ mice correlate with previous findings with isolated islet cells clearly demonstrating

TABLE 6.3
Comparison of Three Studies on the Effect of Streptozocin on Diabetes Induction and Beta Cell Damage in PARP-Deficient Mice

PARP Genotype	PARP Protein Expression	SZ (mg/kg BW)	Effects of Streptozocin Administration				Ref.
			Blood Glucose (mg/dl)	Islet Structure	Reduction of Beta Cell Mass	Beta Cell Damage	
PARP+/+	+++ in islets	160	<50, until d1 to >>250, d2 to d5	Atrophy at d3	≈70%	>90% loss of insulin content	116
	+++ in pancreas	180	>200, d1 to d60	Atrophy beginning d1	>80%	Pyknotic nuclei, eosinophilic cytoplasm	115
	+++ in pancreas	200	>>200, wk1 to wk6	Number and size reduced, structure distorted at wk6	Marked beta cell loss	DNA damage ++	117
PARP−/−	− in islets	160	≈150, d1 to d5	No atrophy at d3	n.s. reduction	n.s. loss of insulin content	116
	− in pancreas	180	<150, d1 to d60, except d21: >200	No atrophy, near normal structure at d60	≈20%	Nearly normal, occasional apoptosis	115
	− in pancreas	200	<150, wk1 to wk6	No atrophy, normal structure at wk6	No detectable reduction	DNA damage +++	117
PARP+/−	− in islets	160	≈150, d1 to d5	No atrophy at d3	n.s. reduction	≈40% loss of insulin content	116
	≈50% of wild-type in pancreas	200	<100, until wk1; >>200, wk2 to wk6	Small, distorted structure at wk6	Marked beta cell loss	DNA damage +++	117

Abbreviations: BW, bodyweight; d, day; n.s., not significant; SZ, streptozocin; wk, week.

that PARP deficiency does not only protect against streptozocin toxicity but also against the beta cell–damaging inflammatory mediators NO and ROI. In these studies radical exposed *PARP*$^{-/-}$ islet cells showed well-preserved NAD$^+$ levels and improved survival in contrast to wild-type islet cells.[118]

Taken together, the studies with PARP-deficient mice and islet cells provided strong evidence for NAD$^+$ depletion caused by PARP activation as the dominant metabolic event in the destruction of beta cells after primary radical-induced DNA damage. Reduction or abrogation of the capacity for a PARP-dependent response to DNA damage will improve the resistance of the beta cell toward the deleterious effects of inflammatory mediators. Future studies with PARP-deficient NOD mice will reveal if PARP activation is also important for the development of spontaneous diabetes.

6.5 POTENTIAL INVOLVEMENT OF PARP IN ISLET CELL DEATH PATHWAYS

The findings outlined above identified extensive PARP activation followed by NAD$^+$ depletion as the dominant metabolic event in inflammatory islet cell destruction and in the development of chemically induced diabetes. Nevertheless, PARP and PARP activation must be regarded as intimate components in a complex series of events leading to cell death. On the one hand, PARP activation triggers processes that may modulate the course and the velocity of cell death. On the other hand, the PARP protein and its enzymatic activity are subjected to modulatory influences exerted by metabolic events involved in cell death. Recent findings provide evidence for PARP interactions with mitochondrial functions and with processes involved in apoptosis or necrosis and further indicate the existence of PARP-independent cell death pathways.

6.5.1 INTERACTION OF PARP ACTIVITY WITH MITOCHONDRIAL FUNCTION

At present, several lines of evidence indicate a potential role for mitochondria in the regulation of PARP activity or in the acceleration of cell death induced by PARP activation. Previous studies clearly show that inflammatory mediators induce mitochondrial dysfunctions in beta cells. After exposure to exogenously generated NO, rat islet cells showed a strong decrease in their respiratory activity.[51,119] In isolated islet cells the IL-1β-induced intracellular formation of NO results in a reduced glucose oxidation rate, defective insulin release after stimulation, and a decrease in ATP production.[34,120] These findings could be explained by the ability of NO to damage the FeS cluster in the citric acid cycle enzyme aconitase[120] or by an impairment of the electron transfer when the radical interacts with the heme-binding site of cytochrome oxidase (complex IV) of the mitochondrial respiratory chain.[121] Recent studies show that not only direct radical effects but also PARP activation can considerably contribute to mitochondrial alterations and cell death. In thymocytes, radical-induced PARP activation was involved in the reduction of the mitochondrial transmembrane potential and the subsequent increase in the secondary production of ROI and signs of mitochondrial membrane damage.[122] Since mitochondria were found to be important regulators of cell death pathways,[123] it could be speculated that radical-mediated mitochondrial dysfunction may trigger a sequence of events that is observed during apoptosis and involves the mitochondrial release of cytochrome C[124,125] and the activation of the PARP-cleaving enzyme caspase 3.[126,127]

However, for the beta cell it still remains to be elucidated whether interactions between PARP and mitochondrial functions contribute to inflammatory cell death. The findings currently available may indicate that the radical-induced impairment of mitochondrial functions, mediated directly or indirectly via PARP activation, will reduce the capacity of the mitochondria to provide the cellular metabolism with sufficient amounts of reduction equivalents and ATP. Therefore, it can be assumed that the induction of mitochondrial dysfunction will exert modulatory effects on the processes of cell death, which are dominated by NAD^+ depletion after PARP is activated by DNA strand breaks.

6.5.2 Potential Role of PARP in Apoptosis or Necrosis of Beta Cells

In the last years, intensive research work focused on the pathways of inflammatory islet cell death. Although the criteria used to characterize the cell death pathways remain to be refined, the studies confirm that beta cells will die via apoptosis and necrosis depending on the experimental conditions and/or the type of inflammatory mediator(s) acting on the cells.[128,129]

Decreased intracellular NAD^+ levels as observed after streptozocin treatment in the pancreas of animals[116] and in NO-exposed isolated islet cells[98] may lead to ATP depletion,[130] which can induce necrosis in various cell types.[131,132] In fact, signs of necrotic cell death could be detected in streptozocin-exposed cells of a murine beta cell line.[128] On the other hand, there is evidence that apoptosis is the preferential type of beta cell death in NOD mice[133-135] and in humans.[136,137]

A common feature of both necrotic and apoptotic cell death is the occurrence of DNA strand breaks. Although the induction, kinetics, and pattern of DNA fragmentation differ in apoptosis and necrosis, the appearance of DNA strand breaks may implicate a role for PARP in the two pathways. Experimental evidence for a potential involvement of PARP comes from the finding of controlled proteolytic cleavage of the enzyme in the course of cell death.[138] In an early stage of apoptosis, caspase 3 specifically cleaves PARP into two fragments of 89 and 24 kDa,[139] and a previous study demonstrated the activation of PARP-cleaving proteases also during the process of necrosis.[140]

However, the role of PARP and its cleavage products in the progression of apoptosis and necrosis are still not fully understood. Moreover, recent studies showed that PARP-deficient cells have normal apoptotic responses indicating that neither activation nor cleavage of PARP is necessary for apoptotic cell death.[141,142] In contrast, in another animal model, PARP-deficient spleen cells even showed an increased susceptibility to death induced by irradiation or alkylating agents.[143]

6.5.3 PARP-Independent Pathways of Beta Cell Death

Recent observations indicate that initial DNA damage may trigger islet cell death pathways which are independent of PARP activation. These pathways are obviously activated after exposure to very high concentrations of beta cell–toxic mediators. Experimental evidence for the existence of these pathways comes from studies with isolated $PARP^{-/-}$ islet cells exposed to reactive radicals and streptozocin. These cells exhibited a pronounced resistance to low concentrations of streptozocin, ROI, and NO, whereas in the presence of high doses of the damaging compounds they showed an increasing sensitivity (streptozocin)[116] or even a complete loss of their resistance (NO, ROI).[118] It is conceivable that high concentrations of radicals may directly cause rapid and extensive damage to vital cellular components, such as mitochondria or intracellular membrane systems, and it could be speculated that these severe damages would then predominate all other events involved in the execution of cell death, irrespective of the extent of DNA damage and the presence of PARP.

In the context of PARP-independent cell death, another interesting new aspect is the possible activation of isoforms of PARP (EC 2.4.2.30), which are still present in $PARP^{-/-}$ mice and which could also exhibit poly(ADP-ribosyl)ating activity in response to the appearance of DNA strand breaks. Among other findings, this speculation is supported by a recent study describing the formation of ADP-ribose polymers also in PARP-deficient cells after exposure to DNA-damaging agents.[144] In fact, recent detailed investigations identified tankyrase (telomeric repeat binding factor-1, or TRF-1, ankyrin-related, ADP-ribose polymerase) as a nuclear protein localized to telomers with several similarities to PARP.[145] Tankyrase exhibits PARP activity *in vitro* and is able to modify itself and TRF-1 by poly(ADP-ribosyl)ation. However, since human tankyrase lacks DNA-binding domains and shows only about 30% amino acid homology to the catalytic domain of PARP, it remains to be elucidated whether tankyrase or other enzymes with "PARP-like" activity play a role in DNA repair and whether their rates of NAD^+ consumption would suffice to cause a critical depletion of intracellular NAD pools, especially in pancreatic islet cells exposed to inflammatory mediators.

The findings outlined above show that despite the continuously increasing knowledge about PARP-related events, understanding of the complex processes associated with PARP activation during inflammatory islet cell death is far from complete. However, from the currently available observations in rodent models it could be assumed that rapid NAD^+ depletion represents the dominant metabolic event in a (distinct) cell death pathway which depends on the presence and activation of PARP. In the absence of PARP the induction of massive DNA strand breaks could result in the activation of alternative pathways of cell death.

Taken together, although *in vitro* and *in vivo* experiments in mice and rats show a central role of PARP in the destruction of beta cells and the development of diabetes, the role of PARP activation and subsequent NAD^+ depletion in the pathogenesis of human type 1 diabetes still remains to be elucidated. Until now, the only observations that may allow one to draw a conclusion on the role of PARP in type 1 diabetes come from studies in which individuals afflicted with diabetes were treated with nicotinamide (outlined below). However, except in some pilot studies, nicotinamide showed only partial preventive or protective effects. Therefore, it cannot be excluded that in humans a more effective approach for the protection of beta cells may comprise the suppression of other diabetogenic mechanisms, such as the inhibition of iNOS (possibly in combination with PARP inhibition).

6.6 CLINICAL TRIALS TO PRESERVE BETA CELL FUNCTION BY PARP INHIBITION

Based on the concept of protecting beta cells from inflammatory damage by PARP inhibition, several clinical studies were initialized in individuals afflicted with type 1 diabetes. Basically, two approaches were performed, evaluating the therapeutic and prophylactic potential of the well-tolerated PARP inhibitor nicotinamide.

6.6.1 PARP Inhibition after Onset of Type 1 Diabetes

To preserve residual beta cell function in patients with newly diagnosed type 1 diabetes, patients were treated with nicotinamide at doses of 20 mg up to 3 g/kg BW for several periods of time within 5 years after diabetes manifestation. A meta-analysis of 10 randomized, controlled clinical trials including 158 nicotinamide treated and 129 control patients indicates a protective effect of nicotinamide on beta cell activity.[146] After an observation period of 1 year, the nicotinamide-treated patients showed a significantly preserved baseline C-peptide level indicative for a preserved insulin secretory capacity.[72] However, an improvement of basic clinical parameters was not achieved by nicotinamide administration as judged by the insulin requirement for metabolic control, which did not differ between the nicotinamide-treated and the control groups.

6.6.2 PARP Inhibition for the Prevention of Type 1 Diabetes

Another approach aimed to prevent diabetes onset by applying nicotinamide during the prediabetic phase in individuals with an increased risk to develop diabetes. Initial trials were performed in children who showed islet cell antibodies (ICA) indicative of an ongoing inflammatory reaction against beta cells. A pilot study including 14 ICA-positive children showed a strong reduction of diabetes incidence in the nicotinamide-treated group compared with untreated controls.[147] A large-scale open trial evaluating the effect of nicotinamide treatment on diabetes incidence in ICA-positive school children showed a significant reduction of diabetes incidence, suggesting a potential protective effect of nicotinamide on beta cells.[148] Based on these observations, two placebo-controlled, double-blind trials have been initiated to evaluate the prophylactic potential of nicotinamide. In the Deutsche Nicotinamid Interventions Studie (DENIS) 3- to 12-year-old, ICA-positive siblings of patients with type 1 diabetes were treated with nicotinamide.[149] In 1997, after 3 years of follow-up, the study was terminated when interim analysis showed that the trial had failed to detect a reduction of the cumulative diabetes incidence from 30 to 6% in the nicotinamide-treated group.[150] However, this study focused on young individuals with a high diabetes risk, which is assumed to be associated with rapid progression of the disease. Therefore, the negative outcome of the DENIS does not exclude the possibility of a protective nicotinamide effect in individuals with lower risk or with a slower progression to type 1 diabetes. The still ongoing European Nicotinamide Diabetes Intervention Trial (ENDIT)[151] includes ICA-positive 5- to 40-year-old first-degree relatives of patients with type 1 diabetes. The study is designed similarly to the

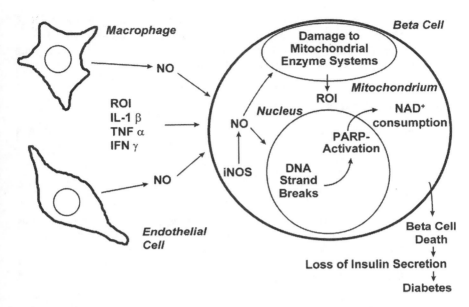

FIGURE 6.3 Hypothetical model on the role of PARP in inflammatory beta cell damage. During insulitis, activated macrophages and cytokine-stimulated capillary endothelial cells release toxic amounts of NO. Inflammatory cytokines such as IL-1β, TNFα, and IFNγ, released from infiltrating immune cells, can induce NO production within the beta cell by iNOS induction. Extracellularly and intracellularly generated NO alone or in combination with ROI rapidly induce widespread damage to nuclear DNA in the beta cell. Subsequent excessive PARP activation results in the critical depletion of NAD$^+$ pools, leading to beta cell death. NO may also impair mitochondrial structures and functions, thereby contributing to beta cell death via an increased formation of ROI. The progressive destruction of the pancreatic beta cell mass will lead to a pathologically reduced insulin secretory capacity and finally to the manifestation of diabetes.

DENIS but will allow one to detect weak effects of nicotinamide treatment on the reduction of the risk to develop type 1 diabetes.

6.7 SUMMARY AND OUTLOOK

The currently available observations from experimental models on the pathogenesis of human type 1 diabetes strongly indicate that inflammatory mediators released during islet infiltration contribute to the destruction of insulin-producing pancreatic beta cells. The radical-induced formation of nuclear DNA strand breaks obviously represents the most prominent primary damage in the beta cell. Recent studies in PARP-deficient animals provided unequivocal evidence for PARP activation and subsequent NAD$^+$ depletion as the dominant metabolic events in the inflammatory destruction of beta cells (Figure 6.3).

The growing knowledge about the processes involved in inflammatory beta cell destruction has allowed the development of strategies aiming at the preservation of beta cell function in individuals at risk to develop diabetes. Although the outcome of these trials, which use nicotinamide for PARP inhibition, may not show strong diabetes-preventive effects, the definite proof of PARP activation as a critical event in beta cell death in animal models of type 1 diabetes will justify the design of future trials with newly developed, more specific PARP inhibitors.

ACKNOWLEDGMENT

The author is grateful to Hubert Kolb for his valuable comments and for critically reading the manuscript.

REFERENCES

1. The Expert Committee on the Diagnosis and Classification of Diabetes Mellitus, Report of the expert committee on the diagnosis and classification of diabetes mellitus, *Diabetes Care*, 20, 1183, 1997.
2. Eisenbarth, G. S., Type I diabetes mellitus: a chronic autoimmune disease, *N. Engl. J. Med.*, 314, 1360, 1986.
3. Atkinson, M. A. and MacLaren, N. K., The pathogenesis of insulin dependent diabetes, *N. Engl. J. Med.*, 331, 1428, 1994.
4. Simone, E. and Eisenbarth, G. S., Chronic autoimmunity of type I diabetes, *Horm. Metab. Res.*, 28, 332, 1996.
5. Schranz, D. B. and Lernmark, A., Immunology in diabetes: an update, *Diabetes Metab. Rev.*, 14, 3, 1998.
6. Palmer, J. P., Asplin, C. M., Clemons, P., Lyen, K., Tatpati, O., Raghu, P. K., and Paguette, T. L., Insulin antibodies in insulin-dependent diabetics before insulin treatment, *Science*, 222, 1337, 1983.
7. Atkinson, M. A., MacLaren, N. K., Riley, W. J., Winter, W. E., Fisk, D. D., and Spillar, R. P., Are insulin autoantibodies markers for insulin-dependent diabetes mellitus? *Diabetes*, 35, 894, 1986.
8. Kuglin, B., Gries, F. A., and Kolb, H., Evidence of IgG autoantibodies against human proinsulin in patients with IDDM before insulin treatment, *Diabetes*, 37, 130, 1988.
9. Lan, M. S., Lu, J., Goto, Y., and Notkins, A. L., Molecular cloning and identification of a receptor-type protein tyrosine phosphatase, IA-2, from human insulinoma, *DNA Cell Biol.*, 13, 505, 1994.
10. Lan, M. S., Wasserfall, C., MacLaren, N. K., and Notkins, A. L., IA-2, a transmembrane protein of the protein tyrosine phosphatase family, is a major autoantigen in insulin-dependent diabetes mellitus, *Proc. Natl. Acad. Sci. U.S.A.*, 93, 6367, 1996.
11. Baekkeskov, S., Aanstoot, H. J., Christgau, S., Reetz, A., Solimena, M., Cascalho, M., Folli, F., Richter-Olesen, H., and DeCamilli, P., Identification of the 64K autoantigen in insulin-dependent diabetes as the GABA-synthesizing enzyme glutamic acid decarboxylase, *Nature*, 347, 151, 1990.
12. Kaufmann, D., Erlander, M., Clare-Salzler, M., Atkinson, M., MacLaren, N. K., and Tobin, A., Autoimmunity to two forms of glutamate decarboxylase in insulin-dependent diabetes mellitus, *J. Clin. Invest.*, 89, 283, 1992.
13. Leslie, R. D. G., Atkinson, M. A., and Notkins, A. L., Autoantigens IA-2 and GAD in type 1 (insulin-dependent) diabetes, *Diabetologia*, 42, 3, 1999.
14. Gepts, W., Pathological anatomy of the pancreas in juvenile diabetes, *Diabetes*, 14, 619, 1965.

15. Foulis, A. K., Liddle, C. N., Farquharson, M. A., Richmond, J. A., and Weir, R. S., The histopathology of the pancreas in type 1 (insulin-dependent) diabetes mellitus: a 25-yr review of deaths in patients under 20 years of age in the United Kingdom, *Diabetologia*, 29, 267, 1986.
16. Itoh, N., Hanafusa, T., Miyazaki, A., Miyagawa, J., Yamagata, K., Yamamoto, K., Waguri, M., Imagawa, A., Tamura, S., Inada, M., Kawata, S., Tarui, S., Kono, N., and Matsuzawa, Y., Mononuclear cell infiltration and its relation to the expression of major histocompatibility complex antigens and adhesion molecules in pancreas biopsy specimens from newly diagnosed insulin-dependent diabetes mellitus patients, *J. Clin. Invest.*, 92, 2313, 1993.
17. Charles, M. A., Suzuki, M., Waldeck, N., Dodson, L. E., Slater, L., Ong, K., Kershnar, A., Buckingham, B., and Golden, M., Immune islet killing mechanisms associated with insulin-dependent diabetes: *in vitro* expression of cellular and antibody-mediated islet cell cytotoxicity in humans, *J. Immunol.*, 130, 1189, 1983.
18. Lohmann, D., Krug, J., Lampeter, E. F., Bierwolf, B., and Verlohren, H.-J., Cell-mediated immune reactions against B cells and defect of suppressor cell activity in type 1 (insulin-dependent) diabetes mellitus, *Diabetologia*, 29, 421, 1986.
19. Kolb, H., Mouse models of insulin-dependent diabetes — low dose streptozotocin induced diabetes and non obese diabetic (NOD) mice, *Diabetes Metab. Rev.*, 3, 751, 1987.
20. Mordes, J. P., Desemone, J., and Rossini, A. A., The BB rat, *Diabetes Metab. Rev.*, 3, 725, 1987.
21. Rerup, C. C. and Tarding, F., Streptozotocin- and alloxan induced diabetes in mice, *Eur. J. Pharmacol.*, 7, 89, 1969.
22. Rerup, C. C., Drugs producing diabetes through damage of the insulin secreting cells, *Pharmacol. Rev.*, 22, 485, 1970.
23. Hanafusa, T. and Tarui, S., Immune pathogenesis of diabetes in the nonobese diabetic mouse: an overview, *Curr. Top. Microbiol. Immunol.*, 156, 15, 1990.
24. Lampeter, E. F., Signore, A., Gale, E. A. M., and Pozzilli, P., Lessons from the NOD mouse for the pathogenesis and immunotherapy of human type 1 (insulin-dependent) diabetes mellitus, *Diabetologia*, 32, 703, 1989.
25. Hanenberg, H., Kolb-Bachofen, V., Kantwerk-Funke, G., and Kolb, H., Macrophage infiltration precedes and is a prerequisite for lymphocytic insulitis in pancreatic islets of prediabetic BB rats, *Diabetologia*, 32, 126, 1989.
26. Appels, B., Burkart, V., Kantwerk-Funke, G., Funda, J., Kolb-Bachofen, V., and Kolb, H., Spontaneous cytotoxicity of macrophages against pancreatic islet cells, *J. Immunol.*, 142, 3803, 1989.
27. Rothe, H., Burkart, V., Faust, A., and Kolb, H., Interleukin-12 gene expression is associated with rapid diabetes development in NOD mice, *Diabetologia*, 39, 119, 1996.
28. Kolb, H. and Kolb-Bachofen, V., Nitric oxide in autoimmune disease: cytotoxic or regulatory mediator? *Immunol. Today*, 19, 556, 1998.
29. Rothe, H., Hausmann, A., Casteels, K., Okamura, H., Kurimoto, M., Burkart, V., Mathieu, C., and Kolb, H., IL-18 inhibits diabetes development in non-obese diabetic mice by counterregulation of Th1 dependent destructive insulitis, *J. Immunol.*, 163, 1230, 1999.
30. Kolb-Bachofen, V., Epstein, S., Kiesel, U., and Kolb, H., Low dose streptozotocin-induced diabetes in mice. Electron microscopy reveals single-cell insulitis before diabetes onset, *Diabetes*, 37, 21, 1988.
31. Mandrup-Poulsen, T., Helqvist, S., Wogensen, L. D., Molvig, J., Pociot, F., Johannesen, J., and Nerup, J., Cytokines and free radicals as effector molecules in the destruction of pancreatic beta cells, *Curr. Top. Microbiol. Immunol.*, 164, 169, 1990.

32. McDaniel, M. L., Kwon, G., Hill, J. R., Marshall, C. A., and Corbett, J. A., Cytokines and nitric oxide in islet inflammation and diabetes, *Proc. Soc. Exp. Biol. Med.*, 211, 24, 1996.
33. Rabinovitch, A., Suarez-Pinzon, W. L., Strynadka, K., Lakey, J. R., and Rajotte, R., Human pancreatic islet beta-cell destruction by cytokines involves oxygen free radicals and aldehyde production, *J. Clin. Endocrinol. Metab.*, 81, 3197, 1996.
34. Southern, C., Schulster, D., and Green, I. G., Inhibition of insulin secretion by interleukin 1β and tumour necrosis factor-α via an L-arginine-dependent nitric oxide generating mechanism, *FEBS Lett.*, 276, 42, 1990.
35. Suarez-Pinzon, W. L., Strynadka, K., Schulz, R., and Rabinovitch, A., Mechanisms of cytokine-induced destruction of rat insulinoma cells: the role of nitric oxide, *Endocrinology*, 134, 1006, 1994.
36. Kröncke, K.-D., Kolb-Bachofen, V., Berschick, B., Burkart, V., and Kolb, H., Activated macrophages kill pancreatic syngeneic islet cells via arginine-dependent nitric oxide generation, *Biochem. Biophys. Res. Commun.*, 175, 752, 1991.
37. Burkart, V., Kröncke, K.-D., Kolb-Bachofen, V., and Kolb, H., Nitric oxide as an inflammatory mediator in insulin-dependent diabetes mellitus. A new therapeutic target? *Clin. Immunother.*, 2, 233, 1994.
38. Kolb, H. and Kolb-Bachofen, V., Nitric oxide: a pathogenetic factor in autoimmunity, *Immunol. Today*, 13, 157, 1992.
39. Kleemann, R., Rothe, H., Kolb-Bachofen, V., Xie, Q., Nathan, C., Martin, S., and Kolb, H., Transcription and translation of inducible nitric oxide synthase in the pancreas of prediabetic BB rats, *FEBS Lett.*, 328, 9, 1993.
40. Suschek, C., Rothe, H., Fehsel, K., Enczmann, J., and Kolb-Bachofen, V., Induction of a macrophage-like nitric oxide synthase in cultured rat aortic endothelial cells, *J. Immunol.*, 151, 3283, 1993.
41. Suschek, C., Fehsel, K., Kröncke, K.-D., Sommer, A., and Kolb-Bachofen, V., Primary cultures of rat islet capillary endothelial cells. Constitutive and cytokine-inducible macrophagelike nitric oxide synthases are expressed and activities are regulated by glucose concentration, *Am. J. Pathol.*, 145, 685, 1994.
42. Steiner, L., Kröncke, K.-D., Fehsel, K., and Kolb-Bachofen, V., Endothelial cells as cytotoxic effector cells: cytokine activated rat islet endothelial cells lyse syngeneic islet cells via nitric oxide, *Diabetologia*, 40, 150, 1997.
43. Corbett, J. A., Lancaster, J. R., Sweetland, M. A., and McDaniel, M. L., Interleukin-1β-induced formation of EPR-detectable iron-nitrosyl-complexes in islets of Langerhans, *J. Biol. Chem.*, 266, 21351, 1991.
44. Welsh, N., Eizirik, D. L., Bendtzen, K., and Sandler, S., Interleukin-1β-induced nitric oxide production in isolated rat pancreatic islets requires gene transcription and may lead to inhibition of the Krebs cycle enzyme aconitase, *Endocrinology*, 129, 3167, 1991.
45. Corbett, J. A., Wang, J. L., Sweetland, M. A., Lancaster, J. R., and McDaniel, M. L., Interleukin-1β induces the formation of nitric oxide by β-cells purified from rodent islets of Langerhans: evidence for the β-cell as a source and site of action of nitric oxide, *J. Clin. Invest.*, 90, 2384, 1992.
46. Welsh, N. and Sandler, S., Interleukin-1β induces nitric oxide production and inhibits the activity of aconitase without decreasing glucose oxidation rates in isolated mouse pancreatic islets, *Biochem. Biophys. Res. Commun.*, 182, 333, 1992.
47. Cetkovic-Cvrlje, M. and Eizirik, D. L., TNF-alpha and IFN-gamma potentiate the deleterious effects of IL-1 beta on mouse pancreatic islets mainly via generation of nitric oxide, *Cytokine*, 6, 399, 1994.

48. Brenner, H.-H., Burkart, V., Rothe, H., and Kolb, H., Oxygen radical production is increased in macrophages from diabetes prone BB rats, *Autoimmunity*, 15, 93, 1993.
49. Heller, B., Bürkle, A., Radons, J., Fengler, E., Jalowy, A., Müller, M., Burkart, V., and Kolb, H., Analysis of oxygen radical toxicity in pancreatic islets at the single cell level, *Biol. Chem. Hoppe-Seyler*, 375, 597, 1994.
50. Fehsel, K., Jalowy, A., Sun, Q., Burkart, V., Hartmann, B., and Kolb, H., Islet cell DNA is a target of inflammatory attack by nitric oxide, *Diabetes*, 42, 496, 1993.
51. Burkart, V., Koike, T., Brenner, H.-H., and Kolb, H., Oxygen radicals generated by the enzyme xanthine oxidase lyse pancreatic islet cells *in vitro*, *Diabetologia*, 35, 1028, 1992.
52. Delaney, C. A., Tyrberg, B., Bouwens, L., Vaghef, H., Hellman, B., and Eizirik, D. L., Sensitivity of human pancreatic islets to peroxynitrite-induced cell dysfunction and death, *FEBS Lett.*, 394, 300, 1996.
53. Corbett, J. A., Sweetland, M. A., Wang, J. L., Lancaster, J. R. J., and McDaniel, M. L., Nitric oxide mediates cytokine-induced inhibition of insulin secretion by human islets of Langerhans, *Proc. Natl. Acad. Sci. U.S.A.*, 90, 1731, 1993.
54. Wilson, G. L., Patton, N. J., and LeDoux, S. P., Mitochondrial DNA is a sensitive target for damage by nitric oxide, *Diabetes*, 46, 1291, 1997.
55. Dunger, A., Cunningham, J. M., Delaney, C. A., Lowe, J. E., Green, M. H., Bone, A. J., and Green, I. C., Tumor necrosis factor-alpha and interferon-gamma inhibit insulin secretion and cause DNA damage in unweaned-rat islets. Extent of nitric oxide involvement, *Diabetes*, 45, 183, 1996.
56. Delaney, C. A., Green, M. H. L., and Green, I. C., Endogenous nitric oxide induced by interleukin-1β in rat islets of Langerhans and HIT-T15 cells causes significant DNA damage as measured by the "comet" assay, *FEBS Lett.*, 333, 291, 1993.
57. Rabinovitch, A., Suarez-Pinzon, W. L., Shi, Y., Morgan, A. R., and Bleackley, R. C., DNA fragmentation is an early event in cytokine-induced islet beta-cell destruction, *Diabetologia*, 37, 733, 1994.
58. Uchigata, Y., Yamamoto, H., Kawamura, A., and Okamoto, H., Protection by superoxide dismutase, catalase, and poly(ADP-ribose)synthetase inhibitors against alloxan- and streptozotocin-induced islet DNA strand breaks and against the inhibition of proinsulin synthesis, *J. Biol. Chem.*, 257, 6084, 1982.
59. Yamamoto, H., Uchigata, Y., and Okamoto, H., Streptozotocin and alloxan induce DNA strand breaks and poly(ADP-ribose)synthetase in pancreatic islets, *Nature*, 294, 284, 1981.
60. Yamamoto, H. and Okamoto, H., Protection by picolinamide, a novel inhibitor of poly(ADP-ribose)synthetase, against both streptozotocin-induced depression of proinsulin synthesis and reduction of NAD content in pancreatic islets, *Biochem. Biophys. Res. Commun.*, 95, 474, 1980.
61. Yamamoto, H., Uchigata, Y., and Okamoto, H., DNA strand breaks in pancreatic islets by *in vivo* administration of alloxan or streptozotocin, *Biochem. Biophys. Res. Commun.*, 103, 1014, 1981.
62. Uchigata, Y., Yamamoto, H., Nagai, H., and Okamoto, H., Effect of poly(ADP-ribose)synthetase inhibitor administration to rats before and after injection of alloxan and streptozotocin on islet proinsulin synthesis, *Diabetes*, 32, 316, 1983.
63. Rakieten, N., Rakieten, M. L., and Nadkarni, M. V., Studies on the diabetogenic action of streptozotocin (NSC-37917), *Cancer Chemother. Rep.*, 29, 91, 1963.
64. Like, A. A. and Rossini, A. A., Streptozotocin-induced pancreatic insulitis: new model of diabetes mellitus, *Science*, 193, 415, 1976.

65. Rossini, A. A., Like, A. A., Chick, W. L., Appel, M. C., and Cahill, G. F., Studies of streptozotocin-insulitis and diabetes, *Proc. Natl. Acad. Sci. U.S.A.*, 74, 2485, 1977.
66. Kolb, H., Kröncke, K.-D., Lessons from the low dose streptozocin model in mice, *Diabetes Rev.*, 1, 116, 1993.
67. Schnedl, W. J., Ferber, S., Johnson, J., and Newgard, C. B., STZ transport and cytotoxicity: specific enhancement in GLUT2-expressing cells, *Diabetes*, 43, 1326, 1994.
68. Wang, Z. and Gleichmann, H., Glucose transporter 2 expression: prevention of streptozotocin-induced reduction in beta-cells with 5-thio-D-glucose, *Exp. Clin. Endocrinol.*, 103, 83, 1995.
69. Turk, J., Corbett, J. A., Ramanadham, S., Bohrer, A., and McDaniel, L., Biochemical evidence for nitric oxide formation from streptozotocin in isolated pancreatic islets, *Biochem. Biophys. Res. Commun.*, 197, 1458, 1993.
70. Kröncke, K.-D., Fehsel, K., Sommer, A., Rodriguez, M.-L., and Kolb-Bachofen, V., Nitric oxide generation during cellular metabolization of the diabetogenic N-methyl-N-nitroso-urea streptozotocin contributes to islet cell DNA damage, *Biol. Chem. Hoppe-Seyler*, 376, 179, 1995.
71. Tsuji, A. and Sakurai, H., Generation of nitric oxide from streptozotocin (STZ) in the presence of copper(II) plus ascorbate: implication for the development of STZ-induced diabetes, *Biochem. Biophys. Res. Commun.*, 245, 11, 1998.
72. Kolb, H. and Burkart, V., Nicotinamide in type 1 diabetes. Mechanism of action revisited, *Diabetes Care*, 22 (Suppl. 2), B16, 1999.
73. Schein, P. S., Cooney, D. A., and Vernon, M. L., The use of nicotinamide to modify the toxicity of streptozotocin diabetes without loss of antitumor activity, *Cancer Res.*, 27, 2324, 1967.
74. Tjälve, H. and Wilander, E., The uptake in the pancreatic islets of nicotinamide, nicotinic acid and tryptophan and their ability to prevent streptozotocin diabetes in mice, *Acta Endocrinol.*, 83, 357, 1976.
75. Schein, P. S., Cooney, D. A., McMenamin, M. G., and Anderson, T., Streptozotocin diabetes — further studies on the mechanism of depression of nicotinamide adenine dinucleotide concentrations in mouse pancreatic islets and liver, *Biochem. Pharmacol.*, 22, 2625, 1973.
76. Lazarus, S. and Shapiro, S. H., Influence of nicotinamide and pyridine nucleotides on streptozotocin and alloxan induced pancreatic B cell cytotoxicity, *Diabetes*, 22, 499, 1973.
77. Dulin, W. E., Wyse, B. M., and Kalamazoo, M. S., Studies on the ability of compounds to block the diabetogenic activity of streptozotocin, *Diabetes*, 18, 459, 1969.
78. Ho, C.-K. and Hashim, S. A., Pyridine nucleotide depletion in pancreatic islets associated with streptozotocin-induced diabetes, *Diabetes*, 21, 789, 1972.
79. Masiello, P., Locci Cubeddu, T., Frosina, G., and Bergamini, E., Protective effect of 3-aminobenzamide, an inhibitor of poly(ADP-ribose)synthetase, against streptozotocin-induced diabetes, *Diabetologia*, 28, 683, 1985.
80. Shima, K., Hirota, M., Sato, M., Numoto, S., and Oshima, I., Effect of poly(ADP-ribose)synthetase inhibitor administration to streptozotocin-induced diabetic rats on insulin and glucagon contents in their pancreas, *Diabetes Res. Clin. Pract.*, 3, 135, 1987.
81. Hu, Y., Wang, Y., Wang, L., Zhang, H., Zhao, B., Zhang, A., and Li, Y., Effects of nicotinamide on prevention and treatment of streptozotocin-induced diabetes mellitus in rats, *Chin. Med. J. Engl.*, 109, 819, 1996.

82. Bouix, O., Reynier, M., Guintrand-Hugret, R., and Orsetti, A., Protective effect of gamma-hydroxybutyrate and nicotinamide on low-dose streptozotocin-induced diabetes in mice, *Horm. Metab. Res.*, 27, 216, 1995.
83. Mendola, G., Wright, J. R., and Lacy, P. E., Oxygen free-radical scavengers and immune destruction of murine islets in allograft rejection and multiple low-dose streptozotocin-induced insulitis, *Diabetes*, 38, 379, 1989.
84. Yamada, K., Nonaka, K., Hanafusa, T., Miyazaki, A., Toyoshima, H., and Tarui, S., Preventive and therapeutic effects of large-dose nicotinamide injections on diabetes associated with insulitis. An observation in nonobese diabetic (NOD) mice, *Diabetes*, 31, 749, 1982.
85. Nakajima, H., Fujino-Kurihara, H., Hanafusa, T., Yamada, K., Miyazaki, A., Miyagawa, J., Nonaka, K., Tarui, S., and Tochino, Y., Nicotinamide prevents the development of cyclophosphamide-induced diabetes mellitus in male non-obese diabetic (NOD) mice, *Biomed. Res.*, 6, 185, 1985.
86. Reddy, S., Bibby, N. J., and Elliott, R. B., Early nicotinamide treatment in the NOD mouse: effects on diabetes and insulitis suppression and autoantibody levels, *Diabetes Res.*, 15, 95, 1990.
87. Kim, J. Y., Chi, J. K., Kim, E. J., Park, S. Y., Kim, Y. W., and Lee, S. K., Inhibition of diabetes in non-obese diabetic mice by nicotinamide treatment for 5 weeks at early age, *J. Korean Med. Sci.*, 12, 293, 1997.
88. Rossini, A. A., Mordes, J. P., Gallina, D. L., and Like, A. A., Hormonal and environmental factors in the pathogenesis of BB rat diabetes, *Metabolism*, 32 (Suppl. 1), 33, 1983.
89. Hermitte, L., Vialettes, B., Atlef, N., Payan, M. J., Doll, N., Scheimann, A., and Vague, P., High dose nicotinamide fails to prevent diabetes in BB rats, *Autoimmunity*, 5, 79, 1989.
90. Kolb, H., Schmidt, M., and Kiesel, U., Immunomodulatory drugs in type 1 diabetes, in *Immunotherapy of Type 1 Diabetes and Selected Autoimmune Disorders*, Eisenbarth, G. S., Ed., CRC Press, Boca Raton, FL, 1989, 111.
91. Sarri, Y., Mendola, J., Ferrer, J., and Gomis, R., Preventive effects of nicotinamide administration on spontaneous diabetes of BB rats, *Med. Sci. Res.*, 17, 987, 1989.
92. Pan, J. Q., Chan, E. K., Cheta, D., Schranz, V., and Charles, M. A., The effect of nicotinamide and glimepiride on diabetes prevention in BB rats, *Life Sci.*, 57, 1525, 1995.
93. Wilson, G. L., Hartig, P. C., Patton, N. J., and LeDoux, S. P., Mechanisms of nitrosourea-induced beta-cell damage. Activation of poly(ADP-ribose)synthetase and cellular distribution, *Diabetes*, 37, 213, 1988.
94. Wilson, G. L., Patton, N. J., McCord, J. M., Mullins, D. W., and Mossman, B. T., Mechanisms of streptozotocin- and alloxan-induced damage in rat B cells, *Diabetologia*, 27, 587, 1984.
95. Hinz, H., Katsilambros, N., Maier, V., Schatz, H., and Pfeiffer, E. F., Significance of streptozotocin induced nicotinamide-adenine-dinucleotide (NAD) degradation in mouse pancreatic islets, *FEBS Lett.*, 30, 225, 1973.
96. Sandler, S., Welsh, M., and Andersson, A., Streptozotocin-induced impairment of islet B-cell metabolism and its prevention by a hydroxyl radical scavenger and inhibitors of poly(ADP-ribose)synthetase, *Acta Pharmacol. Toxicol.*, 53, 392, 1983.
97. LeDoux, S. P., Hall, C. R., Forbes, P. M., Patton, N. J., and Wilson, G. L., Mechanisms of nicotinamide and thymidine protection from alloxan and streptozotocin toxicity, *Diabetes*, 37, 1015, 1988.

98. Radons, J., Heller, B., Bürkle, A., Hartmann, B., Rodriguez, M.-L., Kröncke, K.-D., Burkart, V., and Kolb, H., Nitric oxide toxicity in islet cells involves poly(ADP-ribose) polymerase activation and concomitant NAD$^+$ depletion, *Biochem. Biophys. Res. Commun.*, 199, 1270, 1994.
99. Kallmann, B., Burkart, V., Kröncke, K.-D., Kolb-Bachofen, V., and Kolb, H., Toxicity of chemically generated nitric oxide towards pancreatic islet cells can be prevented by nicotinamide, *Life Sci.*, 51, 671, 1992.
100. Fernandez-Alvarez, J., Tomas, C., Casmitjana, R., and Gomis, R., Nuclear response of pancreatic islets to interleukin-1β, *Mol. Cell. Endocrinol.*, 103, 49, 1994.
101. Buscema, M., Vinci, C., Gatta, C., Rabuazzo, M. A., Vignen, R., and Purrello, F., Nicotinamide partially reverses the interleukin-1β inhibition of glucose-induced insulin release in pancreatic islets, *Metabolism*, 41, 296, 1992.
102. Cetkovic-Cvrlje, M., Sandler, S., and Eizirik, D. L., Nicotinamide and dexamethasone inhibit interleukin-1-induced nitric oxide production by RINm5F cells without decreasing messenger ribonucleic acid expression for nitric oxide synthase, *Endocrinology*, 133, 1739, 1993.
103. Andersen, H. U., Jorgensen, K. H., Egeberg, J., Mandrup-Poulsen, T., and Nerup, J., Nicotinamide prevents interleukin-1 effects on accumulated insulin release and nitric oxide production in rat islets of Langerhans, *Diabetes*, 43, 770, 1994.
104. Eizirik, D. L., Sandler, S., Welsh, M., Bendtzen, K., and Hellerström, C., Nicotinamide decreases nitric oxide production and partially protects human pancreatic islets against the suppressive effects of combinations of cytokines, *Autoimmunity*, 19, 193, 1994.
105. Kolb, H., Burkart, V., Appels, B., Hanenberg, H., Kantwerk-Funke, G., Kiesel, U., Funda, J., Schraermeyer, U., and Kolb-Bachofen, V., Essential contribution of macrophages to islet cell destruction *in vivo* and *in vitro*, *J. Autoimmunity*, 3 (Suppl.), 117, 1990.
106. Kröncke, K.-D., Rodriguez, M.-L., Kolb, H., and Kolb-Bachofen, V., Cytotoxicity of activated rat macrophages against syngeneic islet cells is arginine-dependent, correlates with citrulline and nitrite concentrations and is identical to lysis by the nitric oxide donor nitroprusside, *Diabetologia*, 36, 17, 1992.
107. Burkart, V. and Kolb, H., Protection of islet cells from inflammatory cell death *in vitro*, *Clin. Exp. Immunol.*, 93, 273, 1993.
108. Pellatdeceunynck, C., Wietzerbin, J., and Drapier, J. C., Nicotinamide inhibits nitric oxide synthase mRNA induction in activated macrophages, *Biochem. J.*, 297, 53, 1994.
109. Sandler, S. and Andersson, A., Long-term effects of exposure of pancreatic islets to nicotinamide in vitro on DNA synthesis, metabolism and B-cell function, *Diabetologia*, 29, 199, 1986.
110. Reddy, S., Salari-Lak, N., and Sandler, S., Long-term effects of nicotinamide-induced inhibition of poly(adenosine diphosphate-ribose)polymerase activity in rat pancreatic islets exposed to interleukin-1 beta, *Endocrinology*, 136, 1907, 1995.
111. Hauschildt, S., Scheipers, P., and Bessler, W. G., Inhibitors of poly(ADP-ribose)polymerase suppress lipopolysaccharide-induced nitrite formation in macrophages, *Biochem. Biophys. Res. Commun.*, 179, 865, 1991.
112. Hauschildt, S., Scheipers, P., and Bessler, W. G., Mulsch, A., Induction of nitric oxide synthase in L929 cells by tumour necrosis factor alpha is prevented by inhibitors of poly(ADP-ribose)polymerase, *Biochem. J.*, 288, 255, 1992.
113. Wang, Z.-Q., Auer, B., Stingl, L., Berghammer, H., Haidacher, D., Schwaiger, M., and Wagner, E. F., Mice lacking ADPRT and poly(ADP-ribosyl)ation develop normally but are susceptible to skin disease, *Genes Dev.*, 9, 509, 1995.

PARP in the Pathogenesis of Pancreatic Islet Cell Death and Type 1 Diabetes 127

114. Masutani, M., Nozaki, T., Nishiyama, E., Ochiya, T., Nakagama, H., Wakabayashi, K., Suzuki, H., and Sugimura, T., Establishment of poly(ADP-ribose)polymerase-deficient mouse embryonic stem cell lines, *Proc. Jpn. Acad. Sci. B. Phys. Biol. Sci.*, 74, 233, 1998.
115. Masutani, M., Suzuki, H., Kamada, N., Watanabe, M., Ueda, O., Nozaki, T., Jishage, K.-I., Watanabe, T., Sugimoto, T., Nakagama, H., Ochiya, T., and Sugimura, T., Poly(ADP-ribose)polymerase gene disruption conferred mice resistant to streptozotocin-induced diabetes, *Proc. Natl. Acad. Sci. U.S.A.*, 96, 2301, 1999.
116. Burkart, V., Wang, Z.-Q., Radons, J., Heller, B., Herceg, Z., Stingl, L., Wagner, E. F., and Kolb, H., Mice lacking the poly(ADP-ribose)polymerase gene are resistant to pancreatic beta-cell destruction and diabetes development induced by streptozocin, *Nat. Med.*, 5, 314, 1999.
117. Pieper, A. A., Brat, D. J., Krug, D. K., Watkins, C. C., Gupta, A., Blackshaw, S., Verma, A., Wang, Z.-Q., and Snyder, S. H., Poly(ADP-ribose)polymerase-deficient mice are protected from streptozotocin-induced diabetes, *Proc. Natl. Acad. Sci. U.S.A.*, 96, 3059, 1999.
118. Heller, B., Wang, Z.-Q., Wagner, E. F., Radons, J., Bürkle, A., Fehsel, K., Burkart, V., and Kolb, H., Inactivation of the poly(ADP-ribose)polymerase gene affects oxygen radical and nitric oxide toxicity in islet cells, *J. Biol. Chem.*, 270, 11176, 1995.
119. Burkart, V., Brenner, H.-H., Hartmann, B., and Kolb, H., Metabolic activation of islet cells improves resistance against oxygen radicals or streptozocin, but not nitric oxide, *J. Clin. Endocrinol. Metab.*, 81, 3966, 1996.
120. Sandler, S., Eizirik, D. L., Svensson, C., Strandell, E., Welsh, M., and Welsh, N., Biochemical and molecular actions of interleukin-1 on pancreatic beta-cells, *Autoimmunity*, 10, 241, 1991.
121. Kröncke, K.-D., Fehsel, K., and Kolb-Bachofen, V., Nitric oxide: cytotoxicity versus cytoprotection-how, why, when and where? *Nitric Oxide: Biol. Chem.*, 1, 107, 1997.
122. Virág, L., Salzman, A. L., and Szabó, C., Poly(ADP-ribose)synthetase activation mediates mitochondrial injury during oxidant-induced cell death, *J. Immunol.*, 161, 3753, 1998.
123. Kroemer, G., Zamzami, N., and Susin, S. A., Mitochondrial control of apoptosis, *Immunol. Today*, 18, 44, 1997.
124. Kluck, R. M., Bossy-Wetzel, E., Green, D. R., and Newmeyer, D. D., The release of cytochrome C from mitochondria: primary site for Bcl-2 regulation of apoptosis, *Science*, 275, 1132, 1997.
125. Yang, J., Liu, X., Bhalla, K., Kim, C. N., Ibrado, A. M., Cai, J., Peng, T.-I., Jones, D. P., and Wang, X., Prevention of apoptosis by Bcl-2: release of cytochrome C from mitochondria blocked, *Science*, 275, 1129, 1997.
126. Kaufmann, S. H., Desnoyers, S., Ottaviano, Y., Davidson, N. E., and Poirier, G. G., Specific proteolytic cleavage of poly(ADP-ribose)polymerase: an early marker in chemotherapy induced apoptosis, *Cancer Res.*, 57, 3976, 1997.
127. Lazebnik, Y. A., Kaufmann, S. H., Desnoyers, S., Poirier, G. G., and Earnshaw, W. C., Cleavage of poly(ADP-ribose)polymerase by a proteinase with properties like ICE, *Nature*, 371, 346, 1994.
128. Saini, K. S., Thompson, C., Winterford, C. M., Walker, N. I., Cameron, D. P., Streptozocin at low doses induces apoptosis and at high doses cause necrosis in a murine b cell line, INS-1, *Biochem. Mol. Biol. Int.*, 39, 1229, 1996.
129. Hoorens, A. and Pipeleers, D., Nicotinamide protects human beta cells against chemically induced necrosis, but not against cytokine-induced apoptosis, *Diabetologia*, 42, 55, 1999.

130. Maldonato, A., Trueheart, P. A., Renold, A. E., and Sharp, G. W. G., Effects of streptozotocin *in vitro* on proinsulin biosynthesis, insulin release and ATP content of isolated rat islets of Langerhans, *Diabetologia*, 12, 471, 1976.
131. Eguchi, Y., Shimizu, S., and Tsujimoto, Y., Intracellular ATP levels determine cell death fate by apoptosis or necrosis, *Cancer Res.*, 57, 1835, 1997.
132. Leist, M., Single, B., Castoldi, A. F., Kuhnle, S., and Nicotera, P., Intracellular adenosine triphosphate (ATP) concentration: a switch in the decision between apoptosis and necrosis, *J. Exp. Med.*, 185, 1481, 1997.
133. Kurrer, M. O., Pakala, S. V., Hanson, H. L., and Katz, J. D., Beta cell apoptosis in T cell-mediated autoimmune diabetes, *Proc. Natl. Acad. Sci. U.S.A.*, 94, 213, 1997.
134. O'Brien, B. A., Harmon, B. V., Cameraon, D. P., and Allan, D. J., Apoptosis is the mode of beta-cell death responsible for the destruction of IDDM in the nonobese diabetic (NOD) mouse, *Diabetes*, 46, 750, 1997.
135. Augstein, P., Elefanty, A. G., Allison, J., and Harrison, L. C., Apoptosis and beta-cell destruction in pancreatic islets of NOD mice with spontaneous and cyclophosphamide-accelerated diabetes, *Diabetologia*, 41, 1381, 1998.
136. Stassi, G., DeMaria, R. D., Trucco, G., Rudert, W., Testi, R., Galluzzo, A., Giordano, C., and Trucco, M., Nitric oxide primes pancreatic beta cells for Fas-mediated destruction in insulin-dependent diabetes mellitus, *J. Exp. Med.*, 186, 1193, 1997.
137. Loweth, A. C., Williams, G. T., James, R. F. L., Scarpello, J. H. B., and Morgan, N. G., Human islets of Langerhans express Fas ligand and undergo apoptosis in response to interleukin-1 beta and Fas ligation, *Diabetes*, 47, 727, 1998.
138. Duriez, P. J. and Shah, G. M., Cleavage of poly(ADP-ribose)polymerase: a sensitive parameter to study cell death, *Biochem. Cell. Biol.*, 75, 337, 1997.
139. Shah, G. M., Kaufmann, S. H., and Poirier, G. G., Detection of poly(ADP-ribose)polymerase and its apoptosis-specific fragment by a nonisotopic activity-Western blot technique, *Anal. Biochem.*, 232, 251, 1995.
140. Shah, G. M., Shah, R. G., and Poirier, G. G., Different cleavage pattern for poly(ADP-ribose) polymerase during apoptosis and necrosis in HL-60 cells, *Biochem. Biophys. Res. Commun.*, 229, 838, 1996.
141. Leist, M., Single, B., Künstle, G., Volbracht, C., Hentze, H., and Nicotera, P., Apoptosis in the absence of poly(ADP-ribose)polymerase, *Biochem. Biophys. Res. Commun.*, 233, 518, 1997.
142. Wang, Z.-Q., Stingl, L., Morrison, C., Jantsch, M., Los, M., Schulze-Osthoff, K., and Wagner, E. F., PARP is important for genomic stability but dispensable in apoptosis, *Genes Dev.*, 11, 2347, 1997.
143. Menissier-de Murcia, J., Niedergang, C., Trucco, C., Ricoul, M., Dutrillaux, B., Mark, M., Olivier, F. J., Masson, M., Dierich, A., LeMeur, M., Walztinger, C., Chambon, P., and de Murcia, G., Requirement of poly(ADP-ribose)polymerase in recovery from DNA damage in mice and in cells, *Proc. Natl. Acad. Sci. U.S.A.*, 94, 7303, 1997.
144. Shieh, W. M., Ame, J.-C., Wilson, M. V., Wang, Z.-Q., Koh, D. W., Jacobson, M. K., and Jacobson, E. L., Poly(ADP-ribose)polymerase null mouse cells synthesize ADP-ribose polymers, *J. Biol. Chem.*, 273, 30069, 1998.
145. Smith, S., Giriat, I., Schmitt, A., and de Lange, T., Tankyrase, a poly(ADP-ribose)polymerase at human telomeres, *Science*, 282, 1484, 1998.
146. Pozzilli, P., Browne, P. D., and Kolb, H., Meta-analysis of nicotinamide treatment in patients with recent-onset IDDM. The nicotinamide trialists, *Diabetes Care*, 19, 1357, 1996.
147. Elliott, R. B. and Chase, H. P., Prevention or delay of type 1 (insulin-dependent) diabetes mellitus in children using nicotinamide, *Diabetologia*, 34, 362, 1991.

148. Elliott, R. B. and Pilcher, C. C., Prevention of diabetes in normal school children, *Diabetes Res. Clin. Pract.*, 14 (Suppl. 1), 85, 1991.
149. Lampeter, E. F., Intervention with nicotinamide in pre-type 1 diabetes: the Deutsche Nicotinamid Interventionsstudie-DENIS, *Diabete Metab.*, 19, 105, 1993.
150. Lampeter, E. F., Klinghammer, A., Scherbaum, W. A., Heinze, E., Haastert, B., Giani, G., and Kolb, H., DENIS Group, the Deutsche Nicotinamide Intervention Study: an attempt to prevent type 1 diabetes, *Diabetes*, 47, 980, 1998.
151. Gale, E. A. M. and Bingley, P. J., Can we prevent IDDM? *Diabetes Care*, 17, 339, 1994.

7 Poly(ADP-Ribose) Polymerase in the Immune System: Focus on Reactive Nitrogen Intermediates and Cell Death

László Virág

CONTENTS

7.1 Reactive Nitrogen Intermediate–Induced PARP Activation in the Immune System 132
    7.1.1 Introduction 132
    7.1.2 Intrathymic Selection Processes 133
    7.1.3 Nitric Oxide and Peroxynitrite in the Thymus 133
    7.1.4 Nitric Oxide and Peroxynitrite in Other Lymphoid Organs 134
    7.1.5 Effect of NO and Peroxynitrite on Thymocytes 138
    7.1.6 The Role of PARP Activation in the Regulation of Peroxynitrite-Induced Cell Death *In Vitro:* PARP Activation Shifts Apoptotic Cell Death toward Necrotic Death 139
    7.1.7 The *In Vivo* Significance of Peroxynitrite-Induced PARP Activation in the Negative Selection of Thymocytes 142
    7.1.8 Effect of PARP Activation on Mitochondria 143
    7.1.9 Regulation of Peroxynitrite-Induced PARP Activation and Cytotoxicity 144
        7.1.9.1 Role of Calcium and Zinc 144
        7.1.9.2 Purines: Endogenous PARP Inhibitors? 145
        7.1.9.3 Role of Phosphorylation 146
        7.1.9.4 Role of Bcl-2 146
    7.1.10 What Difference Does PARP Make in Apoptosis? 147
7.2 PARP as Transcriptional Regulator of Inflammation and Lymphocyte Homing 148

7.2.1 Effect of PARP on the Expression of Chemokines,
 Adhesion Molecules, Collagenase, MHC-II Molecules,
 and Inducible Nitric Oxide Synthetase 148
7.2.2 Effect of PARP on Nuclear Receptor Signaling 149
7.3 Role of PARP in the Development of the T- and B-Cell Repertoire 151
7.4 Summary 153
Acknowledgments 153
References 154

7.1 REACTIVE NITROGEN INTERMEDIATE–INDUCED PARP ACTIVATION IN THE IMMUNE SYSTEM

7.1.1 INTRODUCTION

Macrophages were among the first cell types identified as a source of nitric oxide (NO) produced by the inducible nitric oxide synthase (iNOS).[1-5] NO production represents a major cytotoxic mechanism utilized by macrophages to kill certain pathogens and tumor cells.[4-8] Although NO is a free radical capable of inhibiting the respiratory chain,[5,8] inactivating enzymes,[9,10] peroxidating lipids,[11-13] and causing cytotoxicity,[4,5] cell damage associated with increased NO production is more likely to be mediated by NO-derived oxidants such as peroxynitrite than by NO per se.[14] Peroxynitrite (ONOO−) is a binary toxin assembled when NO and superoxide (O_2^-) are produced together.[14] Peroxynitrite is a very potent inducer of DNA single-strand breaks[15-18] and thus poly(ADP-ribose) polymerase (PARP) activation.[18-20] Because activated macrophages upregulate iNOS and produce large amounts of NO in parallel with the overproduction of superoxide in a process called "oxidative burst," macrophage activation provides ideal conditions for peroxynitrite formation. Indeed, activated macrophages have been shown to produce ONOO−.[21] Macrophages and related cell types (e.g., dendritic cells) are present in all lymphoid organs where they sample antigens from the lymph and the bloodstream and present antigens to T lymphocytes. Activated T lymphocytes in turn secrete interferon γ (IFNγ), which is a potent activator of macrophages. Thus, during antigen presentation cross-activation between antigen-presenting cells (APC) and T cells may result in NO production, peroxynitrite formation, and PARP activation. Indeed, NO production by APCs (Langerhans cells, dendritic cells) and modulation of T-cell function by APC-derived NO has been reported in the literature.[22-25] Peroxynitrite-induced PARP activation may occur in the NO-producing cell or in cells localized in the vicinity of the source of NO. We have selected the thymus as a primary lymphoid organ and investigated (1) whether NO is produced in the thymus, (2) whether intrathymic NO production results in peroxynitrite formation in the thymus, and (3) whether PARP activation occurs in thymocytes. Moreover, we have extensively investigated the *in vitro* effect of peroxynitrite on thymocytes as a model cell type and studied how PARP activation affects the mode of peroxynitrite-induced cell death.

As this chapter focuses on the role of PARP in the reactive nitrogen intermediate (RNI)-induced cytotoxicity, a relatively detailed outline of the role of RNI in the immune system is given. The possible effect of PARP on the formation of T and B lymphocyte repertoire and some aspects of PARP as a transcriptional regulator of immune functions are also discussed.

7.1.2 Intrathymic Selection Processes

The thymus is a primary lymphoid organ where thymocytes go through a vigorous selection process. During positive selection, thymocytes capable of recognizing self-histocompatibility (MHC) molecules receive a survival signal from cortical epithelial cells, whereas cells unable to bind to self-MHC undergo apoptosis.[26-29] Positively selected thymocytes, in turn, migrate deeper into the thymus and make contact with dendritic cells at the corticomedullary junction and in the medulla.[27-29] Dendritic cells present self-antigens to thymocytes and induce apoptosis of those carrying T-cell receptors (TCR) with high affinity toward self-antigens (negative selection). As a result of these selection processes, the vast majority of thymocytes (>95%) die within the thymus and only mature T cells with the ability to recognize non-self-antigens in association with self-MHC enter the circulation.[27-29] The predominant death signal provided by dendritic cells during negative selection is transduced via the T-cell receptor. However, signaling through the TCR alone does not necessarily lead to apoptosis.[28,30,31] For instance, cosignaling via the cell-surface Fas molecule, a member of the TNF receptor family, and through CD28 and Thy-1 surface receptors may also contribute to apoptosis induction.[32-36] A role for NO in the negative selection has also been proposed based on findings that *in vivo* TCR stimulation leads to the upregulation of iNOS in the thymus and to a decrease in the number of $CD4^+CD8^+$ thymocytes.[37,38]

7.1.3 Nitric Oxide and Peroxynitrite in the Thymus

We have used NADPH diaphorase (NADPHd) histochemistry to detect NOS activity in the rat thymus. In accordance with previous studies,[39,40] we have found numerous $NADPHd^+$ cells in the rat thymus (Figure 7.1) with the highest density observed at the corticomedullary junction.[41] Based on the distribution and the morphology (long branching processes) of $NADPHd^+$ cells, they may represent dendritic cells; however, their phenotype has not yet been determined.

We have also detected $NADPHd^+$ corticomedullary cells in the mouse thymus (Figure 7.2) where iNOS positivity of $NADPHd^+$ cells was proved by iNOS immunohistochemistry.[42] Furthermore, $NADPHd^+/iNOS^+$ cells were absent from the thymi of iNOS-deficient thymi proving that NADPHd reaction detects iNOS in our system.[42] We have investigated the inducibility of NADPHd/iNOS in the mouse thymus by using staphylococcal enterotoxin B (SEB), a superantigen that cross-links the MHC molecules of APCs (e.g., macrophages and dendritic cells) with the TCR of thymocytes and T lymphocytes. NADPHd reaction revealed a massive increase in the number of positive cells in the SEB-treated mouse thymus as compared with the naive thymus (see Figure 7.2).[42] SEB induced a marked increase in the amount of immunohistochemical staining for iNOS, with a pattern identical to that of NADPHd.[42] In the thymi of iNOS-deficient animals, neither NADPHd nor iNOS staining could be detected after SEB challenge.[42] The lack of these $NADPHd^+$

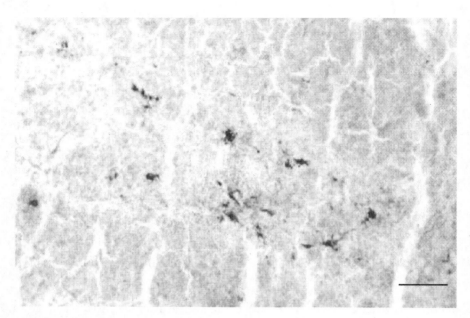

FIGURE 7.1 NADPHd+ positive cells in the rat thymus. Thymi from untreated Sprague–Dawley rats were removed, and frozen sections were stained for NADPHd to detect NOS activity. Strongly stained cells with long branching processes were found in the corticomedullary junction. (Bar: 100 μm; original magnification, 100×.)

medullary cells in the thymocytes of iNOS knockout mice strongly indicates that the NADPHd staining was, indeed, related to iNOS.

Following the proposal of the participation of iNOS-derived NO in the negative selection, the question arose whether intrathymic NO production by iNOS leads to peroxynitrite formation. For the detection of *in vivo* peroxynitrite production, we have taken advantage of the fact that unlike NO, peroxynitrite reacts with tyrosine residues of proteins. Therefore, the presence of nitrated proteins in tissues is regarded as a footprint of peroxynitrite and other oxidized NO products.[14,43,44] We have detected cells containing nitrated proteins in the thymi of untreated mice (Figure 7.3), indicating the *in vivo* "basal" production of peroxynitrite in this organ.[45] The nitrotyrosine-positive cells were predominantly found in the medulla. In addition, nitrotyrosine-positive cells were found in the thymi of iNOS-deficient mice, although they were less abundant than in the wild-type mice and displayed a predominantly perivascular distribution (Figure 7.3).[45]

In summary, NO and peroxynitrite are formed at the site of negative selection (at the corticomedullary junction and in the medulla) and these reactive nitrogen intermediates may provide accessory death signals to thymocytes during negative selection.[46]

7.1.4 Nitric Oxide and Peroxynitrite in Other Lymphoid Organs

Until relatively recently, the thymus appeared to be a unique immune organ where iNOS-derived NO and peroxynitrite are produced without exogenous stimulation.

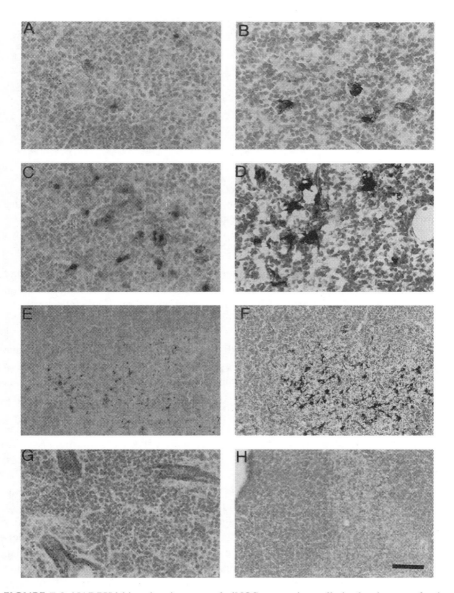

FIGURE 7.2 NADPHd histochemistry reveals iNOS-expressing cells in the thymus of naive and SEB-stimulated mice. (A) NADPHd+ staining in the thymi of naive mice. (B) Immunohistochemical detection of iNOS in the thymi of naive mice. (C, E) NADPHd+ staining in the mouse thymus after SEB stimulation. (D, F) iNOS staining of thymi from SEB-treated mice. (G) NADPHd+ in the thymus of iNOS knockout animals following SEB stimulation. Only endothelial staining could be detected. A similar staining pattern was seen in naive iNOS knockout animals. (H) No iNOS staining could be detected in the thymus of iNOS knockout animals following SEB stimulation. For panels A, B, C, D, G scale bar = 25 μm, (original magnification, 400×); for panels E, F, H scale bar = 100 μm (original magnification, 100×). (From Virág, L. et al., *J. Histochem. Cytochem.*, 46, 787–791, ©1988. The Histochemical Society. With permission.)

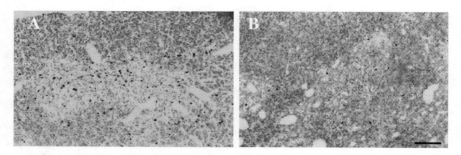

FIGURE 7.3 Immunohistochemical detection of nitrotyrosine in the thymus. Cryostat sections of naive wild-type (A) and iNOS knockout (B) mice were stained for nitrotyrosine. Immunoreactive cells were found predominantly in the medulla. A few nitrotyrosine-positive cells of predominantly perivascular localization could also be detected in the thymi of iNOS-deficient mice (B). (Bar = 100 µm; original magnification, 100×.) (From Virág, L. et al., *Immunology*, 94, 345–355, ©1998. Blackwell Science Ltd. With permission.)

However, Brito et al.[47] have recently demonstrated that nitrotyrosine can be detected in human lymph nodes obtained from routine surgical resections for lung and colon cancers (Figure 7.4). The nitrotyrosine immunostaining was predominantly localized to macrophages; however, lymphocytic and perilymphocytic staining has also been found in the lymph nodes.[47]

Similarly to our hypothesis for intrathymic iNOS activation,[46] Brito et al. have suggested that cross talk between lymphocytes and monocytes/macrophages leads to nitration of proteins in lymphocytes.[47] This scenario is supported by experiments where peripheral blood mononuclear cells (PBMC) containing both lymphocytes and monocytes were stimulated with anti-CD3 antibody (T-cell activator) followed by purification of lymphocytes (Figure 7.5). Under these condition, strong tyrosine nitration was detected by Western blotting.[47] However, when CD3 stimulation was carried out after the separation of lymphocytes (i.e., in the absence of monocytes), no significant increaese in tyrosine nitration could be detected (Figure 7.5).[47]

In summary, these results show that activation of the immune system is accompanied by iNOS expression and peroxynitrite formation due to cross-activation of T lymphocytes and monocytes/macrophages. According to this scenario (a similar one to that is depicted in Figure 7.10), activation of T cells by anti-CD3 or superantigens such as SEB results in cytokine (INFγ) production by activated T cells. INFγ in turn activates monocytes/macrophages/dendritic cells to express iNOS and to switch on the oxidative burst, fueling peroxynitrite production from both sides (NO and superoxide). Peroxynitrite produced by activated monocytes/macrophages triggers tyrosine nitration within the monocytes/macrophages and also diffuses through cell membranes and nitrates tyrosine in lymphocytes.

Peroxynitrite was proposed to inhibit T-cell activation by tyrosine nitration;[47] however, a role for PARP cannot be ruled out in this process. The possibility that peroxynitrite produced during immune reactions activates PARP has not yet been investigated. It would be interesting to know whether PARP becomes activated in this process and how PARP activation may affect the activation of both the T cells and cells of the monocyte/macrophage lineage.

PARP in the Immune System

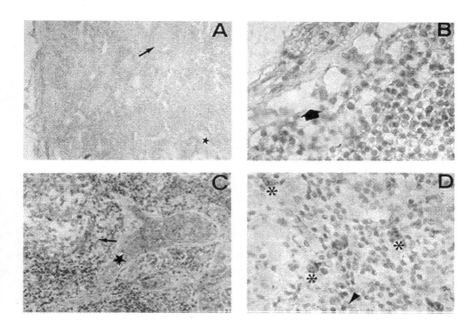

FIGURE 7.4 Nitrotyrosine staining in human lymph nodes. Human lymph nodes were obtained from routine surgical resections for lung and colon cancers, and tyrosine nitration was detected by immunohistochemistry. (A) Subcapsular and medullar sinuses and scanty cortical immunoreactivity with antinitrotyrosine. (B) Nitrotyrosine-positive macrophages surrounding subcapsular sinuses. (C) Dilated sinuses containing large numbers of sinus histiocytes with strong cytoplasmic immunolabeling. (D) Multiple parenchymal macrophages with strong granular reactivity and a poorly defined perilymphocyte and lymphocyte labeling with random distribution. (Original magnifications: A: 40×, B and D: 400×, C: 100×.) (From Brito, C. et al., *J. Immunol.*, 162, 3356–3366, ©1999. The American Association of Immunologists. With permission.)

Another interesting phenomenon is the "burnout" of lymphocytes during sepsis.[48] Prolonged sepsis causes a significant decrease in lymphocyte ATP levels, which correlates with decreased lymphocyte proliferative capacity in response to mitogenic stimulation.[48] Treatment with ATP-$MgCl_2$ at the onset of sepsis significantly increases lymphocyte ATP levels and proliferative response to mitogenic stimuli.[48] Although it has not yet been tested experimentally, PARP appears to be a plausible candidate for being responsible for the lymphocyte burnout. Sepsis is characterized by the overproduction of free radicals, which can activate PARP, and the beneficial effect of ATP supplementation also fits well in this hypothesis.

Free radical/oxidant-induced PARP activation is not the only possible mechanism by which PARP can alter immune activation. Upon recognition of antigen, lymphocytes (both T and B cells) undergo proliferation and express new genes, and B cells differentiate into antibody-secreting plasma cells. All these cellular functions (proliferation,[49-51] differentiation,[51-59] gene expression[60-66]) have been found to be regulated by PARP in certain cell types including lymphocytes, macrophages, and granulocytes.

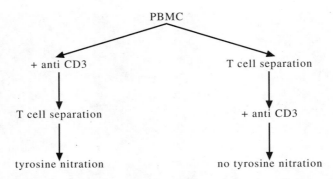

FIGURE 7.5 Monocyte-derived peroxynitrite nitrates proteins in lymphocytes upon T-cell activation. When PBMCs containing both lymphocytes and monocytes were stimulated with anti-CD3 (T-lymphocyte activator) followed by the separation of T cells, nitrated proteins could be detected in T cells. However, when T cells were separated first and T-cell activation was carried out in the absence of monocytes, no tyrosine nitration occurred. This experiment proves that T-cell activation results in the release of a monocyte-activating cytokine (IFNγ ?) that causes the upregulation of iNOS and the increased production of superoxide in monocytes. NO and superoxide form peroxynitrite, which diffuses through the cell membrane and nitrates proteins in T cells.

Besides the thymus and the lymph nodes, Peyer's patches also may represent immune organs where iNOS expression occurs without special stimulation. Peyer's patches are covered by a specialized epithelium called dome epithelium; little is known about its function.[67,68] The dome epithelium is thought to sample the intestinal content and direct "antigen samples" to the lymphocytes localized underneath the epithelial layer.[67,68] With NADPHd histochemistry we have detected strong staining in the dome epithelium of the Peyer's patches in rats and mice (Figure 7.6); however, identification of NADPHd positivity as NOS needs to be done. Nonetheless, the possibility exists that NO and peroxynitrite are produced in the Peyer's patches, which can activate PARP; however, this needs to be confirmed experimentally.

7.1.5 Effect of NO and Peroxynitrite on Thymocytes

NO has been shown to cause apoptotic death in thymocytes[69] as well as several other cell types.[70-73] It has also been demonstrated that peroxynitrite can induce DNA fragmentation (degradation of DNA into oligonucleosomal fragments with the length of the multiple of 180 bp) in thymocytes.[74] We have comprehensively characterized peroxynitrite-induced thymocyte death by means of various flow cytometric techniques, assays of caspase activity and DNA fragmentation, and requirement of intact RNA and DNA synthesis.[45] In summary, we could conclude that peroxynitrite-induced apoptotic death shares the characteristics of apoptotic processes triggered by conventional apoptosis inducers: phosphatidylserine exposure, activation of caspase-3-like but not caspase-1-like proteases, caspase-dependent oligonucleosomal DNA fragmentation, dependence on protein and RNA synthesis.[45]

The central executioners of the apoptotic machinery are thought to be the caspases, a group of cysteinyl aspartate-specific proteases.[75,76] Interestingly, in some

PARP in the Immune System

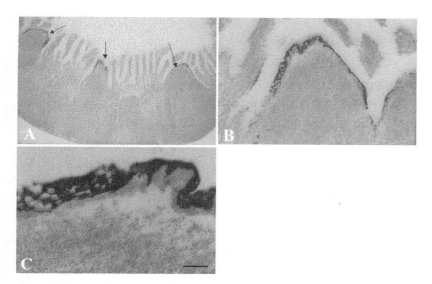

FIGURE 7.6 NADPHd staining of the dome epithelium of Peyer's patches. Frozen sections were cut from untreated Fisher rat Peyer's patches and stained for NADPHd. No staining was found in the intestinal epithelium. However, strong NADPHd staining could be detected in the specialized dome epithelium covering the limphoid tissue. (Original magnifications: A: 40×, bar 100 μm; B: 100×, bar 40 μm; C: 400×, bar 10 μm.)

cell types, NO was found to inhibit apoptosis by inducing the upregulation of the antiapoptotic bcl-2 protein,[77] inhibiting the release of cytochrome c from the mitochondria into the cytoplasm,[78] and by inhibiting caspases by S-nitrosylation.[79-87] In other cell types, however, NO-induced apoptosis was mediated by caspases.[88-92] On the other hand, peroxynitrite has not yet been reported to inhibit caspases, but caused caspase activation in thymocytes[45] and in HL60 cells[93,94] (Figure 7.7). Based on these data, it is tempting to speculate that NO itself inhibits caspases and apoptosis by S-nitrosylation. However, when cellular metabolism (high superoxide leak from the respiratory chain and/or low SOD activity) allows exogenous NO to combine with superoxide to form peroxynitrite, then peroxynitrite causes apoptosis.

7.1.6 THE ROLE OF PARP ACTIVATION IN THE REGULATION OF PEROXYNITRITE-INDUCED CELL DEATH *IN VITRO*: PARP ACTIVATION SHIFTS APOPTOTIC CELL DEATH TOWARD NECROTIC DEATH

PARP was among the first identified caspase targets. Caspase-3 cleaves the intact 116-kb PARP into a large (86-kDa) and a small (30-kDa) fragment.[95] The relevance of PARP cleavage by caspase-3 is unknown, but it has been suggested to serve as an energy-conserving step to prevent the activation of the enzyme by the ensuing DNA damage.[96] This hypothesis is supported by recent findings demonstrating that inserting a mutation into the caspase recognition site of PARP, which renders PARP uncleavable by caspases, switches apoptosis into necrotic death.[97]

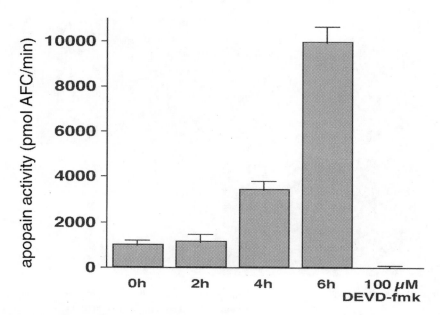

FIGURE 7.7 Peroxynitrite-induced caspase-3 activation in thymocytes. PARP-deficient thymocytes were treated with 50 µM peroxynitrite and DEVD-AFC-cleaving activity of cell lysates prepared 0, 2, 4, and 6 h after peroxynitrite treatment has been measured fluorimetrically. Control samples pretreated with 100 µM apopain inhibitor; DEVD-fmk were also included. AFC liberation was calculated from a standard curve prepared with free AFC. Results are given as DEVD-fmk-inhibitable AFC liberation in picomole AFC/min. Data represent mean ± SD of triplicate samples. (From Virág, L. et al., *Immunology*, 94, 345–355, ©1998. Blackwell Science Ltd. With permission.)

The identification of PARP as a caspase substrate has directed the attention of numerous investigators to the role of PARP in apoptosis. The findings obtained by using pharmacological PARP inhibitors ranged from inhibition[98-102] to lack of effect[45,103-105] and augmentation of apoptosis,[106-109] depending on cell type, culture condition, and apoptosis inducers used. Thus, the eventual role of PARP in apoptosis is still controversial.

We have investigated the effect of PARP on peroxynitrite-induced thymocyte death by comparing the responses of PARP-deficient (PARP$^{-/-}$) and wild-type (PARP$^{+/+}$) thymocytes to peroxynitrite and by using pharmacological PARP inhibitors such as 3-aminobenzamide and 5-iodo-6-amino-1,2-benzopyrone (INH$_2$BP).[45] First, we used a flow cytometry–based technique to detect peroxynitrite-induced apoptosis (based on the binding of Annexin V to phosphatidylserine) and necrosis (uptake of cell membrane impermeable propidium iodide due to the breakdown of membrane integrity). Cytofluorimetric analysis of cells stained with Annexin V-FITC and propidium iodide allows the identification of intact (unstained), apoptotic (Annexin V-FITC-positive), and necrotic (Annexin V-FITC- and propidium-iodide-double-positive) populations.[110]

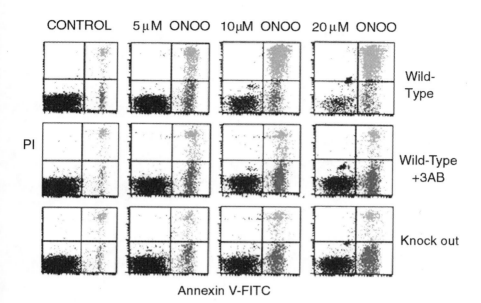

FIGURE 7.8 Cytofluorimetric analysis of thymocyte apoptosis induced by peroxynitrite. Thymocytes were treated for 4 h with peroxynitrite. Cells were then stained with Annexin V-FITC and propidium iodide (PI), and two-color analysis was performed by flow cytometry. An increase in the number of apoptotic (Annexin V-FITC-single-positive) cells was observed in response to 10 μM peroxynitrite treatment, whereas necrotic (stained by both Annexin V-FITC and PI) cells dominate in response to 50 μM peroxynitrite. Wild-type cells pretreated with 3-aminobenzamide as well as PARP-deficient cells were protected against the loss of membrane integrity, as indicated by decreased PI uptake. (From Virág, L. et al., *Immunology*, 94, 345–355, ©1998. Blackwell Science Ltd. With permission.)

As shown in Figure 7.8, peroxynitrite treatment (10 μM) resulted in an increase in the number in apoptotic cells.[45] Higher concentrations (50 μM), however, caused a marked necrosis. By comparison, PARP-deficient cells or 3-aminobenzamide-treated wild-type cells displayed a dramatic decrease in the number of double-positive cells, with an increase of apoptotic (i.e., Annexin V-single-positive) cells. 3-Aminobenzamide had no effect on the apoptosis of PARP-deficient thymocytes.

In line with the flow cytometric data, agarose gel electrophoresis of cellular DNA showed that 10 μM peroxynitrite, but not higher concentrations of the oxidant, induced oligonucleosomal DNA fragmentation indicative of apoptotic death[45] (Figure 7.9). In fact, at higher concentrations of the oxidant, a marked inhibition of the apoptotic process was found, which was reversed by pretreatment with 3-aminobenzamide (Figure 7.9). PARP-deficient thymocytes exposed to peroxynitrite also demonstrated a dose-dependent increase of DNA fragmentation (similar to the results of the wild-type cells treated with 3-aminobenzamide; Figure 7.9). Similarly to peroxynitrite, H_2O_2-induced cell death shifted from necrosis toward apoptosis, whereas cell death induced by non-DNA-damaging stimuli such as dexamethasone

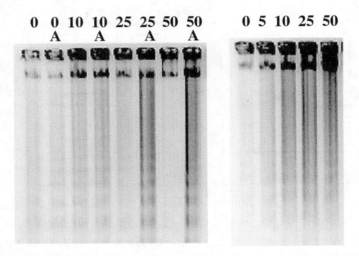

FIGURE 7.9 Peroxynitrite-induced internucleosomal DNA fragmentation in thymocytes. Following a 6-h treatment with the indicated concentrations of peroxynitrite, PARP$^{+/+}$ and PARP$^{-/-}$ thymocytes were analyzed for internucleosomal DNA cleavage with agarose gel electrophoresis. PARP$^{+/+}$ thymocytes were treated with 0, 10, 25, or 50 μM peroxynitrite in the absence or presence of 1 mM 3-aminobenzamide and were then analyzed for DNA laddering. DNA laddering of PARP$^{-/-}$ cells treated with the indicated concentration of peroxynitrite has also been determined. (From Virág, L. et al., *Immunology*, 94, 345–355, ©1998. Blackwell Science Ltd. With permission.)

or anti-Fas antibody was unaffected by the absence or inhibition of PARP as assessed by both flow cytometry and DNA fragmentation.[45]

7.1.7 THE *IN VIVO* SIGNIFICANCE OF PEROXYNITRITE-INDUCED PARP ACTIVATION IN THE NEGATIVE SELECTION OF THYMOCYTES

In summary of the *in vitro* data, we can conclude that peroxynitrite-induced PARP activation results in necrotic-type cell death. Inhibition of PARP activation results in cytoprotection as indicated by an increase of intact or apoptotic phenotype. In light of the necrotic effect of PARP activation, the question arises of how these *in vitro* data translate into the *in vivo* situation. Does PARP have a role in the regulation of negative selection? It is unlikely! Our opinion is based on the following facts:

1. Abnormal thymic size or T-cell maturation defects indicative of a possible failure in negative selection have not been observed in PARP-deficient mice.

2. PARP activation causes necrosis; however, *in vivo* thymocyte death is always apoptotic. Necrosis has never been reported to occur *in vivo* in thymocytes.
3. The effect of "conventional" apoptosis inducers such as dexamethasone or anti-Fas is indistinguishable in PARP$^{+/+}$ and PARP$^{-/-}$ thymocytes.[45,103]
4. Intrathymic peroxynitrite production theoretically could activate PARP and lead to necrosis. The reason we do not think that this would be the case is summarized in Figure 7.10.

According to our scenario (Figure 7.10), the negative selection process is initiated by the high-affinity binding of the T-cell receptors of potentially autoreactive thymocytes to self-antigens presented by dendritic cells. The ligation of TCR switches on the apoptotic machinery (caspase activation, PARP cleavage) and signals thymocytes to produce INFγ.[38] Thymocyte-derived INFγ, in turn, activates stromal cells leading to iNOS expression,[38] NO secretion, and NO/ONOO-mediated apoptosis of thymocytes.[111] According to our proposed model, PARP is dispensable for negative selection because it is already inactivated by caspase-mediated cleavage triggered by TCR signaling when NO and peroxynitrite are produced.

7.1.8 Effect of PARP Activation on Mitochondria

Although peroxynitrite-induced and PARP-mediated necrotic death is not likely to occur *in vivo* in thymocytes, we chose thymocytes as a sensitive model cell type to

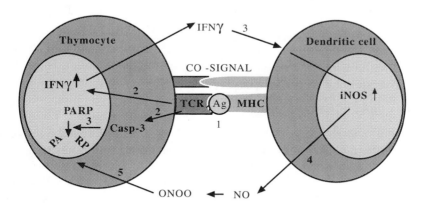

FIGURE 7.10 Hypothetical scenario for the negative selection with the participation of peroxynitrite. Dendritic cells present to thymocytes self-antigenic peptides bound to MHC molecules. Signaling, through TCR and cell surface cosignaling molecules, triggers two pathways: one that switches on the apoptotic machinery including the activation of caspases and another that leads to the upregulation of IFNγ. IFNγ activates dendritic cells resulting in iNOS expression and increased superoxide production. NO and superoxide combine to form peroxynitrite, which provides additional death signals to thymocytes. Peroxynitrite exposure does not activate PARP in thymocytes because PARP is already cleaved by caspase-3 by the time iNOS is upregulated and peroxynitrite production begins. (From Virág, L. et al., *J. Immunol.*, 161, 3752–3759, ©1998. The American Association of Immunologists. With permission.)

investigate the cellular mechanisms of peroxynitrite-induced cell death and the role of PARP in this process.

Mitochondrial dysfunction has been proposed to represent a point of no return during cell death.[112,113] A characteristic sequence of events occurs in the mitocondria during various forms of apoptotic and necrotic cell death. This includes the decrease of mitochondrial membrane potential $\Delta\Psi_m$, overproduction of superoxide, degradation of cardiolipin, and elevation of intracellular calcium level.[113,114] These mitochondrial alterations have also been found to occur during peroxynitrite-induced thymocyte death.[115] Moreover, in this system, PARP activation was found to mediate these events, as evidenced by the potent protection provided by the pharmacological PARP inhibitors and the PARP$^{-/-}$ phenotype from peroxynitrite-induced $\Delta\Psi_m$ decrease, overproduction of superoxide, cardiolipin loss, and elevation of intracellular calcium.[115] Morphological evidence also supports the crucial role of PARP activation in the peroxynitrite-induced necrotic thymocyte death and mitochondrial deterioration.[115] Electronmicroscopic images showed striking differences between PARP$^{+/+}$ and PARP$^{-/-}$ thymocytes exposed to 20 μM peroxynitrite. Wild-type thymocytes displayed a typical necrotic morphology (swollen cytoplasm and organelles and decreased electrondensity) with mitochondrial damage signs ranging from broken cristae to high-amplitude mitochondrial swelling, total disruption of ultrastructure, and appearance of flocculent matrix densities (Figure 7.11). In contrast, mitochondria of the PARP-deficient cells showed no or minor changes (Figure 7.11).

How can PARP activation trigger mitochondrial dysfunction? We hypothesize that PARP activation–mediated depletion of cellular NAD$^+$ and ATP, both of which are important regulators of mitochondrial functions,[116-119] may be responsible for the deleterious effect of PARP activation on the mitochondria. The exact mechanisms, however, require further investigations.

7.1.9 REGULATION OF PEROXYNITRITE-INDUCED PARP ACTIVATION AND CYTOTOXICITY

7.1.9.1 Role of Calcium and Zinc

Calcium is a widely used second messenger mediating a variety of cellular responses, including free radical toxicity. A disturbance of cellular calcium homeostasis is one of the most common signs of apoptotic and necrotic cell death.[120-124] Earlier findings showing that buffering of intracellular calcium protects cells from various lethal challenges[125-131] indicate that elevated Ca$_i^{2+}$ is not simply a passive consequence of the collapse of cellular homeostasis but an active central regulator of cell demise. Zinc accumulation has also been implicated in cell death induced by various stimuli including ROI and RNI.[132-135] Furthermore, intracellular Ca^{2+} and Zn^{2+} chelators had beneficial effects in a series of *in vitro* and *in vivo* conditions such as cerebral ischemia and myocardial reperfusion injury,[125,126,136-138] where PARP was proposed to be a key player.[139-142] Based on these obeservations, we hypothesized that Ca^{2+} and Zn^{2+} chelators may have beneficial effects in peroxynitrite-induced cytotoxicity by interfering with the activation of PARP. Indeed, we have found strong protection provided by cell-permeable Ca^{2+} chelators (BAPTA-AM, QUIN-2-AM, EGTA-AM)[143] and also by TPEN,[144] a cell-permeable Zn^{2+} cheletor with very low affinity

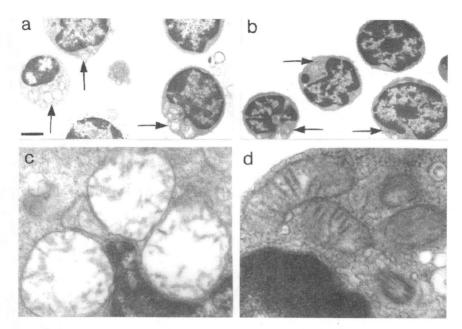

FIGURE 7.11 PARP activation results in mitochondrial destruction. Electron microscopic picture of wild-type thymocytes (a, c) treated with 20 μM peroxynitrite revealed typical necrotic morphology with decreased electrondensity, swollen mitochondria (arrows), disrupted cristae, and flocculant matrix densities. PARP-deficient thymocytes (b, d) exposed to the same concentration of peroxynitrite exhibited markedly reduced mitochondrial injury. (Magnification for panel a, b, c, and d is 6640×, 6640×, 33200×, and 58100×, respectively. Scale bar = 2.5 μm for panel a and b, scale bar = 0.5 μm for panel c, and 0.26 μm for panel d.) (From Virág, L. et al., *J. Immunol.*, 161, 3752–3759, ©1998. The American Association of Immunologists. With permission.)

to calcium against peroxynitrite-induced cytotoxicity. These chelators inhibited peroxynitrite-induced PARP activation in a dose-dependent manner,[143,144] providing further support for the close correlation between PARP activation and cytotoxicity. The mechanism of action of Ca^{2+} and Zn^{2+} chelators, however, is not necessarily the same. Intracellular Ca^{2+} chelators may act by buffering changes of Ca_i^{2+} level. Chelation of Zn^{2+}, however, may have a direct effect on PARP by removing Zn^{2+} from the Zn-finger motifs of the DNA binding domain of PARP. Interestingly, these chelators also provided protection for PARP−/− cells,[143,144] indicating the participation of Ca^{2+} and Zn^{2+} in peroxynitrite-induced PARP-independent cytotoxic pathways. It should be noted, however, that these chelators are not specific for Ca^{2+} or Zn^{2+}; therefore, special care should be taken when the cytoprotective effect of these chelators is attributed to chelation of a particular ion.

7.1.9.2 Purines: Endogenous PARP Inhibitors?

Tissue distruction is accompanied by elevated nucleic acid metabolism. Local concentration of purine–nucleoside metabolites (adenosine, inosine, hypoxanthine) can reach high levels under pathophysiological conditions.[145,146] These metabolites exert diverse biological activities and were implicated in the regulation of neurotransmission, exocrine and endocrine secretory activities, cell growth, differentiation, and proliferation in development and regeneration.[147] Furthermore, purines have been implicated in cardioprotection by early ischemic preconditioning[148-150] and also exert potent anti-inflammatory effects.[147,151-153] Regulation of cytokine production[154-156] can be made responsible in part for the anti-inflammatory effects of purines. A novel anti-inflammatory effect of these compounds may be the inhibition of PARP. Our preliminary results demonstrate that inosine and, more potently, hypoxanthine competitively inhibit PARP activity of peroxynitrite-treated cells and also inhibit the purified enzyme (L. Virág, unpublished observations). The inhibition of PARP by hypoxanthine and inosine may serve as a negative-feedback circle where PARP activation–mediated tissue distruction (e.g., in stroke, myocardial infarction, or colitis) liberates hypoxanthine and inosine from the damaged cells, resulting in downregulation of PARP activity by these metabolites.

7.1.9.3 Role of Phosphorylation

Phosphorylation is a general regulatory mechanism of enzymatic function. PARP has also been shown previously to be phosphorylated by the DNA-dependent protein kinase[157] and by an undefined serine kinase.[158] Earlier reports in the literature indicate that the DNA-binding domain of PARP can serve as substrate for protein kinase C (PKC),[159-161] and thus PARP activation may be regulated by PKC. A multilevel connection between PARP and PKC signaling may exist based on findings that PARP inhibition suppresses PKC activity[162] and that PARP modulates the properties of MARCKS proteins, which are important substrates of PKC.[163,164] Furthermore, apoptosis and/or necrosis induced by various stimuli, including oxidants such as singlet oxygen, has been shown to be suppressed by PKC-activating phorbol esters.[165,166] PKC has also been shown to mediate resistance to myocardial infarction,[167] a potentially fatal condition mediated in part by PARP activation.[139,140] These lines of evidence point to a possible role of PKC in the regulation of different oxidant-induced cytotoxic processes via the inhibition of PARP activation. Indeed, our preliminary data show a 50% reduction in peroxynitrite-induced PARP activation by phorbol esters (L. Virág, unpublished observations).

7.1.9.4 Role of Bcl-2

As detailed above, peroxynitrite-induced cytotoxicity is characterized by features of both apoptosis (at low concentrations of the oxidant) and necrosis (at high peroxynitrite concentrations). Moreover, mitochondria have been demonstrated to play a central role in this cytotoxic process. Thus, members of the Bcl protein family, which are predominantly localized in the mitochondria[168] and are known to protect from both apoptotic and necrotic death,[169-172] appeared likely candidates for being regulators of peroxynitrite cytotoxicity. The Bcl protein family consists of proapoptotic (bax, bak, bad, bik) and antiapoptotic (Bcl-2, Bcl-xL, mcl-1) molecules, each of which is capable of forming homo- or heterodimers with some members of the antagonist group.[173,174] The ratio of pro- and antiapoptotic Bcl proteins is thought to determine the fate of stimulated cells, death vs. survival.[173] The mechanism of action of Bcl-2 involves blocking of the opening of the mitochondrial permeability pore,[170,175] thereby preventing the release of cytochrome c from the mitochondrial intermembrane space to the cytoplasm and thus preventing caspase activation.[176,177] Furthermore, it also interferes with the calcium cycling, inhibiting the calcium overload of the mitochondria.[178-180]

Contrary to our expectations, we found no protection by Bcl-2 overexpression against peroxynitrite-induced and PARP-mediated necrotic death. Consistent with this, Bcl-2 had no effect on PARP activation. PARP-independent apoptotic death, however, was found to be supressed by Bcl-2 at the mitochondrial level (L. Virág, unpublished observations).

7.1.10 What Difference Does PARP Make in Apoptosis?

Based on results generated by our group, the role of PARP in oxidant-induced cytotoxicity can be summarized as follows. Free radicals and oxidants such as superoxide, peroxynitrite, and hydrogen peroxide cause DNA strand breaks, leading to PARP activation. Activation of PARP depletes cellular NAD^+ and ATP pools resulting in necrosis. Pharmacological intervention (3-aminobenzamide, INH_2BP, calcium or zinc chelators, purine metabolites, PKC activators) can protect from the deleterious effect of PARP activation; however, other mechanisms (e.g., Bcl proteins), which do not interfere with PARP activation, have no effect on oxidant-induced necrotic death, however powerfully they work in other PARP-independent systems. At the moment, no intervention is known that could protect from the *consequences* of PARP activation. Once PARP has become activated, NAD^+ and ATP are consumed, and the cell is destined to die by necrosis. This conclusion was drawn from studies carried out on thymocytes, but is likely to be true in most other cell types as well. Macrophages, endothelial cells, fibroblasts, and lung epithelial cells have also been found to be protected by PARP inhibitors (or the $PARP^{-/-}$ phenotype) against the cytotoxic effect of peroxynitrite. Exemption to this rule, also does exist. A promyelocytic leukemic cell line, HL-60, is also a sensitive target of peroxynitrite. As opposed to thymocytes and most other cell types, peroxynitrite-induced cytotoxicity of HL-60 cells was found to be potentiated by PARP inhibitors.[94] The reason HL-60 cells behave differently from most other cells is unknown at present and requires further investigations. These findings, however, emphasize the importance of cell-type-specific differences in the role of PARP in oxidant-induced cytotoxicity.

A much more controversial issue is the role of PARP in apoptotic death. Our results with subnecrotic concentrations of oxidants that do not activate PARP and also with conventional apoptosis inducers such as dexamethasone or anti-Fas revealed no differences between $PARP^{+/+}$ and $PARP^{-/-}$ thymocytes.[45] Other groups investigating the effect of numerous apoptotic stimuli including ceramide, dexamethasone, or etoposide (in thymocytes), tumor necrosis factor α (TNFα), anti-Fas (in hepatocytes), staurosporine, colchicine, potassium withdrawal, peroxynitrite (in primary neurons) to various cell types obtained from $PARP^{+/+}$ and $PARP^{-/-}$ mice were also unable to detect any difference between wild-type and PARP-deficient cells.[103,104] Based on these studies we could conclude that PARP is dispensable for apoptosis. However, a great body of evidence supports the idea that PARP may play a role in apoptosis. During some forms of apoptosis, PARP activation was found to occur, resulting in poly(ADP-ribosyl)ation of nuclear proteins.[181] Furthermore, p53 was also found to be poly(ADP-ribosyl)ated and poly(ADP-ribosyl)ation of p53 led to inhibition of the p53 response.[182,183] Considering the variability of apoptosis systems (p53-dependent or independent, regulated or not by Bcl proteins, dependence or lack of dependence on intact protein and RNA synthesis, etc.), it appears impossible at the moment to explain why PARP is dispensable in some forms of apoptosis, but contributes to other forms of programmed cell death. Furthermore, as our knowledge about the novel PARP enzymes such as PARP2[184] or tankyrase[185] and about their function in different cell types increases, we may get closer to understanding the as yet elusive role of poly(ADP-ribosyl)ation in apoptosis.

In this chapter, we do not aim to discuss the role of PARP in cytotoxicity induced by alkylating agents and ionizing radiation. An increasing body of evidence supports the protective role of PARP in these systems.[186-190] It is not known at present how PARP plays a protective role in ionizing radiation- and alkylating agent-induced cell death on the one hand, and mediates cell death triggered by free radicals and oxidants on the other hand. The solution of this striking dualism may lie in (1) the different nature of DNA damage caused by these genotoxic agents, (2) the different DNA repair mechanisms required to restore DNA structure, and/or (3) the intensity of PARP activation caused.

7.2 PARP AS TRANSCRIPTIONAL REGULATOR OF INFLAMMATION AND LYMPHOCYTE HOMING

7.2.1 Effect of PARP on the Expression of Chemokines, Adhesion Molecules, Collagenase, MHC-II Molecules, and Inducible Nitric Oxide Synthetase

Migration of inflammatory cells (monocytes, granulocytes, and lymphocytes) to the site of inflammation as well as migration of lymphocytes to lymphoid organs (lymph nodes, Peyer's patches, tonsils) where antigens are presented to lymphocytes by APCs is regulated by chemotactic cytokines called chemokines. Chemokines induce the upregulation of adhesion molecules on the surface of endothelial cells and migrating cells (granulocytes, monocytes, lymphocytes) leading to a firm attachment of migrating cells to the endothelium followed by transendothelial migration. At the site of inflammation, movement of leukocytes in the tissues is enhanced by enzymes such as collagenase, hyaluronidase, and elastase, which degrade the intercellular matrix of the connective tissue. Upon exposure to cytokines, inflammatory cells become activated. Activation results in the upregulation of MHC-II molecules and increased production of inflammatory mediators, e.g., NO. Upregulation of MHC-II molecules on the surface of thyroid cells and synovial cells has been implicated in the pathomechanism of autoimmune inflammation (autoimmune thyroiditis, biliary cirrhosis, type 1 diabetes, arthritis).[191-195]

PARP may interfere with the sequence of events at several points. PARP has been shown to enhance activator-dependent transcription.[60] Moreover, PARP has also been found to act as a coactivator for AP-2-mediated transcriptional activation.[61] As synthesis of cytokines and chemokines occurs via activator-dependent transcription of cytokine genes,[196-198] the possibility exists that PARP enhances the production of chemokines and cytokines. Indeed, peritoneal macrophages and embryonal fibroblasts from PARP-deficient mice produced significantly less macrophage inflammatory protein-1α (MIP-1α) a CC chemokine known to be responsible for the recruitment of monocytes to inflammatory environment (Virág, Haskó, and Szabó, unpublished observations). In various inflammatory models ranging from zymozan peritonitis to colitis and arthritis, in animals treated with PARP inhibitors as compared with wild-type mice, we have found a reduced number of neutrophil granulocytes as measured by myeloperoxidase activity of tissue lysates.[199-201] This reduced migration of neutrophil granulocytes could be explained by an inhibition by PARP inhibitors of chemokine expression. However, this hypothesis has not yet been tested in these *in vivo* models. Alternatively, impaired expression of adhesion receptors in the absence of PARP could also be made responsible for the inhibition of granulocyte recruitment. This scenario is supported by findings that human umbilical vein endothelial cells (HUVEC) and fibroblasts respond to cytokines with a reduced expression of intracellular adhesion molecule-1 (ICAM-1) and P-selectin.[202] Furthermore, a decreased expression of these adhesion molecules has been found in the hearts of PARP$^{-/-}$ mice as compared with wild-type (PARP$^{+/+}$) ones after myocardial ischemia reperfusion.[202]

Following transmigration through the endothelial layer, inflammatory cells migrate in the connective tissue toward the source of chemokines. Migration in the connective tissue is facilitated by the production of collagenase and other enzymes degrading the components of extracellular matrix. PARP inhibitors have been shown to inhibit collagenase expression in cultured rabbit synovial fibroblasts.[203] Thus, the beneficial effect of PARP inhibition in reducing the severity of inflammation may be in part due to inhibition of leukocyte migration in the inflammatory environment. The effect of PARP inhibitors on the expression of other enzymes such as elastase or hyaluronidase involved in the degradation of extracellular matrix has not yet been investigated. Activation of inflammatory cells is accompanied by the upregulation of activation markers (e.g., MHC-II molecules) and the production of inflammatory mediators such as NO. PARP has been shown to be involved in the regulation of both MHC[204-205] and iNOS[19,206-208] expression.

Based on these findings, we can conclude that PARP inhibitors exert their potent anti-inflammatory effects by a complex mechanism (Figure 7.12) involving cytoprotection from the necrotic effects of ROIs and RNIs, attenuation of endothelial transmigration of inflammatory cells, and inhibition of proinflammatory cytokines, chemokines, and inflammatory mediators.

7.2.2 Effect of PARP on Nuclear Receptor Signaling

As opposed to activator-dependent transcription, which is enhanced, nuclear receptor signaling is suppressed by PARP. This has been demonstrated by Miyamoto et al.,[209] who showed that PARP bound directly to retinoid X receptor (RXR) and repressed

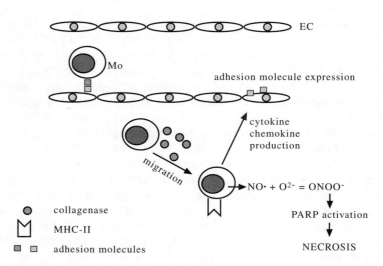

FIGURE 7.12 Involvement of PARP in various steps of the development of inflammation. PARP inhibitors interfere with the migration of inflammatory cells to the sites of inflammation by inhibiting (1) the secretion of chemokines and cytokines and (2) the expression of adhesion molecules. Migration of inflammatory cells at the site of inflammation also may be blocked by PARP inhibitors due to inhibition of collagenase expression. PARP inhibitors attenuate iNOS expression, inhibit peroxynitrite formation, and provide cytoprotection against the necrotic effect of free radicals and oxidants. (From Virág, L. et al., *J. Immunol.*, 161, 3752–3759, ©1998. The American Association of Immunologists. With permission.)

ligand-dependent transcriptional activities mediated by heterodimers of RXR and thyroid hormone receptor.

We have also tested the repressor effect of PARP on the expression of transglutaminase induced by all-trans retinoic acid. In line with Miyamoto's findings, we have found that peritoneal macrophages from PARP[−/−] mice produced more transglutaminase, as measured by Western blotting, both under resting condition and in response to retinoic acid (Figure 7.13). Considering the wide immunomodulatory effects of retinoids,[210] ranging from the potentiation of T-cell activation[211] to induction of differentiation of monocytes,[212] the question arises whether PARP inhibitors could be used as adjuvant therapy to retinoid treatment. It would also be interesting

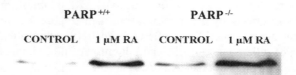

FIGURE 7.13 PARP regulates retinoic acid–induced transglutaminase expression. Peritoneal macrophages from PARP[+/+] and PARP[−/−] mice were treated with 1 μM all-trans retinoic acid (RA) for 24 h. Transglutaminase expression was detected by Western blotting. (From Virág, L. et al., *J. Immunol.*, 161, 3752–3759, ©1998. The American Association of Immunologists. With permission.)

to investigate whether the transcriptional repressor activity of PARP can be extended to other members of the nuclear receptor family with special regard to the steroid receptors. Steroid receptors, comprising a subfamily of the nuclear receptor superfamily, mediate the powerful immunosuppressing activities of glucocorticosteroids, which are widely used potent anti-inflammatory agents. The anti-inflammatory effects of PARP inhibitors could thus be attributed in part to a possible potentiation of the effects of endogenous corticosteroids. This hypothesis also implies that PARP inhibitors may be useful as part of a combined anti-inflammatory treatment together with glucocorticoids.

7.3 ROLE OF PARP IN THE DEVELOPMENT OF THE T- AND B-CELL REPERTOIRE

Lymphocytes recognize antigens by cell surface antigen receptors. The antigen receptor of T lymphocytes is a heterodimer of an α and a β chain, whereas B lymphocytes use immunoglobulins consisting each of two heavy and light chains as antigen receptors. The structural diversity of the antigen receptors of the lymphocytes is achieved in part by the somatic recombination of germ line genes encoding for the V (variable), D (diversity), and J (joining) region of the α and β chains of the TCR or the immunoglobulin receptor of the B cells (BCR). The mammalian genome contains several V, D, and J genes, from which one V, one D, and one J gene will be randomly joined by recombinase enzymes (Figure 7.14). This V(D)J complex is coupled to the C gene encoding the constant region of TCR or BCR, thus forming a full-length antigen receptor gene. In the course of the secondary immune response, when memory B cells are presented the antigen by APCs, the Cμ segment is cleaved off from the original V(D)JC complex and is replaced by another C gene encoding for another type of heavy chain (γ, ε, or α). This process, called isotype switch, results in a change of the isotype of the immunoglobulins from IgM to IgG (or IgA, IgE, etc.).

The V(D)J recombination process requires nicking and joining of DNA. DNA nicks can lead to PARP activation, and thus the possibility exists that PARP activation somehow affects V(D)J recombination and the development of the T- and B-cell repertoire. Some data are available in the literature that support the role of PARP in the regulation of V(D)JC gene rearrangement in both the T- and B-cell lineage.[213] The inhibitory effect of PARP on V(D)J gene rearrangement of T cells is indicated by the finding that the absence of the PARP gene partially restores the ability of DNA-PK-deficient mice to produce T cells.[213] DNA-PK knockout mice suffer from severe combined immune-deficiency (SCID) due to defective antigen receptor V(D)J recombination, which arrests B- and T-lymphocyte development. SCID V(D)J recombination could be rescued partly in T lymphocytes by knocking out the PARP gene. Thymocytes of DNA-PK and PARP double-knockout mice express both CD4 and CD8 coreceptors, bypassing the SCID block and indicating a partial restoration of T-cell development.[213] Double-mutant T cells in the periphery express TCR beta, which is attributable to productive TCR beta joints. A cautionary note for PARP inhibitor therapy is that DNA-PK and PARP double-mutant mice develop a high frequency of T-cell lymphoma; however, no malignancies have so far been linked to the PARP$^{-/-}$ phenotype. Based on the increased recombination activity after the

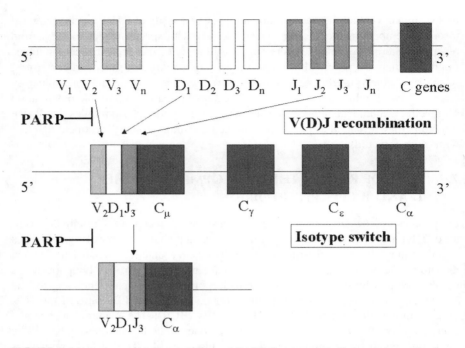

FIGURE 7.14 Antirecombinogenic effect of PARP. Immunoglobulins consist each of two heavy and light chains, whereas TCRs are heterodimers of α and β or γ and δ chains. In the germline configuration, the immunoglobulin and TCR-encoding regions consist of numerous variable (V), diversity (D), joining (J), and constant (C) regions. (There is no D region in the Ig light chain and in the α and γ chain of TCR.) One of each V, D, J, and C gene is randomly recombined by recombinase enzymes to form a functionable immunoglobulin or TCR chain. In the case of immunoglobulins, the immunoglobulin isotype is changed during the secondary immune response from IgM to another isotype (e.g., IgG, IgA, IgE) by replacing the Cμ segment with Cγ, Cϵ, or Cα. This process, called isotype switch, is under the strict control of cytokines. Upon antigen stimulation, Ig genes also undergo somatic hypermutation, further increasing the Ig repertoire. PARP may inhibit the formation of functional Ig or TCR genes from the germ line configuration and may also interfere with the isotype switch.

loss of the antirecombinogenic function of PARP, which rescued V(D)J recombination in SCID mice, authors concluded that PARP and DNA-PK cooperate to minimize genomic damage caused by DNA strand breaks. These findings also implicate that the lack of PARP may affect the development of the B- and T-lymphocyte repertoire. Theoretically, due to the antirecombinogenic effect of PARP, PARP knockout mice may have a larger T- and B-cell repertoire, which must be considered when immune-mediated disease models (e.g., arthritis, experimental allergic encephalomyelitis) are carried out using these mice. No gross anomalies in the immune functions of PARP knockout mice have been reported so far; however, this possibility has not yet been investigated. It should also been noted here, however, that the effect of PARP on V(D)J recombination and isotype switch could not be confirmed by Wang et al.[104] using PARP knockout mice.

In addition to V(D)J recombination, another important source of B-cell diversity is the somatic hypermutation of the immunoglobulin V genes of memory B cells. The possible involvement of DNA repair enzymes including PARP in the somatic hypermutation has been tested, and hypermutation was found unaffected by the lack of PARP and all other repair systems tested.[214]

On the other hand, PARP inhibitors (nicotinamide) have been shown to enhance the endotoxin-induced antibody class switching to IgA in B-cell lymphoma cells.[215] In spleen B cells, nicotinamide increased IgG1 switching by twofold.[215] It would be interesting to know the distribution of different immunoglobulin isotypes in response to different antigen challenges of PARP$^{-/-}$ mice as compared with PARP$^{+/+}$ mice. Considering the differences in the effector functions (complement activation, binding to Fc receptors, transplacental passage) of the individual immunoglobulin isotypes, the lack of PARP may have profound effect on the humoral immune responses.

In summary, due to the antirecombinogenic function of PARP, a wide variety of immunological abnormalities ranging from an altered T- and B-cell repertoire to a different immunoglobulin isotype distribution can be hypothesized to occur in PARP$^{-/-}$ mice. Until the occurrence of this theoretical abnormality of the immune system is either confirmed or ruled out, experimental *in vivo* data obtained in disease models where immune mechanisms play a role must be evaluated with care.

7.4 SUMMARY

On a theoretical basis, a wide variety of immunological abnormalities can be expected to occur in the absence of PARP, but only a few of them have been experimentally investigated. An increasing body of evidence supports the idea that pharmacological PARP inhibitors may offer powerful tools in the treatment of immune-mediated inflammation. Their complex mechanism of action may involve inhibition of chemokine and adhesion molecule expression resulting in decreased recruitment of effector cells to inflammatory sites. Furthermore, production of inflammatory cytokines and inflammatory mediators (NO) can also be reduced. Moreover, proliferation and differentiation of immune effector cells as well as their migration can also be inhibited by PARP inhibitors. In addition, PARP inhibitors may potentiate the effect of immunomodulator drugs (e.g., retinoids) and thereby prove useful in combinational immunotherapy. The protection from necrosis represents a major mechanism by which PARP inhibitors improve disease signs in various animal models. Diversion by PARP inhibitors from oxidant-induced "bad death" (necrosis) to "good death" (apoptosis) attenuates inflammation by reducing the inflammogenic release of cell content characteristic of necrosis.

Anomalies of T- and B-cell development that can be hypothesized to be present in PARP$^{-/-}$ mice may explain possible discrepancies between the results of pharmacological studies and studies using PARP$^{-/-}$ mice when immune-mediated diseases are investigated.

ACKNOWLEDGMENTS

This work was supported by a grant from the National Institutes of Health (R01GM60915) and the Hungarian Ministry of Health (ETT104/99).

REFERENCES

1. Stuehr, D. J., Gross, S. S., Sakuma, I., Levi, R., and Nathan, C. F., Activated murine macrophages secrete a metabolite of arginine with the bioactivity of endothelium-derived relaxing factor and the chemical reactivity of nitric oxide. *J. Exp. Med.,* 169:1011–1020, 1989.
2. Stuehr, D. J., Kwon, N. S., Gross, S. S., Thiel, B. A., Levi, R., and Nathan, C. F., Synthesis of nitrogen oxides from L-arginine by macrophage cytosol: requirement for inducible and constitutive components. *Biochem. Biophys. Res. Commun.,* 161:420–426, 1989.
3. Marletta, M. A., Yoon, P. S., Iyengar, R., Leaf, C. D., and Wishnok, J. S., Macrophage oxidation of L-arginine to nitrite and nitrate: nitric oxide is an intermediate. *Biochemistry,* 27:8706–8711, 1988.
4. Stuehr, D. J. and Nathan, C. F., Nitric oxide. A macrophage product responsible for cytostasis and respiratory inhibition in tumor target cells. *J. Exp. Med.,* 169:1543–1555, 1989.
5. Hibbs, J. B. J., Taintor, R. R., Vavrin, Z., and Rachlin, E. M., Nitric oxide: a cytotoxic activated macrophage effector molecule. *Biochem. Biophys. Res. Commun.,* 157:87–94, 1988.
6. Nathan, C. F. and Hibbs, J. B. J., Role of nitric oxide synthesis in macrophage antimicrobial activity. *Curr. Opin. Immunol.,* 3:65–70, 1991.
7. Adams, L. B., Franzblau, S. G., Vavrin, Z., Hibbs, J. B. J., and Krahenbuhl, J. L., L-Arginine-dependent macrophage effector functions inhibit metabolic activity of *Mycobacterium leprae. J. Immunol.,* 147:1642–1646, 1991.
8. Stadler, J., Billiar, T. R., Curran, R. D., Stuehr, D. J., Ochoa, J. B., and Simmons, R. L., Effect of exogenous and endogenous nitric oxide on mitochondrial respiration of rat hepatocytes. *Am. J. Physiol.,* 260:C910–C916, 1991.
9. Drapier, J. C. and Hibbs, J. B. J., Aconitases: a class of metalloproteins highly sensitive to nitric oxide synthesis. *Methods Enzymol.,* 269:26–36, 1996.
10. Gardner, P. R., Costantino, G., Szabó, C., and Salzman, A. L., Nitric oxide sensitivity of the aconitases. *J. Biol. Chem.,* 272:25071–25076, 1997.
11. Hogg, N. and Kalyanaraman, B., Nitric oxide and lipid peroxidation. *Biochim. Biophys. Acta,* 1411:378–384, 1999.
12. van der Veen, R. C. and Roberts, L. J., Contrasting roles for nitric oxide and peroxynitrite in the peroxidation of myelin lipids. *J. Neuroimmunol.,* 95:1–7, 1999.
13. O'Donnell, V. B., Eiserich, J. P., Bloodsworth, A., Chumley, P. H., Kirk, M., Barnes, S., Darley-Usmar, V. M., and Freeman, B. A., Nitration of unsaturated fatty acids by nitric oxide-derived reactive species. *Methods Enzymol.,* 301:454–470, 1999.
14. Beckman, J. S. and Koppenol, W. H., Nitric oxide, superoxide, and peroxynitrite: the good, the bad, and the ugly. *Am. J. Physiol.,* 271:C1424–C1437, 1996.
15. Inoue, S. and Kawanishi, S., Oxidative DNA damage induced by simultaneous generation of nitric oxide and superoxide. *FEBS Lett.,* 371:86–88, 1995.
16. Salgo, M. G., Bermudez, E., Squadrito, G. L., and Pryor, W. A., Peroxynitrite causes DNA damage and oxidation of thiols in rat thymocytes. *Arch. Biochem. Biophys.,* 322:500–505, 1995.
17. Salgo, M. G., Stone, K., Squadrito, G. L., Battista, J. R., and Pryor, W. A., Peroxynitrite causes DNA nicks in plasmid pBR322. *Biochem. Biophys. Res. Commun.,* 210:1025–1030, 1995.
18. Szabó, C. and Ohshima, H., DNA damage induced by peroxynitrite: subsequent biological effects. *Nitric Oxide,* 1:373–385, 1997.

19. Szabó, C., Zingarelli, B., O'Connor, M., and Salzman, A. L., DNA strand breakage, activation of poly(ADP-ribose) synthetase, and cellular energy depletion are involved in the cytotoxicity of macrophages and smooth muscle cells exposed to peroxynitrite. *Proc. Natl. Acad. Sci. U.S.A.*, 93:1753–1758, 1996.
20. Szabó, C., DNA strand breakage and activation of poly-ADP ribosyltransferase: a cytotoxic pathway triggered by peroxynitrite. *Free Radical Biol. Med.*, 21:855–869, 1996.
21. Ischiropoulos, H., Zhu, L., and Beckman, J. S., Peroxynitrite formation from macrophage-derived nitric oxide. *Arch. Biochem. Biophys.*, 298:446–451, 1992.
22. Lu, L., Induction of nitric oxide synthase in mouse dendritic cells by IFN-gamma, endotoxin, and interaction with allogeneic T cells: nitric oxide production is associated with dendritic cell apoptosis. *J. Immunol.*, 157:3577–3586, 1996.
23. Qureshi, A. A., Hosoi, J., Xu, S., Takashima, A., Granstein, R. D., and Lerner, E. A., Langerhans cells express inducible nitric oxide synthase and produce nitric oxide. *J. Invest. Dermatol.*, 107:815–821, 1996.
24. Yamamoto, K., Increased nitric oxide (NO) production by antigen-presenting dendritic cells is responsible for low allogeneic mixed leucocyte reaction (MLR) in primary biliary cirrhosis (PBC). *Clin. Exp. Immunol.*, 114:94–101, 1998.
25. Bonham, C. A., Nitric oxide production by dendritic cells is associated with impairment of T cell responses. *Transplant. Proc.*, 29:1116–1117, 1997.
26. Marrack, P. and Kappler, J., Positive selection of thymocytes bearing alpha beta T cell receptors. *Curr. Opin. Immunol.*, 9:250–255, 1997.
27. Anderson, G., Moore, N. C., Owen, J. J., and Jenkinson, E. J., Cellular interactions in thymocyte development. *Annu. Rev. Immunol.*, 14:73–99, 1996.
28. Reza, J. N. and Ritter, M. A., Positive and negative selection in the thymus and the thymic paradox. *Dev. Immunol.*, 5:161–168, 1998.
29. Saito, T. and Watanabe, N., Positive and negative thymocyte selection. *Crit. Rev. Immunol.*, 18:359–370, 1998.
30. Page, D. M., Kane, L. P., Allison, J. P., and Hedrick, S. M., Two signals are required for negative selection of CD4+CD8+ thymocytes. *J. Immunol.*, 151:1868–1880, 1993.
31. Amsen, D. and Kruisbeek, A. M., Thymocyte selection: not by TCR alone. *Immunol. Rev.*, 165:209–229, 1998.
32. Castro, J. E., Listman, J. A., Jacobson, B. A., Wang, Y., Lopez, P. A., Ju, S., Finn, P. W., and Perkins, D. L., Fas modulation of apoptosis during negative selection of thymocytes. *Immunity*, 5:617–627, 1996.
33. Kishimoto, H., Surh, C. D., and Sprent, J., A role for Fas in negative selection of thymocytes *in vivo*. *J. Exp. Med.*, 187:1427–1438, 1998.
34. Noel, P. J., Alegre, M. L., Reiner, S. L., and Thompson, C. B., Impaired negative selection in CD28-deficient mice. *Cell Immunol.*, 187:131–138, 1998.
35. Page, D. M., Tokugawa, Y., Silver, J., and Stewart, C. L., Role of Thy-1 in T cell development. *J. Immunol.*, 159:5285–5292, 1997.
36. Hueber, A. O., Raposo, G., Pierres, M., and He, H. T., Thy-1 triggers mouse thymocyte apoptosis through a bcl-2-resistant mechanism. *J. Exp. Med.*, 179:785–796, 1994.
37. Tai, X. G., Saitoh, Y., Satoh, T., Yamamoto, N., Kita, Y., Takenaka, H., Yu, W. G., Zou, J. P., Hamaoka, T., and Fujiwara, H., Thymic stromal cells eliminate T cells stimulated with antigen plus stromal Ia molecules through their cross-talk involving the production of interferon-gamma and nitric oxide. *Thymus*, 24:41–56, 1994.
38. Tai, X. G., Toyo, O. K., Yamamoto, N., Yashiro, Y., Mu, J., Hamaoka, T., and Fujiwara, H., Expression of an inducible type of nitric oxide (NO) synthase in the thymus and involvement of NO in deletion of TCR-stimulated double-positive thymocytes. *J. Immunol.*, 158:4696–4703, 1997.

39. Gulati, P., Chan, A. S., and Leong, S. K., NADPH-diaphorase positive cells in the chick and rat thymus. *Thymus,* 22:117–124, 1993.
40. Downing, J. E., Multiple nitric oxide synthase systems in adult rat thymus revealed using NADPH diaphorase histochemistry. *Immunology,* 82:659–664, 1994.
41. Downing, J. E., Virág, L., and Jones, I. W., NADPH diaphorase-positive dendritic profiles in rat thymus are discrete from autofluorescent cells, immunoreactive for inducible nitric oxide synthase, and show strain-specific abundance differences. *Immunology,* 95:148–155, 1998.
42. Virág, L., Hasko, G., Salzman, A. L., and Szabó, C., NADPH diaphorase histochemistry detects inducible nitric oxide synthetase activity in the thymus of naive and staphylococcal enterotoxin B-stimulated mice. *J. Histochem. Cytochem.,* 46:787–791, 1998.
43. van der Vliet, A., O'Neill, C. A., Halliwell, B., Cross, C. E., and Kaur, H., Aromatic hydroxylation and nitration of phenylalanine and tyrosine by peroxynitrite. Evidence for hydroxyl radical production from peroxynitrite. *FEBS Lett.,* 339:89–92, 1994.
44. Crow, J. P. and Beckman, J. S., Reactions between nitric oxide, superoxide, and peroxynitrite: footprints of peroxynitrite *in vivo. Adv. Pharmacol.,* 34:17–43, 1995.
45. Virág, L., Scott, G. S., Cuzzocrea, S., Marmer, D., Salzman, A. L., and Szabó, C., Peroxynitrite-induced thymocyte apoptosis: the role of caspases and poly (ADP-ribose) synthetase (PARS) activation. *Immunology,* 94:345–355, 1998.
46. Downing, J. E., Virág, L., and Perry, M. E., Nitrergic mechanism of DC-mediated T-cell elimination [letter; comment]. *Immunol. Today,* 19:190–191, 1998.
47. Brito, C., Naviliat, M., Tiscornia, A. C., Vuillier, F., Gualco, G., Dighiero, G., Radi, R., and Cayota, A. M., Peroxynitrite inhibits T lymphocyte activation and proliferation by promoting impairment of tyrosine phosphorylation and peroxynitrite-driven apoptotic death. *J. Immunol.,* 162:3356–3366, 1999.
48. Meldrum, D. R., Ayala, A., and Chaudry, I. H., Energetics of lymphocyte "burnout" in late sepsis: adjuvant treatment with ATP-MgCl$_2$ improves energetics and decreases lethality. *J. Surg. Res.,* 56:537–542, 1994.
49. Ittel, M. E., Jongstra-Bilen, J., Rochette-Egly, C., and Mandel, P., Involvement of polyADP-ribose polymerase in the initiation of phytohemagglutinin induced human lymphocyte proliferation. *Biochem. Biophys. Res. Commun.,* 116:428–434, 1983.
50. Greer, W. L. and Kaplan, J. G., Early nuclear events in lymphocyte proliferation. The role of DNA strand break repair and ADP ribosylation. *Exp. Cell Res.,* 166:399–415, 1986.
51. Chabert, M. G., Niedergang, C. P., Hog, F., Partisani, M., and Mandel, P., Poly(ADPR) polymerase expression and activity during proliferation and differentiation of rat astrocyte and neuronal cultures. *Biochim. Biophys. Acta,* 1136:196–202, 1992.
52. Bhatia, M., Kirkland, J. B., and Meckling-Gill, K. A., Modulation of poly(ADP-ribose) polymerase during neutrophilic and monocytic differentiation of promyelocytic (NB4) and myelocytic (HL-60) leukaemia cells. *Biochem. J.,* 308:131–137, 1995.
53. Bhatia, M., Kirkland, J. B., and Meckling-Gill, K. A., Overexpression of poly(ADP-ribose) polymerase promotes cell cycle arrest and inhibits neutrophilic differentiation of NB4 acute promyelocytic leukemia cells. *Cell Growth Differ.,* 7:91–100, 1996.
54. Damji, N., Khoo, K. E., Booker, L., and Browman, G. P., Influence of the poly (ADP-ribose) polymerase inhibitor 3-aminobenzamide on macrophage and granulocyte differentiation of HL-60 cells. *Am. J. Hematol.,* 21:67–78, 1986.
55. Kanai, M., Miwa, M., Kondo, T., Tanaka, Y., Nakayasu, M., and Sugimura, T., Involvement of poly(ADP-ribose) metabolism in induction of differentiation of HL-60 promyelocytic leukemia cells. *Biochem. Biophys. Res. Commun.,* 105:404–411, 1982.

56. Kun, E., Poly (ADP-ribose) polymerase, a potential target for drugs: Part II. Regulation of differentiation by the poly ADP-ribose system. *Int. J. Mol. Med.*, 2:591–592, 1998.
57. Menegazzi, M., Suzuki, H., Carcereri, D. P., Tommasi, M., Miwa, M., Gandini, G., and Gerosa, F., Increase of poly(ADP-ribose) polymerase mRNA levels during TPA-induced differentiation of human lymphocytes. *FEBS Lett.*, 297:59–62, 1992.
58. Smulson, M. E., Kang, V.H., Ntambi, J. M., Rosenthal, D. S., Ding, R., and Simbulan, C. M., Requirement for the expression of poly(ADP-ribose) polymerase during the early stages of differentiation of 3T3-L1 preadipocytes, as studied by antisense RNA induction. *J. Biol. Chem.*, 270:119–127, 1995.
59. Ueda, K., Banasik, M., Nakajima, S., Yook, H. Y., and Kido, T., Cell differentiation induced by poly(ADP-ribose) synthetase inhibitors. *Biochimie*, 77:368–373, 1995.
60. Meisterernst, M., Stelzer, G., and Roeder, R. G., Poly(ADP-ribose) polymerase enhances activator-dependent transcription *in vitro*. *Proc. Natl. Acad. Sci. U.S.A.*, 94:2261–2265, 1997.
61. Kannan, P., Yu, Y., Wankhade, S., and Tainsky, M. A., Poly(ADP-ribose) polymerase is a coactivator for AP-2-mediated transcriptional activation. *Nucl. Acids Res.*, 27:866–874, 1999.
62. Butler, A. J. and Ordahl, C. P., Poly(ADP-ribose) polymerase binds with transcription enhancer factor 1 to MCAT1 elements to regulate muscle-specific transcription. *Mol. Cell Biol.*, 19:296–306, 1999.
63. Le Page, C., Sanceau, J., Drapier, J. C., and Wietzerbin, J., Inhibitors of ADP-ribosylation impair inducible nitric oxide synthase gene transcription through inhibition of NF kappa B activation. *Biochem. Biophys. Res. Commun.*, 243:451–457, 1998.
64. Oei, S. L., Griesenbeck, J., Ziegler, M., and Schweiger, M., A novel function of poly(ADP-ribosyl)ation: silencing of RNA polymerase II-dependent transcription. *Biochemistry*, 37:1465–1469, 1998.
65. Rawling, J. M. and Alvarez-Gonzalez, R., TFIIF, a basal eukaryotic transcription factor, is a substrate for poly(ADP-ribosyl)ation. *Biochem. J.*, 324:249–253, 1997.
66. Yamagoe, S., Kohda, T., and Oishi, M., Poly(ADP-ribose) polymerase inhibitors suppress UV-induced human immunodeficiency virus type 1 gene expression at the posttranscriptional level. *Mol. Cell Biol.*, 11:3522–3527, 1991.
67. Heel, K. A., McCauley, R. D., Papadimitriou, J.M., and Hall, J. C., Review: Peyer's patches. *J. Gastroenterol. Hepatol.*, 12:122–136, 1997.
68. Gebert, A., Rothkotter, H. J., and Pabst, R., M cells in Peyer's patches of the intestine. *Int. Rev. Cytol.*, 167:91–159, 1996.
69. Fehsel, K., Kroncke, K. D., Meyer, K. L., Huber, H., Wahn, V., and Kolb-Bachofen, V., Nitric oxide induces apoptosis in mouse thymocytes. *J. Immunol.*, 155:2858–2865, 1995.
70. Kayahara, M., Felderhoff, U., Pocock, J., Hughes, M. N., and Mehmet, H., Nitric oxide (NO·) and the nitrosonium cation (NO$^+$) reduce mitochondrial membrane potential and trigger apoptosis in neuronal PC12 cells. *Biochem. Soc. Trans.*, 26:S340, 1998.
71. Nishikawa, M., Sato, E. F., Kuroki, T., Utsumi, K., and Inoue, M., Macrophage-derived nitric oxide induces apoptosis of rat hepatoma cells *in vivo*. *Hepatology*, 28:1474–1480, 1998.
72. Palluy, O. and Rigaud, M., Nitric oxide induces cultured cortical neuron apoptosis. *Neurosci. Lett.*, 208:1–4, 1996.
73. Smith, J. D., McLean, S. D., and Nakayama, D. K., Nitric oxide causes apoptosis in pulmonary vascular smooth muscle cells. *J. Surg. Res.*, 79:121–127, 1998.

74. Salgo, M. G., Squadrito, G. L., and Pryor, W. A., Peroxynitrite causes apoptosis in rat thymocytes. *Biochem. Biophys. Res. Commun.,* 215:1111–1118, 1995.
75. Nicholson, D. W. and Thornberry, N. A., Caspases: killer proteases. *Trends Biochem. Sci.,* 22:299–306, 1997.
76. Thornberry, N. A. and Lazebnik, Y., Caspases: enemies within. *Science,* 281:1312–1316, 1998.
77. Suschek, C. V., Krischel, V., Bruch-Gerharz, D., Berendji, D., Krutmann, J., Kröncke, K. D., and Kolb-Bachofen, V., Nitric oxide fully protects against UVA-induced apoptosis in tight correlation with Bcl-2 up-regulation. *J. Biol. Chem.,* 274:6130–6137, 1999.
78. Leist, M., Single, B., Naumann, H., Fava, E., Simon, B., Kuhnle, S., and Nicotera, P., Nitric oxide inhibits execution of apoptosis at two distinct ATP- dependent steps upstream and downstream of mitochondrial cytochrome c release. *Biochem. Biophys. Res. Commun.,* 258:215–221, 1999.
79. Li, J., Bombeck, C. A., Yang, S., Kim, Y. M., and Billiar, T. R., Nitric oxide suppresses apoptosis via interrupting caspase activation and mitochondrial dysfunction in cultured hepatocytes. *J. Biol. Chem.,* 274:17325–17333, 1999.
80. Mannick, J. B., Asano, K., Izumi, K., Kieff, E., and Stamler, J. S., Nitric oxide produced by human B lymphocytes inhibits apoptosis and Epstein-Barr virus reactivation. *Cell,* 79:1137–1146, 1994.
81. Mannick, J. B., Miao, X. Q., and Stamler, J. S., Nitric oxide inhibits Fas-induced apoptosis. *J. Biol. Chem.,* 272:24125–24128, 1997.
82. Kim, Y. M., Talanian, R. V., and Billiar, T. R., Nitric oxide inhibits apoptosis by preventing increases in caspase-3-like activity via two distinct mechanisms. *J. Biol. Chem.,* 272:31138–31148, 1997.
83. Kim, Y. M., Kim, T. H., Seol, D. W., Talanian, R. V., and Billiar, T. R., Nitric oxide suppression of apoptosis occurs in association with an inhibition of Bcl-2 cleavage and cytochrome c release. *J. Biol. Chem.,* 273:31437–31441, 1998.
84. Dimmeler, S., Haendeler, J., Nehls, M., and Zeiher, A. M., Suppression of apoptosis by nitric oxide via inhibition of interleukin-1beta-converting enzyme (ICE)-like and cysteine protease protein (CPP)-32-like proteases. *J. Exp. Med.,* 185:601–607, 1997.
85. Melino, G., Bernassola, F., Knight, R. A., Corasaniti, M. T., Nistico, G., and Finazzi-Agro, A., S-Nitrosylation regulates apoptosis [letter]. *Nature,* 388:432–433, 1997.
86. Tenneti, L., D'Emilia, D. M., and Lipton, S. A., Suppression of neuronal apoptosis by S-nitrosylation of caspases. *Neurosci. Lett.,* 236:139–142, 1997.
87. Fiorucci, S., Santucci, L., Federici, B., Antonelli, E., Distrutti, E., Morelli, O., Renzo, G. D., Coata, G., Cirino, G., Soldato, P. D., and Morelli, A., Nitric oxide-releasing NSAIDs inhibit interleukin-1beta converting enzyme-like cysteine proteases and protect endothelial cells from apoptosis induced by TNFalpha. *Aliment Pharmacol. Ther.,* 13:421–435, 1999.
88. Messmer, U. K., Reimer, D. M., Reed, J. C., and Brune, B., Nitric oxide induced poly(ADP-ribose) polymerase cleavage in RAW 264.7 macrophage apoptosis is blocked by Bcl-2. *FEBS Lett.,* 384:162–166, 1996.
89. Leist, M., Volbracht, C., Kuhnle, S., Fava, E., Ferrando-May, E., and Nicotera, P., Caspase-mediated apoptosis in neuronal excitotoxicity triggered by nitric oxide. *Mol. Med.,* 3:750–764, 1997.
90. Messmer, U. K., Reimer, D. M., and Brune, B., Protease activation during nitric oxide-induced apoptosis: comparison between poly(ADP-ribose) polymerase and U1-70 kDa cleavage. *Eur. J. Pharmacol.,* 349:333–343, 1998.

91. Uehara, T., Kikuchi, Y., and Nomura, Y., Caspase activation accompanying cytochrome c release from mitochondria is possibly involved in nitric oxide-induced neuronal apoptosis in SH-SY5Y cells. *J. Neurochem.*, 72:196–205, 1999.
92. Chlichlia, K., Peter, M. E., Rocha, M., Scaffidi, C., Bucur, M., Krammer, P. H., Schirrmacher, V., and Umansky, V., Caspase activation is required for nitric oxide-mediated, CD95(APO-1/Fas)-dependent and independent apoptosis in human neoplastic lymphoid cells. *Blood,* 91:4311–4320, 1998.
93. Lin, K. T., Xue, J. Y., Lin, M. C., Spokas, E. G., Sun, F. F., and Wong, P. Y., Peroxynitrite induces apoptosis of HL-60 cells by activation of a caspase-3 family protease. *Am. J. Physiol.,* 274:C855–C860, 1998.
94. Virág, L., Marmer, D. J., and Szabó, C., Crucial role of apopain in the peroxynitrite-induced apoptotic DNA fragmentation. *Free Radical Biol. Med.,* 25:1075–1082, 1998.
95. Kaufmann, S. H., Desnoyers, S., Ottaviano, Y., Davidson, N. E., and Poirier, G. G., Specific proteolytic cleavage of poly(ADP-ribose) polymerase: an early marker of chemotherapy-induced apoptosis. *Cancer Res.,* 53:3976–3985, 1993.
96. Earnshaw, W. C., Nuclear changes in apoptosis. *Curr. Opin. Cell Biol.,* 7:337–343, 1995.
97. Herceg, Z. and Wang, Z. Q., Failure of poly(ADP-ribose) polymerase cleavage by caspases leads to induction of necrosis and enhanced apoptosis. *Mol. Cell Biol.,* 19:5124–5133, 1999.
98. Nosseri, C., Coppola, S., and Ghibelli, L., Possible involvement of poly(ADP-ribosyl) polymerase in triggering stress-induced apoptosis. *Exp. Cell Res.,* 212:367–373, 1994.
99. Tanaka, Y., Yoshihara, K., Tohno, Y., Kojima, K., Kameoka, M., and Kamiya, T., Inhibition and downregulation of poly(ADP-ribose) polymerase results in a marked resistance of HL-60 cells to various apoptosis-inducers. *Cell Mol. Biol. (Noisy.-le.-grand.),* 41:771–781, 1995.
100. Shiokawa, D., Maruta, H., and Tanuma, S., Inhibitors of poly(ADP-ribose) polymerase suppress nuclear fragmentation and apoptotic-body formation during apoptosis in HL-60 cells. *FEBS Lett.,* 413:99–103, 1997.
101. Guo, T. L., Miller, M. A., Datar, S., Shapiro, I. M., and Shenker, B. J., Inhibition of poly(ADP-ribose) polymerase rescues human T lymphocytes from methylmercury-induced apoptosis. *Toxicol. Appl. Pharmacol.,* 152:397–405, 1998.
102. Hivert, B., Cerruti, C., and Camu, W., Hydrogen peroxide-induced motoneuron apoptosis is prevented by poly ADP ribosyl synthetase inhibitors. *Neuroreport,* 9:1835–1838, 1998.
103. Leist, M., Single, B., Kunstle, G., Volbracht, C., Hentze, H., and Nicotera, P., Apoptosis in the absence of poly-(ADP-ribose) polymerase. *Biochem. Biophys. Res. Commun.,* 233:518–522, 1997.
104. Wang, Z. Q., Stingl, L., Morrison, C., Jantsch, M., Los, M., Schulze-Osthoff, K., and Wagner, E. F., PARP is important for genomic stability but dispensable in apoptosis. *Genes Dev.,* 11:2347–2358, 1997.
105. Watson, A. J., Askew, J. N., and Benson, R. S., Poly(adenosine diphosphate ribose) polymerase inhibition prevents necrosis induced by H_2O_2 but not apoptosis. *Gastroenterology,* 109:472–482, 1995.
106. Ghibelli, L., Nosseri, C., Coppola, S., Maresca, V., and Dini, L., The increase in H_2O_2-induced apoptosis by ADP-ribosylation inhibitors is related to cell blebbing. *Exp. Cell Res.,* 221:470–477, 1995.
107. Tentori, L., Orlando, L., Lacal, P. M., Benincasa, E., Faraoni, I., Bonmassar, E., D'Atri, S., and Graziani, G., Inhibition of O_6-alkylguanine DNA-alkyltransferase or poly(ADP-ribose) polymerase increases susceptibility of leukemic cells to apoptosis induced by temozolomide. *Mol. Pharmacol.,* 52:249–258, 1997.

108. Payne, C. M., Crowley, C., Washo-Stultz, D., Briehl, M., Bernstein, H., Bernstein, C., Beard, S., Holubec, H., and Warneke, J., The stress-response proteins poly(ADP-ribose) polymerase and NF-kappaB protect against bile salt-induced apoptosis. *Cell Death Differ.*, 5:623–636, 1998.
109. Tentori, L., Turriziani, M., Franco, D., Serafino, A., Levati, L., Roy, R., Bonmassar, E., and Graziani, G., Treatment with temozolomide and poly(ADP-ribose) polymerase inhibitors induces early apoptosis and increases base excision repair gene transcripts in leukemic cells resistant to triazene compounds. *Leukemia,* 13:901–909, 1999.
110. Vermes, I., Haanen, C., Steffens-Nakken, H., and Reutelingsperger, C., A novel assay for apoptosis. Flow cytometric detection of phosphatidylserine expression on early apoptotic cells using fluorescein-labelled Annexin V. *J. Immunol. Methods,* 184:39–51, 1995.
111. Lu, L., Bonham, C. A., Chambers, F. G., Watkins, S. C., Hoffman, R. A., Simmons, R. L., and Thomson, A. W., Induction of nitric oxide synthase in mouse dendritic cells by IFN-gamma, endotoxin, and interaction with allogeneic T cells: nitric oxide production is associated with dendritic cell apoptosis. *J. Immunol.*, 157:3577–3586, 1996.
112. Kroemer, G., Dallaporta, B., and Resche-Rigon, M., The mitochondrial death/life regulator in apoptosis and necrosis. *Annu. Rev. Physiol.*, 60:619–642, 1998.
113. Zamzami, N., Marchetti, P., Castedo, M., Decaudin, D., Macho, A., Hirsch, T., Susin, S. A., Petit, P. X., Mignotte, B., and Kroemer, G., Sequential reduction of mitochondrial transmembrane potential and generation of reactive oxygen species in early programmed cell death. *J. Exp. Med.*, 182:367–377, 1995.
114. Zamzami, N., Marchetti, P., Castedo, M., Zanin, C., Vayssiere, J. L., Petit, P. X., and Kroemer, G., Reduction in mitochondrial potential constitutes an early irreversible step of programmed lymphocyte death in vivo. *J. Exp. Med.*, 181:1661–1672, 1995.
115. Virág, L., Salzman, A. L., and Szabó, C., Poly(ADP-ribose) synthetase activation mediates mitochondrial injury during oxidant-induced cell death. *J. Immunol.*, 161:3753–3759, 1998.
116. Lee, A. C., Xu, X., and Colombini, M., The role of pyridine dinucleotides in regulating the permeability of the mitochondrial outer membrane. *J. Biol. Chem.*, 271:26724–26731, 1996.
117. Zizi, M., Forte, M., Blachly-Dyson, E., and Colombini, M., NADH regulates the gating of VDAC, the mitochondrial outer membrane channel. *J. Biol. Chem.*, 269:1614–1616, 1994.
118. Costantini, P., Chernyak, B. V., Petronilli, V., and Bernardi, P., Modulation of the mitochondrial permeability transition pore by pyridine nucleotides and dithiol oxidation at two separate sites. *J. Biol. Chem.*, 271:6746–6751, 1996.
119. Devin, A., Guerin, B., and Rigoulet, M., Cytosolic NAD+ content strictly depends on ATP concentration in isolated liver cells. *FEBS Lett.*, 410:329–332, 1997.
120. Orrenius, S. and Nicotera, P., The calcium ion and cell death. *J. Neural Transm. Suppl.*, 43:1–11, 1994.
121. Orrenius, S., Ankarcrona, M., and Nicotera, P., Mechanisms of calcium-related cell death. *Adv. Neurol.*, 71:137–149, 1996.
122. Nicotera, P. and Orrenius, S., The role of calcium in apoptosis. *Cell Calcium,* 23:173–180, 1998.
123. Lipton, S. A. and Nicotera, P., Calcium, free radicals and excitotoxins in neuronal apoptosis. *Cell Calcium,* 23:165–171, 1998.
124. Leist, M. and Nicotera, P., Calcium and neuronal death. *Rev. Physiol. Biochem. Pharmacol.*, 132:79–125, 1998.

125. Tymianski, M., Wallace, M. C., Spigelman, I., Uno, M., Carlen, P. L., Tator, C. H., and Charlton, M. P., Cell-permeant Ca^{2+} chelators reduce early excitotoxic and ischemic neuronal injury *in vitro* and *in vivo*. *Neuron*, 11:221–235, 1993.
126. Tymianski, M., Neuroprotection *in vitro* and *in vivo* by cell membrane-permeant Ca^{2+} chelators. *Clin. Exp. Pharmacol. Physiol.*, 22:299–300, 1995.
127. Tymianski, M., Spigelman, I., Zhang, L., Carlen, P. L., Tator, C. H., Charlton, M. P., and Wallace, M. C., Mechanism of action and persistence of neuroprotection by cell-permeant Ca^{2+} chelators. *J. Cereb. Blood Flow Metab.*, 14:911–923, 1994.
128. Tymianski, M., Charlton, M. P., Carlen, P. L., and Tator, C. H., Properties of neuroprotective cell-permeant Ca^{2+} chelators: effects on $[Ca^{2+}]_i$ and glutamate neurotoxicity *in vitro*. *J. Neurophysiol.*, 72:1973–1992, 1994.
129. Frandsen, A. and Schousboe, A., Excitatory amino acid-mediated cytotoxicity and calcium homeostasis in cultured neurons. *J. Neurochem.*, 60:1202–1211, 1993.
130. Schousboe, A., Belhage, B., and Frandsen, A., Role of Ca^{2+} and other second messengers in excitatory amino acid receptor mediated neurodegeneration: clinical perspectives. *Clin. Neurosci.*, 4:194–198, 1997.
131. Zager, R. A. and Burkhart, K., Myoglobin toxicity in proximal human kidney cells: roles of Fe, Ca^{2+}, H_2O_2, and terminal mitochondrial electron transport. *Kidney Int.*, 51:728–738, 1997.
132. Tatsumi, T. and Fliss, H., Hypochlorous acid mobilizes intracellular zinc in isolated rat heart myocytes. *J. Mol. Cell Cardiol.*, 26:471–479, 1994.
133. Kim, C. H., Kim, J. H., Xu, J., Hsu, C. Y., and Ahn, Y. S., Pyrrolidine dithiocarbamate induces bovine cerebral endothelial cell death by increasing the intracellular zinc level. *J. Neurochem.*, 72:1586–1592, 1999.
134. Cuajungco, M. P. and Lees, G. J., Nitric oxide generators produce accumulation of chelatable zinc in hippocampal neuronal perikarya. *Brain Res.*, 799:118–129, 1998.
135. Adler, M., Dinterman, R. E., and Wannemacher, R. W., Protection by the heavy metal chelator N,N,N',N'-tetrakis (2-pyridylmethyl)ethylenediamine (TPEN) against the lethal action of botulinum neurotoxin A and B. *Toxicon*, 35:1089–1100, 1997.
136. Ferdinandy, P., Appelbaum, Y., Csonka, C., Blasig, I. E., and Tosaki, A., Role of nitric oxide and TPEN, a potent metal chelator, in ischaemic and reperfused rat isolated hearts. *Clin. Exp. Pharmacol. Physiol.*, 25:496–502, 1998.
137. Karck, M., Appelbaum, Y., Schwalb, H., Haverich, A., Chevion, M., and Uretzky, G., TPEN, a transition metal chelator, improves myocardial protection during prolonged ischemia. *J. Heart Lung Transplant.*, 11:979–985, 1992.
138. Chevion, M., Protection against free radical-induced and transition metal-mediated damage: the use of "pull" and "push" mechanisms. *Free Radical Res. Commun.*, 12–13 Pt 2:691–696, 1991.
139. Zingarelli, B., Cuzzocrea, S., Zsengeller, Z., Salzman, A. L., and Szabó, C., Protection against myocardial ischemia and reperfusion injury by 3-aminobenzamide, an inhibitor of poly(ADP-ribose) synthetase. *Cardiovasc. Res.*, 36:205–215, 1997.
140. Thiemermann, C., Bowes, J., Myint, F. P., and Vane, J. R., Inhibition of the activity of poly(ADP ribose) synthetase reduces ischemia-reperfusion injury in the heart and skeletal muscle. *Proc. Natl. Acad. Sci. U.S.A.*, 94:679–683, 1997.
141. Eliasson, M. J., Sampei, K., Mandir, A. S., Hurn, P. D., Traystman, R. J., Bao, J., Pieper, A., Wang, Z. Q., Dawson, T. M., Snyder, S. H., and Dawson, V. L., Poly(ADP-ribose) polymerase gene disruption renders mice resistant to cerebral ischemia. *Nat. Med.*, 3:1089–1095, 1997.

142. Endres, M., Wang, Z. Q., Namura, S., Waeber, C., and Moskowitz, M. A., Ischemic brain injury is mediated by the activation of poly(ADP-ribose)polymerase. *J. Cereb. Blood Flow Metab.*, 17:1143–1151, 1997.
143. Virág, L., Scott, G. S., Antal-Szalmas, P, O'Connor, M., Ohshima, H., and Szabó, C., Requirement of intracellular calcium mobilization for peroxynitrite-induced poly(ADP-ribose) synthetase (PARS) activation and cytotoxicity. *Mol. Pharmacol.*, 56:824–833, 1999.
144. Virág, L. and Szabó, C., Inhibition of poly(ADP-ribose) synthetase (PARS) and protection against peroxynitrite-induced cytotoxicity by zinc chelation. *Br. J. Pharmacol.*, 126:769–777, 1999.
145. Rodriguez-Nunez, A., Camina, F., Lojo, S., Rodriguez-Segade, S., and Castro-Gago, M., Concentrations of nucleotides, nucleosides, purine bases and urate in cerebrospinal fluid of children with meningitis. *Acta Paediatr.*, 82:849–852, 1993.
146. Gudbjornsson, B., Zak, A., Niklasson, F., and Hallgren, R., Hypoxanthine, xanthine, and urate in synovial fluid from patients with inflammatory arthritides. *Ann. Rheum. Dis.*, 50:669–672, 1991.
147. Abbracchio, M. P. and Burnstock, G., Purinergic signalling: pathophysiological roles. *Jpn. J. Pharmacol.*, 78:113–145, 1998.
148. Liu, G. S., Thornton, J., Van Winkle, D. M., Stanley, A. W., Olsson, R. A., and Downey, J. M., Protection against infarction afforded by preconditioning is mediated by A1 adenosine receptors in rabbit heart. *Circulation*, 84:350–356, 1991.
149. de Jonge, R., Bradamante, S., and de Jong, J. W., Cardioprotection by ischemic preconditioning. Role of adenosine and glycogen. *Adv. Exp. Med. Biol.*, 431:279–282, 1998.
150. Miura, T. and Tsuchida, A., Adenosine and preconditioning revisited. *Clin. Exp. Pharmacol. Physiol.*, 26:92–99, 1999.
151. Cronstein, B. N., Adenosine, an endogenous anti-inflammatory agent. *J. Appl. Physiol.*, 76:5–13, 1994.
152. Schrier, D. J., Lesch, M. E., Wright, C. D., and Gilbertsen, R. B., The antiinflammatory effects of adenosine receptor agonists on the carrageenan-induced pleural inflammatory response in rats. *J. Immunol.*, 145:1874–1879, 1990.
153. Marak, G. E. J., de Kozak, Y., Faure, J. P., Rao, N. A., Romero, J. L., Ward, P. A., and Till, G. O., Pharmacologic modulation of acute ocular inflammation. I. Adenosine. *Ophthalmic Res.*, 20:220–226, 1988.
154. Hasko, G., Szabó, C., Nemeth, Z. H., Kvetan, V., Pastores, S. M., and Vizi, E. S., Adenosine receptor agonists differentially regulate IL-10, TNF-alpha, and nitric oxide production in RAW 264.7 macrophages and in endotoxemic mice. *J. Immunol.*, 157:4634–4640, 1996.
155. Hasko, G. and Szabó, C., Regulation of cytokine and chemokine production by transmitters and co-transmitters of the autonomic nervous system. *Biochem. Pharmacol.*, 56:1079–1087, 1998.
156. Szabó, C., Scott, G. S., Virág, L., Egnaczyk, G., Salzman, A. L., Shanley, T. P., and Hasko, G., Suppression of macrophage inflammatory protein (MIP)-1alpha production and collagen-induced arthritis by adenosine receptor agonists. *Br. J. Pharmacol.*, 125:379–387, 1998.
157. Ruscetti, T., Lehnert, B. E., Halbrook, J. et al., Stimulation of the DNA-dependent protein kinase by poly(ADP-ribose) polymerase. *J. Biol. Chem.*, 273:14461–14467, 1998.
158. Ariumi, Y., Ueda, K., Masutani, M., Copeland, T. D., Noda, M., Hatanaka, M., and Shimotohno, K., *In vivo* phosphorylation of poly(ADP-ribose) polymerase is independent of its activation. *FEBS Lett.*, 436:288–292, 1998.

159. Tanaka, Y., Koide, S. S., Yoshihara, K., and Kamiya, T., Poly (ADP-ribose) synthetase is phosphorylated by protein kinase C *in vitro*. *Biochem. Biophys. Res. Commun.*, 148:709–717, 1987.
160. Bauer, P. I., Farkas, G., Buday, L., Mikala, G., Meszaros, G., Kun, E., and Farago, A., Inhibition of DNA binding by the phosphorylation of poly ADP-ribose polymerase protein catalysed by protein kinase C. *Biochem. Biophys. Res. Commun.*, 187:730–736, 1992.
161. Bauer, P. I., Farkas, G., Mihalik, R., Kopper, L., Kun, E., and Farago, A., Phosphorylation of poly(ADP-ribose)polymerase protein in human peripheral lymphocytes stimulated with phytohemagglutinin. *Biochim. Biophys. Acta,* 1223:234–239, 1994.
162. Ricciarelli, R., Palomba, L., Cantoni, O., and Azzi, A., 3-Aminobenzamide inhibition of protein kinase C at a cellular level. *FEBS Lett.,* 431:465–467, 1998.
163. Schmitz, A. A., Pleschke, J. M., Kleczkowska, H. E., Althaus, F. R., and Vergeres, G., Poly(ADP-ribose) modulates the properties of MARCKS proteins. *Biochemistry,* 37:9520–9527, 1998.
164. Chao, D., Severson, D. L., Zwiers, H., and Hollenberg, M. D., Radiolabelling of bovine myristoylated alanine-rich protein kinase C substrate (MARCKS) in an ADP-ribosylation reaction. *Biochem. Cell Biol.,* 72:391–396, 1994.
165. Zhuang, S., Lynch, M. C., and Kochevar, I. E., Activation of protein kinase C is required for protection of cells against apoptosis induced by singlet oxygen. *FEBS Lett.,* 437:158–162, 1998.
166. Mansat, V., Laurent, G., Levade, T., Bettaieb, A., and Jaffrezou, J. P., The protein kinase C activators phorbol esters and phosphatidylserine inhibit neutral sphingomyelinase activation, ceramide generation, and apoptosis triggered by daunorubicin. *Cancer Res.,* 57:5300–5304, 1997.
167. Joyeux, M., Baxter, G. F., Thomas, D. L., Ribuot, C., and Yellon, D. M., Protein kinase C is involved in resistance to myocardial infarction induced by heat stress. *J. Mol. Cell Cardiol.,* 29:3311–3319, 1997.
168. Hockenbery, D., Nunez, G., Milliman, C., Schreiber, R. D., and Korsmeyer, S. J., Bcl-2 is an inner mitochondrial membrane protein that blocks programmed cell death. *Nature,* 348:334–336, 1990.
169. Kane, D. J., Ord, T., Anton, R., and Bredesen, D. E., Expression of bcl-2 inhibits necrotic neural cell death. *J. Neurosci. Res.,* 40:269–275, 1995.
170. Shimizu, S., Eguchi, Y., Kamiike, W., Waguri, S., Uchiyama, Y., Matsuda, H., and Tsujimoto, Y., Retardation of chemical hypoxia-induced necrotic cell death by Bcl-2 and ICE inhibitors: possible involvement of common mediators in apoptotic and necrotic signal transductions. *Oncogene,* 12:2045–2050, 1996.
171. Subramanian, T., Tarodi, B., and Chinnadurai, G., p53-independent apoptotic and necrotic cell deaths induced by adenovirus infection: suppression by E1B 19K and Bcl-2 proteins. *Cell Growth Differ.,* 6:131–137, 1995.
172. Tsujimoto, Y., Shimizu, S., Eguchi, Y., Kamiike, W., and Matsuda, H., Bcl-2 and Bcl-xL block apoptosis as well as necrosis: possible involvement of common mediators in apoptotic and necrotic signal transduction pathways. *Leukemia,* 11 Suppl. 3:380–382, 1997.
173. Chao, D. T. and Korsmeyer, S. J., BCL-2 family: regulators of cell death. *Annu. Rev. Immunol.,* 16:395–419, 1998.
174. Adams, J. M. and Cory, S., The Bcl-2 protein family: arbiters of cell survival. *Science,* 281:1322–1326, 1998.

175. Marzo, I., Brenner, C., Zamzami, N., Susin, S. A., Beutner, G., Brdiczka, D., Remy, R., Xie, Z. H., Reed, J. C., and Kroemer, G., The permeability transition pore complex: a target for apoptosis regulation by caspases and bcl-2-related proteins. *J. Exp. Med.*, 187:1261–1271, 1998.
176. Kluck, R. M., Bossy-Wetzel, E., Green, D. R., and Newmeyer, D. D., The release of cytochrome c from mitochondria: a primary site for Bcl-2 regulation of apoptosis [see comments]. *Science*, 275:1132–1136, 1997.
177. Yang, J., Liu, X., Bhalla, K., Kim, C. N., Ibrado, A. M., Cai, J., Peng, T. I., Jones, D. P., and Wang, X., Prevention of apoptosis by Bcl-2: release of cytochrome c from mitochondria blocked [see comments]. *Science*, 275:1129–1132, 1997.
178. Magnelli, L., Cinelli, M., Turchetti, A., and Chiarugi, V. P., Bcl-2 overexpression abolishes early calcium waving preceding apoptosis in NIH-3T3 murine fibroblasts. *Biochem. Biophys. Res. Commun.*, 204:84–90, 1994.
179. Murphy, A. N., Bredesen, D. E., Cortopassi, G., Wang, E., and Fiskum, G., Bcl-2 potentiates the maximal calcium uptake capacity of neural cell mitochondria. *Proc. Natl. Acad. Sci. U.S.A.*, 93:9893–9898, 1996.
180. Wei, H., Wei, W., Bredesen, D. E., and Perry, D. C., Bcl-2 protects against apoptosis in neuronal cell line caused by thapsigargin-induced depletion of intracellular calcium stores. *J. Neurochem.*, 70:2305–2314, 1998.
181. Simbulan-Rosenthal, C. M., Rosenthal, D. S., Iyer, S., Boulares, A. H., and Smulson, M. E., Transient poly(ADP-ribosyl)ation of nuclear proteins and role of poly(ADP-ribose) polymerase in the early stages of apoptosis. *J. Biol. Chem.*, 273:13703–13712, 1998.
182. Kumari, S. R., Mendoza-Alvarez, H., and Alvarez-Gonzalez, R., Functional interactions of p53 with poly(ADP-ribose) polymerase (PARP) during apoptosis following DNA damage: covalent poly(ADP-ribosyl)ation of p53 by exogenous PARP and noncovalent binding of p53 to the M(r) 85,000 proteolytic fragment. *Cancer Res.*, 58:5075–5078, 1998.
183. Simbulan-Rosenthal, C. M., Rosenthal, D. S., Luo, R., and Smulson, M. E., Poly(ADP-ribosyl)ation of p53 during apoptosis in human osteosarcoma cells. *Cancer Res.*, 59:2190–2194, 1999.
184. Ame, J. C., Rolli, V., Schreiber, V., Niedergang, C., Apiou, F., Decker, P., Muller, S., Hoger, T., Ménissier-de Murcia, J., and de Murcia, G., PARP-2, a novel mammalian DNA damage-dependent Poly(ADP-ribose) polymerase. *J. Biol. Chem.*, 274:17860–17868, 1999.
185. Smith, S., Giriat, I., Schmitt, A., and de Lange, T., Tankyrase, a poly(ADP-ribose) polymerase at human telomeres. *Science*, 282:1484–1487, 1998.
186. Chatterjee, S., Cheng, M. F., and Berger, N. A., Hypersensitivity to clinically useful alkylating agents and radiation in poly(ADP-ribose) polymerase-deficient cell lines. *Cancer Commun.*, 2:401–407, 1990.
187. Chatterjee, S., Cheng, M. F., Berger, S. J., and Berger, N. A., Alkylating agent hypersensitivity in poly(adenosine diphosphate-ribose) polymerase deficient cell lines. *Cancer Commun.*, 3:71–75, 1991.
188. Witmer, M. V., Aboul-Ela, N., Jacobson, M. K., and Stamato, T. D., Increased sensitivity to DNA-alkylating agents in CHO mutants with decreased poly(ADP-ribose) polymerase activity. *Mutat. Res.*, 314:249–260, 1994.
189. de Murcia, J. M., Niedergang, C., Trucco, C., Ricoul, M., Dutrillaux, B., Mark, M., Oliver, F. J., Masson, M., Dierich, A., LeMeur, M., Walztinger, C., Chambon, P., and de Murcia, G., Requirement of poly(ADP-ribose) polymerase in recovery from DNA damage in mice and in cells. *Proc. Natl. Acad. Sci. U.S.A.*, 94:7303–7307, 1997.

190. Masutani, M., Nozaki, T., Nishiyama, E., Shimokawa, T., Tachi, Y., Suzuki, H., Nakagama, H., Wakabayashi, K., and Sugimura, T., Function of poly(ADP-ribose) polymerase in response to DNA damage: gene-disruption study in mice. *Mol. Cell Biochem.,* 193:149–152, 1999.
191. Hanafusa, T., Pujol-Borrell, R., Chiovato, L., Russell, R. C., Doniach, D., and Bottazzo, G. F., Aberrant expression of HLA-DR antigen on thyrocytes in Graves' disease: relevance for autoimmunity. *Lancet,* 2:1111–1115, 1983.
192. Ballardini, G., Mirakian, R., Bianchi, F. B., Pisi, E., Doniach, D., and Bottazzo, G. F., Aberrant expression of HLA-DR antigens on bileduct epithelium in primary biliary cirrhosis: relevance to pathogenesis. *Lancet,* 2:1009–1013, 1984.
193. Foulis, A. K., Class II major histocompatibility complex and organ specific autoimmunity in man. *J. Pathol.,* 150:5–11, 1986.
194. Pujol-Borrell, R., Todd, I., Londei, M., Foulis, A. K., Feldmann, M., and Bottazzo, G. F., Inappropriate major histocompatibility complex class II expression by thyroid follicular cells in thyroid autoimmune disease and by pancreatic beta cells in type I diabetes. *Mol. Biol. Med.,* 3:159–165, 1986.
195. Feldmann, M., Londei, M., Kissonerghis, M., Portillo, G., Leech, Z., Ziegler, A., and Grubeck-Loebenstein, B., Regulation of HLA class II expression and the pathogenesis of autoimmunity. *Concepts Immunopathol.,* 5:44–56, 1988.
196. Li, Q., Vaingankar, S. M., Green, H. M., and Martins-Green, M., Activation of the 9E3/cCAF chemokine by phorbol esters occurs via multiple signal transduction pathways that converge to MEK1/ERK2 and activate the Elk1 transcription factor. *J. Biol. Chem.,* 274:15454–15465, 1999.
197. Roebuck, K. A., Carpenter, L. R., Lakshminarayanan, V., Page, S. M., Moy, J. N., and Thomas, L. L., Stimulus-specific regulation of chemokine expression involves differential activation of the redox-responsive transcription factors AP-1 and NF-kappaB. *J. Leukoc. Biol.,* 65:291–298, 1999.
198. Park, J. H., Kaushansky, K., and Levitt, L., Transcriptional regulation of interleukin 3 (IL3) in primary human T lymphocytes. Role of AP-1– and octamer-binding proteins in control of IL3 gene expression. *J. Biol. Chem.,* 268:6299–6308, 1993.
199. Szabó, C., Lim, L. H., Cuzzocrea, S., Getting, S. J., Zingarelli, B., Flower, R. J., Salzman, A. L., and Perretti, M., Inhibition of poly(ADP-ribose) synthetase attenuates neutrophil recruitment and exerts antiinflammatory effects. *J. Exp. Med.,* 186:1041–1049, 1997.
200. Zingarelli, B., Szabó, C., and Salzman, A. L., Blockade of poly(ADP-ribose) synthetase inhibits neutrophil recruitment, oxidant generation, and mucosal injury in murine colitis. *Gastroenterology,* 116:335–345, 1999.
201. Cuzzocrea, S., Zingarelli, B., Costantino, G., Szabó, A., Salzman, A. L., Caputi, A. P., and Szabó, C., Beneficial effects of 3-aminobenzamide, an inhibitor of poly(ADP-ribose) synthetase in a rat model of splanchnic artery occlusion and reperfusion. *Br. J. Pharmacol.,* 121:1065–1074, 1997.
202. Zingarelli, B., Salzman, A. L., and Szabó, C., Genetic disruption of poly(ADP-ribose) synthetase inhibits the expression of P-selectin and intercellular adhesion molecule-1 in myocardial ischemia/reperfusion injury. *Circ. Res.,* 83:85–94, 1998.
203. Ehrlich, W., Huser, H., and Kroger, H., Inhibition of the induction of collagenase by interleukin-1 beta in cultured rabbit synovial fibroblasts after treatment with the poly(ADP-ribose)-polymerase inhibitor 3-aminobenzamide. *Rheumatol. Int.,* 15:171–172, 1995.

204. Taniguchi, T., Ota, K., Qu, Z., and Morisawa, K., Effect of poly(ADP-ribose) synthetase on the expression of major histocompatibility complex (MHC) class II genes. *Biochimie,* 77:472–479, 1995.
205. Otsuka, A., Hanafusa, T., Miyagawa, J., Kono, N., and Tarui, S., Nicotinamide and 3-aminobenzamide reduce interferon-gamma-induced class II MHC (HLA-DR and -DP) molecule expression on cultured human endothelial cells and fibroblasts. *Immunopharmacol. Immunotoxicol.,* 13:263–280, 1991.
206. Fujimura, M., Tominaga, T., and Yoshimoto, T., Nicotinamide inhibits inducible nitric oxide synthase mRNA in primary rat glial cells. *Neurosci. Lett.,* 228:107–110, 1997.
207. Hauschildt, S., Scheipers, P., Bessler, W. G., and Mulsch, A., Induction of nitric oxide synthase in L929 cells by tumour-necrosis factor alpha is prevented by inhibitors of poly(ADP-ribose) polymerase. *Biochem. J.,* 288:255–600, 1992.
208. Le Page, C., Sanceau, J., Drapier, J. C., and Wietzerbin, J., Inhibitors of ADP-ribosylation impair inducible nitric oxide synthase gene transcription through inhibition of NF kappa B activation. *Biochem. Biophys. Res. Commun.,* 243:451–457, 1998.
209. Miyamoto, T., Kakizawa, T., and Hashizume, K., Inhibition of nuclear receptor signalling by poly(ADP-ribose) polymerase. *Mol. Cell Biol.,* 19:2644–2649, 1999.
210. Evans, T. R. and Kaye, S. B., Retinoids: present role and future potential. *Br. J. Cancer,* 80:1–8, 1999.
211. Jiang, X. L., Everson, M. P., and Lamon, E. W., A mechanism of retinoid potentiation of murine T-cell responses: early upregulation of interleukin-2 receptors. *Int. J. Immunopharmacol.,* 15:309–317, 1993.
212. Botling, J., Oberg, F., Torma, H., Tuohimaa, P., Blauer, M., and Nilsson, K., Vitamin D3- and retinoic acid-induced monocytic differentiation: interactions between the endogenous vitamin D3 receptor, retinoic acid receptors, and retinoid X receptors in U-937 cells. *Cell Growth Differ.,* 7:1239–1249, 1996.
213. Morrison, C., Smith, G. C., Stingl, L., Jackson, S. P., Wagner, E. F., and Wang, Z. Q., Genetic interaction between PARP and DNA-PK in V(D)J recombination and tumorigenesis. *Nat. Genet.,* 17:479–482, 1997.
214. Jacobs, H., Fukita, Y., van der Horst, G. T., de Boer, J., Weeda, G., Essers, J., de Wind, N., Engelward, B. P., Samson, L., Verbeek, S., de Murcia, J. M., de Murcia, G., Riele H., and Rajewsky, K., Hypermutation of immunoglobulin genes in memory B cells of DNA repair- deficient mice. *J. Exp. Med.,* 187:1735–1743, 1998.
215. Shockett, P. and Stavnezer, J., Inhibitors of poly(ADP-ribose) polymerase increase antibody class switching. *J. Immunol.,* 151:6962–6976, 1993.

8 Protective Effect of Poly(ADP-Ribose) Polymerase Inhibitors against Cell Damage Induced by Antiviral and Anticancer Drugs

Balázs Sümegi, György Rablóczky, Ildikó Rácz, Kálmán Tory, Sándor Bernáth, Gábor Várbiró, Ferenc Gallyas, Jr., and Péter Literati Nagy

CONTENTS

8.1 Introduction 167
8.2 Role of PARP Activation in the Cytotoxicity of Deoxynucleoside
    Analogues and Dideoxynucleoside Antiviral Drugs 168
8.3 ROS-Mediated Cytotoxicity of Antitumor Drugs 171
8.4 Protective Effect of BGP-15, a Novel PARP Inhibitor, against
    Antitumor Drug–Induced Cytotoxicity 173
8.5 Conclusion 177
References 177

8.1 INTRODUCTION

Antitumor and antiviral agents have limited selectivity. They kill tumor cells and suppress the proliferation of viruses but can also damage normal tissues. The molecular mechanism of antitumor and antiviral effects relies mainly on specific interactions with the replication process of the tumor cells and the viruses. However, they can induce reactive oxygen species (ROS) formation in the well-oxygenated healthy tissues and activate poly(ADP-ribose) polymerase (PARP). In many cases, oxidative cell damage and PARP activation provide the molecular basis of the dose-limiting side effects of these drugs. The different mechanism of action for the therapeutic and the side effects raises the possibility that nontoxic PARP inhibitors, by protecting normal tissues from the side effect of anticancer and antiviral drugs, may increase the effectiveness and safety of these drugs in a combination therapy.

This chapter provides new experimental evidence and supporting literature data showing that PARP inhibitors can diminish the side effects of 3′-azido-3′-deoxythymidine (AZT) and cisplatin without interfering with their primary antitumor and antiviral effects. It suggests that PARP inhibitors can specifically protect vital tissues from antiviral and antitumor drug–induced side effects, and can further extend the use of these drugs in antitumor and antiviral therapy.

The currently available antitumor and antiviral agents have limited selectivity. The antitumor effect of several such drugs is based on intercalation into the DNA, DNA alkylation and cross-linking, interference with DNA unwinding, inhibition of topoisomerase II, and so on. The antiviral effect of nucleoside and dideoxynucleoside analogues mainly rely on reverse transcriptase inhibition. However, the dose-limiting side effects of these drugs in many cases rely on ROS-mediated cytotoxicity, raising the possibility of designing compounds that can protect vital tissues from the side effects without interfering with the primary antiviral and antitumor action. The permanently high ROS level in several tissues (e.g., heart, kidney, and the nervous system) can induce the formation of single-strand DNA breaks, which in turn activate nuclear PARP, a chromatin-bound enzyme involved in the repair of DNA strand breaks.[1,2] The activation of the nuclear PARP accelerates NAD^+ catabolism resulting in NAD^+ depletion and compromised mitochondrial energy metabolism.[3] Mitochondria have an important role in ROS-induced cell damage, because ROS can induce a massive influx of Ca^{2+} into the mitochondrial matrix and specifically inhibit the electron transport between NADH and ubiquinone.[4] The inhibition of mitochondrial electrontransport at complex I induces excessive ROS formation,[5,6] which in turn can induce mitochondrial permeability transition (MPT). MPT can lead to intramitochondrial NAD^+ loss and the inhibition of mitochondrial NAD^+-linked substrate oxidation,[7] further damaging the mitochondrial energy production and finally resulting in apoptotic or necrotic cell death.[8,9]

PARP inhibitors can diminish these unfavorable processes because PARP inhibitors significantly decrease the rate of ROS-activated NAD^+ catabolism, reduce the ATP utilization for the resynthesis of NAD^+, decrease the ROS-induced mitochondrial NAD^+ loss, and under certain conditions[10,11] decrease the endogenous, predominantly mitochondrial, ROS formation. Since the antiviral and antitumor effects of the abovementioned drugs rely on other characteristics of the drugs (reverse transcriptase inhibition, telomerase inhibition induction of apoptosis, etc.) and not on specific oxidative cell damages, it is likely that PARP inhibitors may protect normal tissues (heart, kidneys, and the nervous system) from antiviral and anticancer drug–induced oxidative side effects without decreasing the therapeutic effectiveness of these drugs.

8.2 ROLE OF PARP ACTIVATION IN THE CYTOTOXICITY OF DEOXYNUCLEOSIDE ANALOGUES AND DIDEOXYNUCLEOSIDE ANTIVIRAL DRUGS

AZT and dideoxynucleoside analogues are effective inhibitors of the reverse transcriptase of the human immunodeficiency virus (HIV), so they are used beneficially in the therapy of acquired immune deficiency syndrome (AIDS).[12-14] Even after the discovery of protease inhibitors, they have remained potent anti-HIV agents exerting a synergistic antiviral effect[15,16] in combination therapy. Previous studies demonstrated that these drugs have several toxic side effects, depending on the length of the application time. After a short treatment, proliferating tissues are affected (hemopoetic and gastrointestinal systems) resulting in anemia and gastrointestinal disorders.[17,18] Furthermore, AZT and dideoxynucleoside analogues, in addition to the inhibition of the HIV reverse transcriptase, inhibit the mitochondrial DNA polymerase gamma,[14,19-21] so these drugs inhibit mitochondrial DNA replication, causing depletions of mtDNA.[22] The mammalian mitochondrial genome encodes components of the respiratory chain; therefore, any damage of mtDNA will cause defective mitochondrial protein synthesis, which results in abnormal respiratory complexes and impaired oxidative energy production. In addition, AZT (or its metabolites) has an inhibitory effect on mitochondrial oxidative energy production that cannot be associated with depletion of mtDNA. The inhibition by AZT is localized to NADH:citochrome c oxidoreductase, while succinate:citochrome c oxidoreductase is not affected. It is also known that AZT and 2′,3′-dideoxycytidine (ddC) treatment increases the ROS level in several tissues, which can lead to oxidative damages of lipid membranes, proteins, and DNA in the affected cells.

We showed previously that AZT and ddC treatment in rats increased ROS production, lipid peroxidation, protein oxidation, and single-strand DNA breaks in rat heart leading to abnormal cardiac function.[23,24] It was also demonstrated that the increased PARP activity in rat heart muscle was accompanied by a significant decrease in NAD+ level, and this resulted in an abnormal energy status of the hearts.[23,24] These data demonstrate that the molecular mechanism of the cytotoxicity of ddC and AZT was indistinguishable from ROS-induced cell damage. Since PARP inhibitors can prevent or decrease ROS-induced cell damages, it is likely that PARP inhibitors can decrease the side effects of anitviral nucleoside analogues and dideoxynucleosides.

Previously, we showed that BGP-15, a novel PARP inhibitor, protects the heart from postischemic injury.[24] Therefore, we used BGP-15 in combination with AZT to investigate whether PARP inhibitors can protect the heart from AZT-induced cardiac damages. AZT treatment for 2 weeks significantly increased the RR, PR, and QT intervals, and caused a significant change in J point depressions in leads I and aVL that correspond to the main muscle mass of the left ventricle (Table 8.1). When the rats were treated with AZT in combination with BGP-15, the heart abnormalities were much lighter and no significant differences were seen in the electrocardiographic (ECG) data of the control group and the AZT plus BGP-15–treated group. These data show that BGP-15, a novel PARP inhibitor, protected the heart function from AZT-induced abnormalities.

We also reported earlier that AZT-induced oxidative damage in cardiomyocytes caused a decrease in the activity of the respiratory complexes and abnormalities in the mitochondrial membrane systems.[25] Using AZT in combination with BGP-15, we found BGP-15 to protect rat hearts from AZT-induced decreases in the activity of the respiratory complexes (Table 8.2). That is, BGP-15 prevented the AZT-induced inactivation of NADH:cytochrome c oxidoreductase and cytochrome oxidase, while AZT treatment alone or in combination with BGP-15 had no effect on

TABLE 8.1
Effect of BGP-15 on the AZT Treatment–Induced Changes of RR, PR, QT Intervals and J Point Depression in Rat Heart

Treatments	RR	PR	QT (ms)	J (mm)
Control group	174 ± 12	53 ± 2	70 ± 2	−0.1 ± 0.1
AZT treatment	284 ± 16*	82 ± 3*	112 ± 9*	−1.1 ± 0.1*
AZT treatment plus BGP-15	161 ± 24**	47 ± 6**	76 ± 5**	−0.2 ± 0.14**

Note: Rats (80 to 100 g) were treated daily with AZT (50 mg/kg) alone or in combination with BGP-15 (100 mg/kg) intraperitoneally for up to 14 days. Cardiac function was recorded by Schiller AT-6 ECG. RR, PR, QT intervals and J point values were determined by standard methods.[74] Data represent mean ± SEM of five animals.

* Values are different from the corresponding values of age-matched control rats at the significance level of $p < 0.01$.
** Values are different from the corresponding AZT-treated group values at the significance level of $p < 0.01$.

the activity of malate dehydrogenase and mitochondrial carnitine acetyltransferase (see Table 8.2). Data presented in Table 8.1 and 8.2 indicate that a novel PARP inhibitor, BGP-15,[25] can normalize the heart function and help minimize the defect of respiratory complexes without interfering with the reverse transcriptase inhibitory effect of this drug.

AZT-induced ROS formation can explain why AZT treatment causes inflammation.[26] It is well known that viral or retroviral infections alone cause inflammation, which is also associated with elevated ROS and peroxynitrite production.[27,28] That is, both the viral infection–induced inflammation and the side effect of antiviral

TABLE 8.2
Effect of BGP-15 on the AZT Treatment–Caused Inhibition of Respiratory Complexes in Rat Heart

Treatment	NADH:Cytochrome c Oxidoreductase	Cytochrome Oxidase	Malate Dehydrogenase	Carnitine Acetyltransferase
Control	3.5 ± 0.3	10.7 ± 0.8	1257 ± 138	3.6 ± 0.7
AZT treated	1.6 ± 0.4*	7.8 ± 0.7*	1202 ± 78	3.1 ± 0.4
AZT treated + BGP-15	3.1 ± 0.2	10.2 ± 0.8	1233 ± 101	3.5 ± 0.4

Note: The treated rats were sacrificed 3 days after the completion of AZT treatment to avoid the possible toxic effect of AZT metabolites. Cytochrome oxidase,[75] NADH:cytochrome c oxidoreductase,[75] and carnitine acetyltransferase[76] were measured as described earlier. Data represent mean ± SEM of five animals. Values are unit/g wet tissue.

* Values are different from the corresponding control and AZT treated + BGP-15 values at the significance level of $p < 0.05$.

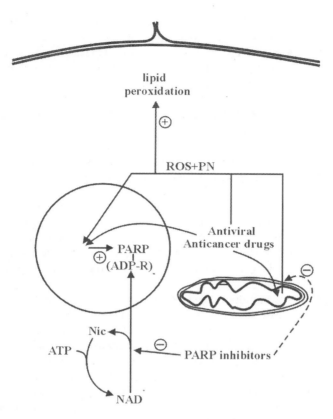

FIGURE 8.1 Antiviral and anticancer drug–induced ROS-mediated signaling. PN, peroxynitrite; Nic, nicotinamide; ADP-R, ADP-ribose; ⊕, induction or activation, −, inhibition.

drugs can lead to PARP activation and oxidative cell damage (Figure 8.1). Therefore, it is expected that PARP inhibitors may protect different tissues from the inflammatory effect of viral infection and from the side effects of antiviral drugs. There are data suggesting that antioxidants can attenuate the side effects of AZT in patients with HIV,[29,30] which also indicate the importance of oxidative cell damage in the side effects of antiviral drugs. It is likely that nontoxic PARP inhibitors, such as BGP-15, can decrease the inflammation caused by viral infection and antiviral drug–induced oxidative cell damages, as well. Therefore, PARP inhibitors may increase the safety of antiviral therapy.

8.3 ROS-MEDIATED CYTOTOXICITY OF ANTITUMOR DRUGS

Involvement of free radicals in the mediation of side effects is well documented in the case of cisplatin and adriamycin and may be implicated in the toxicity of other cytostatics as well. High-dose cisplatin therapy produces untoward side effects of nephrotoxicity, bone marrow toxicity, gastrointestinal toxicity, ototoxicity, and peripheral neuropathy.[31,32] The mechanism of nephrotoxicity is still not understood completely, but generation of free oxygen radicals by the proximal tubular cells has been proposed as a major pathogenic mechanism.[33] Cisplatin treatment resulted in depletion of glutathion (GSH) and protein thiols[34] in the kidney. Studies performed on renal cortical slices revealed that depletion of mitochonrial GSH was an early event after cisplatin treatment that was followed by increased lipid peroxidation (measured by the amount of thiobarbituric acid reactive substance) and decreased mitochondrial protein concentration.[35] The imbalance of the antioxidant system was partly caused by decreased activity of the antioxidant enzymes, as decreased levels of superoxide dismutase (SOD), catalase, and GSH peroxidase activity (GSH-Px) were detected *in vivo* in cisplatin-treated kidney.[36] The dysfunction of mitochondria caused by oxidative damage may maintain a prolonged increase of oxygen free radical production. The role of oxidative damage in the toxicity of cisplatin was further supported by the observations that buthionine sulfoximine, which causes GSH depletion, potentiated the nephrotoxicity of cisplatin,[33] whereas treatment with antioxidants (GSH and ebselen) attenuated cisplatin toxicity. Data suggest that the antioxidant system is involved in the pathomechanism of cisplatin toxicity in other organs, as well.[37]

The anthracyclin antibiotics, like doxorubicin, have been in clinical use for more than 30 years. The extensive research of the mechanism of action of these drugs revealed several biological activities, but the exact mechanisms of the antitumor and toxic effect are still debated. The widely supported mechanism of anthracyclin activity include intercalation into the DNA, DNA alkylation and cross-linking, interference with DNA unwinding, inhibition of topoisomerase II, and free radical formation.[38] The quinone structure allows anthracyclines to accept electrons in reactions mediated by oxidoreductive enzymes including xanthine oxidase, cytochrome P-450 reductase, and NADH dehydrogenase. The formed semiquinone is a free radical by itself or after interacting with molecular oxygen,[39] and can damage DNA. Doxorubicin is capable of generating free radicals through formation of complexes with iron, as well.[40] The well-documented protective effect of the iron-chelating agent ICRF-187 against doxorubicin-induced cardiotoxicity[41] clearly implicates this mechanism in the toxicity of this drug.

ROS plays a significant role in the cytotoxicity of bleomycins since bleomycins form a complex with Fe^{2+} and molecular oxygen, and produce ROS to cleave DNA.[42,43] In addition, metallo-bleomycins can be activated by NADPH–cytochrome P-450 reductase,[44] which results in activated ROS formation in the cells. Bleomycin-induced lung fibrosis is caused by elevated ROS level, and can be moderated by antioxidants.[45,46] Furthermore, it is also known that mitomycin C and 5-fluorouracil increase ROS levels in different tissues,[47,48] which can contribute to their side effects. These data show that several anticancer drugs activate ROS formation both in normal tissues and tumors, and elevated ROS level can cause serious damage in well-oxygenated normal tissues. Since elevated ROS level can cause single-strand DNA breaks, anticancer drug treatment can activate PARP and, therefore, induce NAD^+ catabolism, defective energy metabolism, and finally cell death. The PARP activation and PARP-related cell damage can be an important pathway in the development of the cytotoxic side effect of anticancer drugs in vital tissues, raising the possibility of introducing nontoxic PARP inhibitors as protective agents against the toxic side effects of anticancer drugs.

8.4 PROTECTIVE EFFECT OF BGP-15, A NOVEL PARP INHIBITOR, AGAINST ANTITUMOR DRUG–INDUCED CYTOTOXICITY

It is well documented that cisplatin increases ROS formation, single-strand DNA breaks, and activates PARP,[49] so PARP activation and PARP-dependent cell damage can be important factors in the cytotoxicity of cisplatin. Since PARP inhibitors significantly decrease ROS-induced cell damage in normal tissue,[50-53] it is to be expected that PARP inhibitors may decrease the cisplatin-induced damage of normal tissues. Previously, we found that BGP-15 effectively protected Langendorff perfused heart from ischemia–reperfusion-induced injury and that BGP-15 had low toxicity (LD_{50} = 1203.4 mg/kg i.p. in rats). Therefore, we investigated whether BGP-15 can decrease the mortality caused by cisplatin treatment in mice. Figure 8.2 shows that cisplatin alone caused 67% mortality while BGP-15 reduced the mortality rate to 40% (Figure 8.2). That is, the PARP inhibitor partially protected the mice from cisplatin toxicity. In addition, we found that combined treatment with BGP-15 caused a faster regain of cisplatin-induced body weight loss in mice (data not shown).

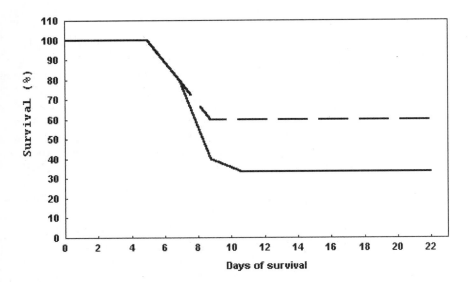

FIGURE 8.2 Effect of BGP-15 on the survival of cisplatin-treated mice. Groups of 15 NMRI mice were treated for 5 consecutive days with cisplatin (3.8 mg/kg i.p.) alone (– – –) or in combination with BGP-15 (200 mg/kg p.o.) (- - - -). Body weight changes and survival times were recorded. Chi square = 1.268; df = 1; p = 0.06.

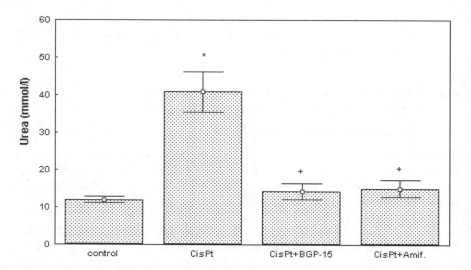

FIGURE 8.3 Effect of BGP-15 and amifostine on cisplatin-induced nephrotoxicity. NMRI mice were treated with cisplatin (3.8 mg/kg i.p.) for 5 consecutive days. In addition to cisplatin, BGP-15 (200 mg/kg p.o.) and amifostine (200 mg/kg i.p.) were administered daily to groups of animals until the termination of the experiment. On the seventh day, serum urea levels were evaluated.[77] The data were analyzed by multiway analysis of variance (ANOVA) and post hoc least significant dose (LSD) test. Data represent mean value ± SEM, ANOVA; $F(8.55) = 12.25$; $p < 0.0001$. Significant difference * = compared to the control, + = compared with the cisplatin-treated group. LSD; $p < 0.05$.

The nephrotoxicity of cisplatin can be a limiting factor in clinical usage. The serum urea level, a good parameter of kidney function, was used to monitor cisplatin-induced nephrotoxicity. Cisplatin treatment caused a significant increase in serum urea levels (Figure 8.3) indicating defective kidney function. When cisplatin was administrated in combination with either BGP-15 or amifostine (a known chemoprotective drug), urea levels remained close to control levels, showing that both substances effectively protected kidneys from cisplatin-induced damage (see Figure 8.2). These data show that BGP-15, a novel PARP inhibitor, effectively protects normal cells from cisplatin-induced damage. The following presents data that BGP-15 does not inhibit the antitumor activity of cisplatin.

BGP-15 treatment alone did not inhibit the growth of Du-145 or A549 cell lines up to the 1-mg/ml concentration (data not shown). Cisplatin treatment in the concentration range of up to 15 mg/l effectively inhibited the growth of both tumor cell lines (Figure 8.4). Combined treatment with BGP-15 did not significantly change the growth inhibitory curve of cisplatin on Du-145 (Figure 8.4), and a similar result was obtained on A549 cells (data not shown). When S-180 mouse sarcoma were transplanted subcutaneously into BDF1 mice, cisplatin treatment caused a significant delay in tumor growth (Figure 8.5). Combined treatment with BGP-15 resulted in a somewhat more persistent tumor suppression than cisplatin treatment alone (Figure 8.5), showing that BGP-15 did not diminish the antitumor activity of cisplatin. These data are in accord with the view that PARP binds to DNA breaks and is involved

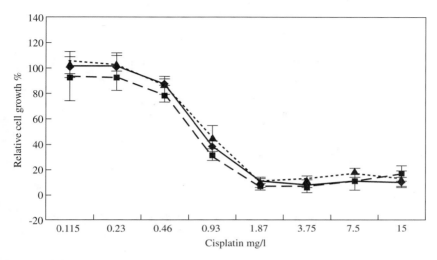

FIGURE 8.4 Effect of BGP-15 on the *in vitro* cytotoxicity of cisplatin. Effect of BGP-15 on the *in vitro* cytotoxicity of cisplatin was evaluated in a 4-day assay. Du-145 and A549 human tumor cell lines were maintained in RPMI 1640 medium supplemented with 10% fatal calf serum in humidified air containing 5% CO_2; 10^4 cells were plated into the wells of a 96-well plate in the same medium. On the following day, cells were exposed to a series of concentrations of cisplatin and BGP-15 (10; 30 µg/ml) either separately or in combination. Cell growth was evaluated after 3 days by MTT assays.[78] Dose–response curves (mean ± SEM) were evaluated. ■ = cisplatin, ▲ = cisplatin + BGP-15 (10 mg/l), ♦ = cisplatin + BGP-15 (30 mg/l). Data represent mean value ± SEM.

in DNA repair; therefore, it is assumed that PARP inhibitors increase the antitumor activity of DNA-damaging drugs. Potentiation of the antitumor effect of monofunctional alkylating agents (MNNG, MNU, temozolomide) and bleomycin was observed.[54,55] In some cases, PARP inhibitors had synergistic or no effect on the antitumor efficiency of cisplatin.[56-58] In a recent comprehensive study,[59] it was found that PARP inhibitors *in vitro* did not change the cytotoxicity of cisplatin in cultured tumor cell lines. However, there are data showing that PARP inhibitors can reverse acquired cisplatin resistance.[60] These effects correlate well with the repair mechanisms involved in the removal of DNA damage caused by different antitumor agents. DNA modifications achieved by monofunctional alkylating agents are repaired by PARP-dependent base excision,[61] whereas DNA lesions caused by cisplatin are removed by PARP-independent nucleotide excision repair.[62] The increased sensitivity of PARP knockout mice to alkylating agents and ionizing radiation also confirms this view.[63] Our data showing that BGP-15 did not decrease the antitumor efficacy of cisplatin *in vitro* in cell culture and *in vivo* in transplantable tumor in mice are in accord with the previous observations.[55-57] However, data showing that BGP-15 cotreatment results in a higher survival rate of cisplatin-treated mice and BGP-15 significantly protects kidneys from cisplatin nephrotoxicity indicate that PARP inhibitors, like BGP-15, can protect normal tissues from the toxic side effects of cisplatin without interfering with its antitumor effect.

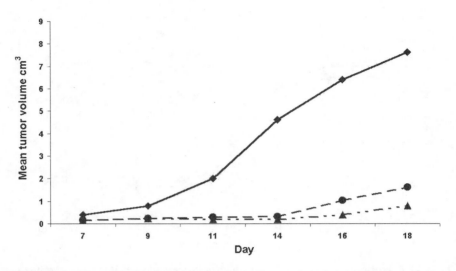

FIGURE 8.5 Effect of BGP-15 on the antitumor efficacy of cisplatin on S-180 mouse sarcoma. BDF1 mice were implanted subcutaneously with 25 mg tumor homogenate of S-180 mouse sarcoma. On the following day, groups of seven animals were treated as follows: control (untreated), cisplatin (10 mg/kg i.p.), cisplatin (10 mg/kg i.p.) + BGP-15 (200 mg/kg p.o.). BGP-15 was given 30 min before the application of cisplatin. Tumor size was calculated from the perpendicular diameters. Treatment: ● = cisplatin (10 mg/kg i.p.), ▲ = cisplatin (10 mg/kg i.p.) plus BGP-15 (200 mg/kg p.o.), ♦ = untreated.

It is well documented that cisplatin kills tumor cells by apoptosis,[63-65] which is not directly related to PARP activation. The cisplatin-induced apoptosis of tumor cells can be mediated by p53-activated programmed cell death or, in p53-defective cells, by the cAbl–p73 (mismatch-repair-dependent) apoptosis pathway. At an early stage of apoptotic cell death, apoptotic proteases (e.g., caspase 3) cleave and inactivate PARP, which can contribute to the effective chromatin degradation.[66-68] The inactivation of PARP can be advantageous for apoptosis because during apoptosis, numerous single-strand DNA breaks are formed with which PARP interacts and helps to recruit base excision repair (BER) proteins to repair damaged DNA.[67,69] With the same logic, the inhibition of PARP could also facilitate apoptosis in tumor cells. In solid tumors, the hypoxic environment can decrease cisplatin-induced ROS formation; therefore, cisplatin toxicity is mediated via a ROS-independent apoptotic pathway. On the other hand, normal tissues are well oxygenated (like kidneys, heart, and nervous system); therefore, cisplatin can induce severe oxidative damage (Figure 8.1), which can explain the severe nephrotoxicity and neurotoxicity of cisplatin. The cisplatin-induced oxidative damage, including the formation of single-strand DNA breaks, can activate PARP, which catabolizes NAD^+ and compromises energy metabolism (see Figure 8.1). This can lead to predominantly necrotic cell death in the kidney. Therefore, antitumor drug–induced cell damage to normal tissues is very similar to the ROS- and peroxynitrite-induced damage that is mediated via PARP activation.[1,2] Consequently, PARP inhibitors can effectively protect normal tissues under various pathological conditions, as was described for neuronal damages in stroke,[70] myocardial infarction,[51] cytokine or oxidant damage to pancreatic islet cells and hepatocytes,[2,71] and the oxidant damage to pulmonary epithelium, macrophages, and smooth muscle cells.[51,72] The significance of PARP[70,73] and the effectiveness of PARP inhibitors in protecting cells against oxidative damage are well documented.[50-53] Therefore, it is understandable that BGP-15, a new PARP inhibitor, can protect kidneys from the toxicity of cisplatin and can decrease the mortality of mice (see Figure 8.2).

8.5 CONCLUSION

This chapter has drawn attention to the fact that toxic side effects of certain antitumor and antiviral drugs are often mediated by ROS or single-strand DNA breaks, suggesting that PARP, which is activated by these mechanisms, may play a significant role in the toxicity of these drugs. The microenvironment of solid tumors is significantly different from that of normal tissues, because most of the cells in solid tumors are under hypoxic conditions, whereas normal tissues are well oxygenated. Therefore, antitumor and antiviral drugs can induce severe oxidative damage in normal tissues (like kidneys, heart, and nervous system), which can explain the severe nephrotoxicity, cardiotoxicity, and neurotoxicity of antitumor and antiviral drugs. These drugs induce oxidative damage, cause single-strand DNA break formation, and activate PARP, which in turn catabolizes NAD^+ and compromises energy metabolism. This can lead predominantly to necrotic cell death, while the antitumor and antiviral actions of these drugs progress through different mechanisms (reverse transcriptase inhibition by antiviral drugs, and induction of apoptosis or other specific action by antitumor drugs). Therefore, nontoxic PARP inhibitors, like BGP-15, provide a new specific means to protect normal tissues from the toxic side effects of antitumor and antiviral drugs without interfering with their therapeutic effect. It is to be expected that PARP inhibitors can improve the effectiveness and the safety of antitumor and antiviral drugs in a combination therapy.

REFERENCES

1. Lindahl, T., Satoh, M. S., Poirier, G. G., and Klungland, A., Post-translational modification of poly(ADP-ribose) polymerase induced by DNA strand breaks, *Trends Biochem. Sci.,* 20, 405, 1995.
2. Radons, J., Heller, B., Burkle, A., Hartmann, B., Rodriguez, M. L,. Kröncke, K. D., Burkart, V., and Kolb, H., Nitric oxide toxicity in islet cells involves poly(ADP-ribose) polymerase activation and concomitant NAD^+ depletion, *Biochem. Biophys. Res. Commun.,* 199, 1270, 1994.
3. Janero, D. R., Hreniuk, D., Sharif, H. M., and Prout, K. C., Hydroperoxide-induced oxidative stress alters pyridine nucleotide metabolism in neonatal heart muscle cells, *Am. J. Physiol.,* 264, C1401, 1993.
4. Turrens, J. F., Beconi, M., Barilla, J., Chavez, U. B., and McCord, J. M., Mitochondrial generation of oxygen radicals during reoxygenation of ischemic tissues, *Free Radical Res. Commun.,* 12–13, 681, 1991.

5. Ambrosio, G., Zweier, J. L., Duilio, C., Kuppusamy, P., Santoro, G., Elia, P. P., Tritto, I., Cirillo, P., Condorelli, M., and Chiariello., M., Evidence that mitochondrial respiration is a source of potentially toxic oxygen free radicals in intact rabbit hearts subjected to ischemia and reflow, *J. Biol. Chem.,* 268, 18532, 1993.
6. Vanden-Hoek, T. L., Becker, L. B., Shao, Z., Li, C., and Schumacker, P. T., Reactive oxygen species released from mitochondria during brief hypoxia induce preconditioning in cardiomyocytes, *J. Biol. Chem.,* 273, 18092, 1998.
7. Takeyama, N., Matsuo, N., and Tanaka, T., Oxidative damage to mitochondria is mediated by the $Ca^{(2+)}$-dependent inner-membrane permeability transition, *Biochem. J.,* 294, 719, 1993.
8. Halestrap, A. P., Kerr, P. M., Javadov, S., and Woodfield, K. Y., Elucidating the molecular mechanism of the permeability transition pore and its role in reperfusion injury of the heart, *Biochim. Biophys. Acta,* 1366, 79, 1998.
9. Lemasters, J. J., Nieminen, A. L., Qian, T., Trost, L. C., Elmore, S. P., Nishimura, Y., Crowe, R. A., Cascio, W. E., Bradham, C. A., Brenner, D. A., and Herman, B., The mitochondrial permeability transition in cell death: a common mechanism in necrosis, apoptosis and autophagy, *Biochim. Biophys. Acta,* 1366, 177, 1998.
10. Cuzzocrea, S., Zingarelli, B., and Caputi, A. P., Peroxynitrate-mediated DNA strand breakage activates poly(ADP-ribose) synthetase and causes cellular energy depletion in a nonseptic shock model induced by zymosan in the rat, *Shock,* 9, 336, 1998.
11. Szabó, C., Virág, L., Cuzzocrea, S., Scott, G. S., Hake, P., O'Connor, M. P., Zingarelli, B., Salzman, A. L., and Kun, E., Protection against peroxynitrite-induced fibroblast injury and arthritis development by inhibition of poly(ADP-ribose) synthase, *Proc. Natl. Acad. Sci. U.S.A.,* 95, 3867, 1998.
12. Copeland, W. C., Cheng, M. S., and Wang, T. S. F., Human DNA polymerase are able to incorporate anti-HIV deoxynucleotides into DNA, *J. Biol. Chem.,* 267, 21459, 1992.
13. Meng, T. C., Fischl, M. A., Boota, A. M., Spector, S. A., Bennett, D., Bassiakos, Y., Lai, S. H., Wright, B., and Richman, D. D., Combination therapy with zidovudine and dideoxycytidine in patients with advanced human immundeficiency virus infection: a phase I/II study, *Ann. Intern. Med.,* 116, 13, 1992.
14. Lipsky, J. J., Zalcitabine and didanosine, *Lancet,* 341, 30, 1993.
15. Von der Helm, K., Retroviral proteases: structure, function and inhibition from a non-anticipated viral enzyme to the target of a most promising HIV therapy, *Biol. Chem.,* 377, 765, 1996.
16. Deminie, C. A., Bechtold, C. M., Stock, D., Alam, M., Djang, F., Balch, A. H., Chou, T. C., Prichard, M., Colonno, R. J., and Lin, P. F., Evaluation of reverse transcriptase and protease inhibitors in two-drug combinations against human immundeficiency virus replication, *Antimicrob. Agents Chemother.,* 40, 1346, 1996.
17. Indorf, A. M. and Pergam, P. S., Esophageal ulceration related to zalcitabine(ddC), *Ann. Intern. Med.,* 117, 133, 1992.
18. Lerza, R., Castello, G., Mela, G. S., Arboscello, E., Cerruti, A., Bogliolo, G., Menconi, M., Ballarino, P., and Pannacciulli, I., In vitro synergistic inhibition of human bone marrow hemopoetic progenitor growth by a 3'-azido-3'deoxy-thymidine, 2',3'-dideoxycytidine combination, *Exp. Hematol.,* 25, 252, 1997.
19. Izuta, S., Saneyoshi, M., Sakurai, T., Suzuki, M., Kojima, K., and Yoshida, S., The 5'-triphosphates of azido-3'-dezoxythymidine and 2',3'-dideoxynucleosides inhibit DNA polymerase gamma by different mechanisms, *Biochem. Biophys. Res. Commun.,* 179, 776, 1991.

20. Chen, C. H., Vazquez Padua, M., and Cheng, Y. C., Effect of anti-human immundeficiency virus nucleoside analogs on mitochondrial DNA and its implication for delayed toxicity, *Mol. Pharmacol.,* 39, 625, 1991.
21. Herskowitz, A., Willoughby, S. B., Baughman, K. L., Schulman, S. P., and Bartlett, J. D., Cardiomyopathy associated with antiretroviral therapy in patients with HIV infection, *Ann. Intern. Med.,* 116, 311, 1992.
22. Lewis, L. D., Hamzeh, F. M., and Lietman, P. S., Ultrastructural changes associated with reduced mitochondrial DNA and impaired mitochondrial function in the presence of 2′,3′-dideoxycytidine, *Antimicrob. Agents Chemother.,* 36, 2061, 1992.
23. Skuta, G., Fischer, M. G., Janaky, T., Kele, Z., Szabó, P., Tozser, J., and Sumegi, B., Molecular mechanism of the short term cardiotoxicity caused by 2′,3′dideoxycytidine (ddC): modulation of reactive oxygen species levels and ADP ribosylation reactions, *Biochem. Pharmacol.,* 59, 937, 2000.
24. Szabados, E., Fischer, M. G., Toth, K., Csete, B., Nemeti, B., Trombitas, K., Habon, T., Endrei, D., and Sumegi, B., Role of reacrive oxygen species and poly-ADP-ribose polymerase in the development of AZT-induced cardiomyopathy in rat, *Free Radical Biol. Med.,* 26, 309, 1999.
25. Szabados, E., Literati-Nagy, P., Farkas, B., and Sumegi, B., BGP-15, a nicotinic amidoxime derivate protecting heart from ischemia-reperfusion injury through modulation of poly(ADP-ribose) polymerase activity, *Biochem. Pharmacol.,* 58, 1915, 1999.
26. Raymond, P., Blais, C., Jr., Decarie, A., Morais, R., and Adam, A., Zidovudine potentiates local and systemic inflammatory responses in the rat, *J. Acquired Immune Deficiency Syndr. Hum. Retrovirol.,* 14, 399, 1997.
27. Ensoli, F., Fiorelli, V., DeCristofaro, M., Santini Muratori, D., Novi, A., Vannelli, B., Thiele, C. J., Luzi, G., and Aiuti, F., Inflammatory cytokines and HIV-1-associated neurodegeneration: oncostatin-M produced by mononuclear cells from HIV-1-infected individuals induces apoptosis of primary neurons, *J. Immunol.,* 162, 6268, 1999.
28. Elbim, C., Pillet, S., Prevost, M. H., Preira, A., Girard, P. M., Rogine, N., Matusani, H., Hakim, J., Israel, N., and Gougerot Pocidalo, M. A., Redox and activation status of monocytes from human immunodeficiency virus–infected patients: relationship with viral load, *J. Virol.,* 73, 4561, 1999.
29. Wang, Y. and Watson, R. R., Is vitamin E supplementation a useful agent in AIDS therapy? *Prog. Food Nutr. Sci.,* 17, 351, 1993.
30. de la Asuncion, J. G., del Olmo, M. L., Sastre, J., Millan, A., Pellin, A., Pallardo, F. V., and Vina, J., AZT treatment induces molecular and ultrastructural oxidative damage to muscle mitochondria. Prevention by antioxidant vitamins, *J. Clin. Invest.,* 102, 4, 1998.
31. Cozzaglio, L., Doci, R., Colla, G., Zumino, F., Casciarri, G., and Genarri, L., A feasibility study of high-dose cisplatin and 5-fluorouracil with glutathione protection in the treatment of advanced colorectal cancer, *Tumori,* 76, 590, 1990.
32. Gandara, D. R., Perez, E. A., Weibe, V., and DeGregorio, M. W., Cisplatin chemoprotection and rescue: pharmacological modulation of toxicity, *Semin. Oncol.,* 18, 49, 1991.
33. Ishikawa, M., Takayanagi, Y., and Sasaki, K. I., Enhancement of cisplatin toxicity by buthionine sulfoximine, a glutathion-depleting agent in mice, *Res. Commun. Chem. Pathol. Pharmacol.,* 67, 131, 1990.
34. Levi, J., Jacobs, C., Kalman, S. B., Mctigue, M., and Weiner, M. W., Mechanism of cisplatinum nephrotoxicity: I. Effect of sulfhydryl groups in the rat kidneys, *J. Pharmacol. Exp. Ther.,* 213, 545, 1980.

35. Zhang, J. and Lindup, E., Role of mitochondria in cisplatin-induced oxidative damage exhibited by rat renal cortical slices, *Biochem. Pharmacol.*, 45, 2215, 1990.
36. Husain, K., Morris, C., Whithworth, C., Trammel, G. L., Rybak, L. P., and Somani, S. M., Protection by ebselen against cisplatin-induced nephrotoxicity: antioxidant system, *Mol. Cell. Biochem.*, 178, 127, 1998.
37. Rybak, L. P., Husain, K., Whitworth, C., and Somani, S. M., Dose-dependent protection by lipoic acid against cisplatin-induced ototoxicity in rats: antioxidant defense system, *Toxicol. Sci.*, 47, 195, 1999.
38. Gewirtz, D. A., A critical evaluation of the mechanism of action proposed for the antitumor effects of the anthracyclin antibiotics adriamycin and daunorubicin, *Biochem. Pharmacol.*, 57, 727, 1999.
39. Bachur, N. R., Gordon, S. L., and Gee, M. V., Anthracyclin antibiotic augmentation of microsomal electron transport and free radical formation, *Mol. Pharmacol.*, 13, 901, 1977.
40. Eliot, H., Gianni, L., and Myers, C., Oxidative destruction of DNA by the adriamycin–iron complex, *Biochemistry*, 23, 928, 1994.
41. Wexler, L. H., Andrich, M. P., Venzon, D., Berg, S. L., Weaver-McClure, L., Chen, C. C., Dilsizian, V., Avila, N., Jarosinski, P., Balis, F. M., Poplack, D. G., and Horowitz, M. E., Randomized trials of the cardioprotective agent ICR-187 in pediatric sarcoma patients treated with doxorubicin, *J. Clin. Oncol.*, 14, 362, 1996.
42. Parizada, B., Werber, M. M., and Nimrod, A., Protective effects of human recombinant MnSOD in adjuvant arthritis and bleomycin-induced lung fibrosis, *Free Radical Res. Commun.*, 15, 297, 1991.
43. Lopez-Larraza, D., De-Luca, J. C., and Bianchi, N. O., The kinetics of DNA damage by bleomycin in mammalian cells, *Mutat. Res.*, 232, 57, 1990.
44. Nakamura, M., Lactoferrin-mediated formation of oxygen radicals by NADPH-cytochrome P-450 reductase system, *J. Biochem. Tokyo*, 107, 395, 1990.
45. Kilinc, C., Ozcan, O., Karaoz, E. Sunguroglu, K., Kutluay, T., and Karaca, L., Vitamin E reduces bleomycin-induced lung fibrosis in mice: biochemical and morphological studies, *J. Basic Clin. Physiol. Pharmacol.*, 4, 249, 1993.
46. Nici, L., Santos-Moore, A., Kuhn, C., and Calabresi, P., Modulation of bleomycin-induced pulmonary toxicity in the hamster by the antioxidant amifostine, *Cancer*, 83, 2008, 1998.
47. Sinha, B. K. and Mimnaugh, E. G., Free radicals and anticancer drug resistance: oxygen free radicals in the mechanisms of drug cytotoxicity and resistance by certain tumors (adriamycin and mitomycin C), *Free Radical Biol. Med.*, 8, 567, 1990.
48. Millart, H., Kantelip, J. P., Platonoff, N., Descous, I., Trenque, T., Lamiable, D., and Choisy, H., Increased iron content in rat myocardium after 5-fluorouracil chronic administration, *Anticancer Res.*, 13, 779, 1993.
49. Burkle, A., Chen, G., Kupper, J. H., Grube, K., and Zeller, W. J., Increased poly(ADP-ribosyl)ation in intact cells by cisplatin treatment, *Carcinogenesis*, 14, 559, 1993.
50. Thiemermann, C., Bowes, J., Myint, F. P., and Vane, J. R., Inhibition of the activity of poly(ADP-ribose) synthetase reduces ischemia–reperfusion injury in the heart and skeletal muscle, *Proc. Natl. Acad. Sci. U.S.A.*, 94, 679, 1997.
51. Szabó, C., Zingarelli, B., O'Connor, M., and Salzman, A. L., DNA strand breakage, activation of poly(ADP-ribose) synthetase, and cellular energy depletion are involved in the cytotoxicity of macrophages and smooth muscle cells exposed to peroxynitrite, *Proc. Natl. Acad. Sci. U.S.A.*, 93, 1753, 1996.
52. Szabó, C. and Dawson, V. L., Role of poly(ADP-ribose) synthetase in inflammation and ischaemia–reperfusion, *Trends Pharmacol. Sci.*, 19, 287, 1998.

53. Zingarelli, B., Cuzzocrea, S., Zsengeller, Z., Salzman, A. L., and Szabó, C., Protection against myocardial ischemia and reperfusion injury by 3-aminobenzamide, an inhibitor of poly(ADP-ribose) synthetase, *Cardiovasc. Res.,* 36, 205, 1997.
54. Althaus, E. M. and Richter, C., ADP-ribosylation of proteins. Enzymology and biological significance, *Mol. Biol. Biochem. Biophys.,* 37, 1, 1987.
55. Bowman, K. J., White, A., Golding, B. T., Griffin, R. J., and Curtin, N. J., Potentiation of anti-cancer agent cytotoxicity by the potent poly(ADP-ribose)polymerase inhibitors NU1025 and NU1064, *Br. J. Cancer,* 78, 1269,1998.
56. Boike, G. M., Petru, E., Sevin, B. U. et al., Chemical enhancement of cisplatin cytotoxicity in a human ovarian and cervical cancer cell line, *Gynecol. Oncol.,* 38, 315, 1990.
57. Chen, G. and Pan, Q. C., Potentiation of the antitumor activity of cisplatin in mice by 3-aminobenzamide, *Cancer Chemother. Pharmacol.,* 22, 303, 1988.
58. Boorstein, R. J. and Pardee, A. B., Factors modifying 3-aminobenzamide cytotoxicity in normal and repair-deficient human fibroblast, *J. Cell. Physiol.,* 120, 335, 1984.
59. Bernges, F. and Zeller, W. J., Combination effects of poly(ADP-ribose)polymerase inhibitors and DNA-damaging agents in ovarian tumor cell lines — with special reference to cisplatin, *J. Cancer Res. Clin. Oncol.,* 122, 665, 1996.
60. Chen, G. and Zeller, W. J., Reversal of acquired cisplatin resistance by nicotinamide *in vitro* and *in vivo, Cancer Chemother. Pharmacol.,* 33, 157, 1993.
61. Dantzer, S. V., Niedergang, C., Trucco, C., Flatter, E., de la Rubia, G., Oliver, F. J., Rolli, V., and Menissier-de Murcia, J., Role of poly(ADP-ribose)polymerase in base excision repair, *Biochimie,* 81, 69, 1991.
62. Wang, Z. and Wu, X., Nucleotide excision repair of DNA in cell-free extract of the yeast *Saccaromyces cerevisiae, Proc. Natl. Acad. Sci. U.S.A.,* 90, 4907, 1993.
63. Henkels, K. M. and Turchi, J. J., Cisplatin-induced apoptosis proceeds by caspase-3-dependent and -independent pathways in cisplatin-resistant and -sensitive human ovarian cancer cell lines, *Cancer Res.,* 59, 3077, 1999.
64. Zhan, Y., Van de Water, B., Wang, Y., and Stevens, J. L., The roles of caspase-3 and bcl-2 in chemically-induced apoptosis but necrosis of renal epithelial cells, *Oncogene,* 18, 6505, 1999.
65. Gong, J. G., Costanzo, A., Yang, H. Q., Melino, G., Kaelin, W. G., Jr., Levrero, M., and Wang, J. Y., The tyrosine kinase c-Abl regulates p73 in apoptotic response to cisplatin-induced DNA damage, *Nature,* 399, 806, 1999.
66. Basu, A. and Akkaraju, G. R., Regulation of caspase activation and cis-diamminedichloroplatinum(II)-induced cell death by protein kinase C, *Biochemistry,* 38, 4245, 1999.
67. Oliver, F. J., Menissier-de Murcia, J., and de Murcia, G., Poly(ADP-ribose) polymerase in the cellular response to DNA damage, apoptosis, and disease, *Am. J. Hum. Genet.,* 64, 1282, 1999.
68. Casciola-Rosen, L., Nicholson, D. W., Chong, T., Rowan, K. R., Thornberry, N. A., Miller, D. K., and Rosen, A., Apopain/CPP32 cleaves proteins that are essential for cellular repair: a fundamental principle of apoptotic death, *J. Exp. Med.,* 183, 1957, 1996.
69. Le Cam, F., Fack, F., Menissier-de Murcia, J., Cognet, J. A., Barbin, A., Sarantoglu, V., and Revet, B., Conformational analysis of a 139 base-pair DNA fragment containing a single-stranded break and its interaction with human poly(ADP-ribose) polymerase, *J. Mol. Biol.,* 235, 1062, 1994.
70. Eliasson, M. J. L., Sampei, K., Mandir, A. S., Hurn, P. D., Traystman, R. J., Bao, J., Pieper, A., Wang, Z. Q., Dawson, T. M., Snyder, S. H., and Dawson, V. L., Poly(ADP-ribose) polymerase gene disruption renders mice resistant to celebral ischaemia, *Nat. Med.,* 3, 1089, 1994.

71. Mizumoto, K., Glascott, P. A., Jr., and Farber, J. L., Roles of oxidative stress and poly(ADP-ribosyl)ation in the killing of cultured hepatocytes by methyl methanesulfonate, *Biochem. Pharmacol.,* 46, 1811, 1993.
72. Said, S. I., Berisha, H. I., and Pakbaz, H., Excitotoxicity in the lung: *N*-Methyl-D-asparate-induced, nitric oxide-dependent, pulmonary edema is attenuated by vasoactive intestinal peptide and by inhibitors of poly(ADP-ribose) polymerase, *Proc. Natl. Acad. Sci. U.S.A.,* 93, 4688, 1996.
73. Wang, Z. Q., Auer, B., Stingl, L., Berghammer, H., Haidacher, D., Schweiger, M., and Wagner, E. F., Mice lacking ADPRT and poly-ADP-ribosylation develop normally but are susceptible to skin disease, *Genes Dev.,* 9, 509, 1995.
74. MacFarlane, P. W. and Lawrie, T. D. V., *Comprehensive Electrocardiology,* Pergamon Press, Oxford, 1989.
75. Sumegi, B., Melegh, B., Adamovich, K., and Trombitas, K., Cytochrome oxidase deficiency affecting the structure of the myofiber and the shape of cristal membrane, *Clin. Chim. Acta,* 192, 9, 1990.
76. Sumegi, B., Gilbert, H. F., and Srere, P. A., Interaction between citrate synthase and thiolase, *J. Biol. Chem.,* 260, 188, 1985.
77. Abul-Ezz, S. R., Walker, P. D., and Shah, S. V., Role of glutathione in an animal model of myoglobinuric acute renal failure, *Proc. Natl. Acad. Sci. U.S.A.,* 88, 9833, 1991.
78. Scudiero, D. A., Shoemaker, R. H., Paull, K. D., Monks, A., Tierney, S., Nofzinger, T. H., Currens, M. J., Seniff, D., and Boyd, M. R., Evaluation of a soluble tetrezolium/formazan assay for cell growth and drug sensitivity in culture using human and other tumor cell lines, *Cancer Res.,* 48, 4827, 1988.

9 Involvement of Poly(ADP-Ribose) Polymerase in the Cellular Response to DNA Damage

Claude Niedergang, F. Javier Oliver, Josiane Ménissier-de Murcia, and Gilbert de Murcia

CONTENTS

9.1 Introduction 184
9.2 The Modular Structure of Poly(ADP-Ribose) Polymerase 184
9.3 Poly(ADP-Ribose) Polymerase is a Survival Factor 188
    9.3.1 PARP-Depleted Cell Lines and *trans*-Dominant Negative Mutants 188
    9.3.2 PARP Knockout Mice are Sensitive to Ionizing Radiation and Alkylating Agents 189
    9.3.3 PARP Knockout Cells are Severely Affected in the Base Excision Repair Pathway 190
    9.3.4 Some Partners of PARP are Involved in DNA Repair and Replication 193
9.4 Poly(ADP-Ribose) Polymerase and Cell Death 195
    9.4.1 Cleavage of PARP during Apoptosis 195
    9.4.2 Apoptosis in PARP-Deficient Cells 197
9.5 Future Prospects 197
Abbreviations 199
Acknowledgments 199
References 200

9.1 INTRODUCTION

To ensure accurate transmission of genetic information in dividing cells, specific biochemical pathways maintain integrity. Fundamental to these pathways is the recognition of genomic lesions by specific proteins, which signals the presence of DNA damage to nuclear factors involved in DNA repair and cell survival. DNA strand breaks, generated either directly by genotoxic agents (oxygen radicals, ionizing radiation, monofunctional alkylating agents) or indirectly following enzymatic incision of a DNA-base lesion, trigger the synthesis of poly(ADP-ribose) by the enzyme poly(ADP-ribose) polymerase (PARP). PARP (E.C. 2.4.2.30) is a nuclear zinc-finger DNA-binding protein that detects DNA strand breaks. At a site of breakage, PARP catalyzes the transfer of the ADP-ribose moiety from the respiratory coenzyme NAD^+ to a limited number of protein acceptors (Figure 9.1) involved in chromatin architecture (histones H1, H2B, HMG proteins, lamin B) or in DNA metabolism (DNA replication factors, topoisomerases including PARP itself).[1,2] Because of the negative charges on ADP-ribose polymers, poly(ADP-ribosyl)ated proteins lose their affinity for DNA and in many cases their biological activities. PARP and other modified proteins may be restored to their native state following the action of poly(ADP-ribose) glycohydrolase (Figure 9.1). Poly(ADP-ribosyl)ation is therefore an immediate and transient post-translational modification of nuclear proteins induced by DNA lesions (DNA nicks and base damage–generating nicks) mainly repaired by the base excision repair (BER) pathway. Detection, translation of signals emanating from DNA interruptions, and signal amplification by poly(ADP-ribose) formation are the main characteristics of this enzymatic activity, catalyzed by a highly conserved protein.[1,3]

The physiological role of PARP has been much debated this last decade, but molecular and genetic approaches have only recently been exploited to study the role of the immediate poly(ADP-ribose) synthesis that occurs in response to DNA strand breaks. This review summarizes the most recent results, which define unambiguously its role in the cell response to DNA damage and repair, including cell death by apoptosis.

9.2 THE MODULAR STRUCTURE OF POLY(ADP-RIBOSE) POLYMERASE

From the title of their initial paper in 1963, "Nicotinamide Mononucleotide Activation of a New DNA-Dependent Polyadenylic Acid Synthesizing Nuclear Enzyme," Chambon et al.[4] already anticipated correctly the major property of PARP, that is, the stimulation of its enzymatic activity by DNA strand breaks. The cloning of the human PARP cDNA 24 years later, simultaneously by the groups of Shizuta,[5] Miwa,[6] Smulson,[7] and Schweiger,[8] opened the field of poly(ADP-ribosyl)ation reactions to various fruitful molecular and genetic approaches. We know now that the PARP mRNA of 3042 nucleotides encodes a multifunctional and highly conserved protein of 113,135 Da made up of 1014 amino acids (Figure 9.2A).

Involvement of PARP in the Cellular Response to DNA Damage

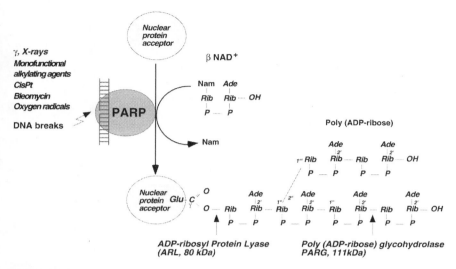

FIGURE 9.1 The poly(ADP-ribosyl)ation reaction during DNA damage and repair induced by various genotoxins.

The PARP DNA-binding domain, DBD, is located in the amino-terminal region (Figure 9.2A). It contains a repeated sequence (amino acids, aa 2 to 97 and 106 to 207) within which a zinc-finger motif of the form $Cx_2Cx_{28-30}Hx_2C$ is evolutionarily conserved, suggesting that the DBD might have arisen by sequence divergence of a primordial element, duplicated in two independently folded zinc-containing minidomains (FI, FII). This region, alone, acts as a detector of single-strand breaks and interacts with one and a half turns of the double helix, independently of the sequence.[9] The interaction between PARP and a single-strand DNA break, visualized by electron microscopy, clearly shows that PARP induces a bent conformation to the DNA, leading to a characteristic V-shaped conformation of the complex.[10] Strikingly, a similar bent conformation of DNA has been observed recently in DNA polymerase β cocrystallized with a nicked duplex.[11] These structures may constitute strong roadblocks for ongoing replication forks.

It has been shown recently that the N-terminal region of PARP not only binds to DNA but also constitutes an interface for protein–protein interaction with several partners including PARP, histones,[12] DNA polymerase α,[13] X-ray cross-complementing factor-1 (XRCC1),[14] and the transcription factors TEF-1[15] and RXRα[16] (Figure 9.2A). Interestingly, the integrity of the zinc fingers is also important both for DNA binding[17,18] and protein binding.[13] Clearly, this dual function of the PARP zinc-finger domain, also demonstrated for several other zinc-finger proteins,[19] needs to be investigated further to be understood fully.

A module of 38 amino acids (Figure 9.2A) containing a bipartite motif of the form $KRK-X_{11}-KKKSKK$, constitutes the nuclear homing sequence of PARP.[20] This short region is not only recognized by the nuclear transport machinery, but also contains several proteolytic cleavage sites of caspase-3 (Figure 9.2B) localized in

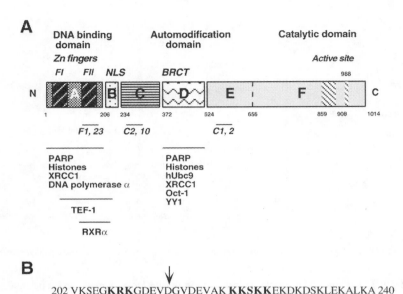

FIGURE 9.2 (A) Schematic of the modular organization of human PARP (F1, F2: zinc-fingers 1, 2; NLS: nuclear localization sequence; BRCT: sequence homologous to the BRCT motif). The functional domains are indicated with their respective monoclonal antibodies (F1, 23; C2, 10; C1, 2)[129] and their protein partners (see also text). (B) PARP nuclear localization domain. The amino acids belonging to the bipartite NLS are in bold. The proteolytic cleavage sequence is underlined and the cleavage site is indicated by an arrow.

the sequence 211-DEVD-214,[21] plasmin,[22] and granzyme A at position K221 (J. Tchopp, personal communication) confirming that it is an exposed and regulatory region of the molecule, particularly during apoptosis (see Section 9.3).

The central region of PARP comprises an auto-poly(ADP-ribosyl)ation domain (aa 372 to 524, Figure 9.2A). PARP appears to be the main acceptor of ADP-ribose polymers, both *in vitro*[23] and *in vivo*, under DNA damage conditions.[24] The auto-poly(ADP-ribosyl)ation of PARP, which takes place mostly at glutamate residues, interferes with DNA binding due to charge repulsion between covalently bound polymers and nicked DNA.[25] Interestingly, the automodification domain contains a BRCT motif (for *b*reast *ca*ncer susceptibility protein, BRCA1, *C-t*erminus) recently identified in a number of DNA damage response proteins and in DNA repair and cell cycle checkpoint proteins.[26,27] This module of about 100 residues (aa 372 to 476 in human PARP) appears to be typical of specific protein–protein contacts in several pairs of DNA repair proteins (XRCC1/DNA Ligase III;[28] XRCC4/DNA Ligase IV[29]). In human PARP, the BRCT module is involved in the interface with multiple protein partners such as XRCC1,[14] hUbc9,[30] histones,[12,22] and transcription factors such as Oct-1[31] and YY1.[32] The BRCT domain of the human PARP contains two highly conserved glutamate residues (E 407 and E 413) that might be implicated as polymer acceptor sites, in a negative regulation of the interactions between PARP and modified species.

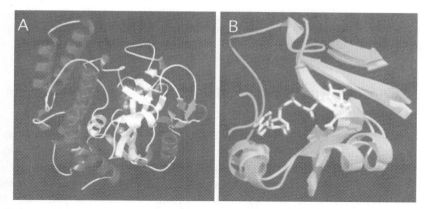

FIGURE 9.3 Crystal structure of chicken PARP catalytic domain (domain F, aa 654 to 1014). (A) The active site of PARP constituted by the structural elements (β–α-loop/β–α; aa 859 to 908) is light gray.[38,39,41] (B) Superposition of the active site of chicken PARP (medium gray) to the catalytic site of diphtheria toxin (light gray).[40]

The carboxy-terminal region of PARP (Figure 9.2A, domain F) bears all the different catalytic activities associated with the full-length enzyme: NAD^+ hydrolysis and initiation, elongation, branching, and termination of ADP-ribose polymers.[33] This basal activity of the enzyme is independent of the presence of DNA breaks. Domain F is by far the most evolutionarily conserved region. It contains a block of 50 amino acids (aa 859 to 908) representing "the PARP signature" virtually unchanged from human to plants[34,35] (M. Katzmaier, personal communication), which turned out to be the PARP catalytic site. The crystal structure of the chicken PARP catalytic domain, overproduced from the Sf9/baculovirus system[36,37] has recently been solved at 2.4 Å resolution (Figure 9.3A) in the presence and in the absence of the nicotinamide analogue PD 128763 (3,4-dihydro-5-methyl-isoquinolinone; Parke-Davis).[38] It comprises (1) an α-helical N-terminal region (aa 662 to 784) and (2) a C-terminal region made of two β-sheets surrounded by five α-helices (aa 785 to 1014) containing the NAD^+-binding cavity.[39] Interestingly, this region, which includes the PARP signature, is folded in a β–α-loop/β–α motif (Figure 9.3B) structurally similar to the NAD^+-binding fold of several ADP-ribosylating bacterial toxins like diphtheria toxin (DT),[40] heat labile enterotoxin from *Escherichia coli*, exotoxin A from *Pseudomonas aeruginosa*, and pertussis toxin from *Bordetella pertussis*.[38] The NAD^+-binding site was finally derived from cocrystal structures and homology modeling.[39] A strong conservation was found among the residues forming the NAD^+-binding pocket (donor site) in DT and PARP. Therefore, PARP and ADP-ribosylating toxins define a new superfamily of ADP-ribosyl transferases (ARTases) containing a NAD^+ fold different from the Rossman fold present in dehydrogenases.

The ADP-ribosyl acceptor-binding site (polymer site) was recently identified by chance in a cocrystallization experiment using carba-NAD.[41] This second site, located close to the NAD^+-binding pocket, is unique to PARP and not present in the crystal structure of other ARTases. Site-directed and random mutagenesis confirmed the existence of the acceptor site, revealing its dual function for both elongation and branching.[41,42]

9.3 POLY(ADP-RIBOSE) POLYMERASE IS A SURVIVAL FACTOR

Much of the knowledge of the *in vivo* role of PARP has been based on the utilization of competitive inhibitors of the 3-substituted benzamide family which are in fact now regarded as "benchmark" inhibitors. Although their specificity has been questioned many times, it has been repeatedly demonstrated that treatment of damaged cells with PARP inhibitors delays DNA strand breaks rejoining and potentiates the cytotoxicity of DNA-damaging agents that trigger the BER of DNA (for a review see Reference 54). The following sections summarize the various contributions based on several different and independent molecular and genetic approaches, to the better understanding of the physiological role of PARP.

9.3.1 PARP-Depleted Cell Lines and *trans*-Dominant Negative Mutants

In spite of the long half-life of the PARP molecule, antisense RNA expressed in human or in rodent cells has been used to limit the endogenous translation of PARP mRNA. These cells showed alterations in cell morphology and chromatin architecture and performed very limited DNA repair during early time periods of recovery from MMS (methylmethanesulfonate)-induced DNA strand breaking.[43] Moreover, gene amplification, genomic stability, and cell survival were strongly affected.[44,45]

Parallel to this approach, rodent cell lines deficient in PARP activity were generated by random mutagenesis.[46-49] These mutants displayed a drastically reduced PARP activity, an increased frequency of sister chromatid exchanges (SCEs), and a reduced cell survival in response to alkylating agents.

Alternatively, overexpression in mammalian cells of the PARP DBD, either transiently[50-52] or constitutively,[53] was used to inhibit the endogenous PARP activity in *trans*. As a consequence of PARP inhibition, DBD expression increased sensitivity to alkylating agents and ionizing radiation but not to UV-C. This was shown by an increased level of N-methyl-N'nitro-N-nitrosoguanidine (MNNG)-induced gene amplification, SCEs, doubling time and apoptotic response, a G2/M accumulation, a marked reduction in cell survival, and a strongly decreased level of MNNG-induced DNA repair synthesis.

Most of the results obtained from these studies are generally in agreement with those obtained with PARP inhibitors (for a review see Reference 54) and favor the implication of PARP in the maintenance of genomic integrity during BER. To clarify the role of PARP in the mammalian BER process, we and others have developed animal and cellular models deficient in PARP.[55-57]

9.3.2 PARP Knockout Mice are Sensitive to Ionizing Radiation and Alkylating Agents

To establish an animal model and derived cell lines deficient in PARP activity, the PARP gene has been disrupted in exon 1,[57] exon 2,[55] or exon 4[56] by homologous recombination. PARP knockout (KO) animals are viable and fertile, indicating that PARP is dispensable for mice development and tissue differentiation. However, we found that the average litter size was smaller than that of the wild-type mice. Moreover, the adult body weight of the homozygous mice was significantly reduced by about 23% compared with their wild-type littermates, suggesting that perhaps PARP KO mice lose cells by spontaneous apoptosis.[56]

In contrast to Wang and collaborators,[55] who first showed that a null mutation in the PARP gene had no influence on repair of MNNG-damaged DNA, we have reported that PARP KO mice are hypersensitive to genotoxic agents, like monofunctional alkylating agents or γ rays.[56] Injection of a single dose of 75 mg/kg per body weight (BW) of N-methyl-N-nitrosourea (MNU) at 6 weeks of age resulted in 100% mortality in PARP KO and only 43% in PARP wild-type mice during more than 8 weeks of observation (Figure 9.4A). Similarly, following a whole-body irradiation of 8 Gy at 6 weeks of age, 62% of PARP KO mice died 4 days postirradiation and all mutant mice were dead by 9 days postirradiation. In contrast, only 50% of wild-type mice were dead by 15 days postirradiation (Figure 9.4B). Sensitivity to ionizing radiation was confirmed using PARP KO mice developed by Wang and colleagues.[58]

To understand the cause of this precocious lethality of the PARP KO mice, wild-type and mutant mice were irradiated with 8 Gy, and an autopsy was performed 3 days postirradiation. In all KO mice, the small intestine was distended by intraluminal fluid accumulation, and histological sections through the duodenum indicated that the villi of irradiated PARP KO mice were considerably shortened, together with a necrosis of their epithelial cells (Figure 9.5). These data demonstrated that

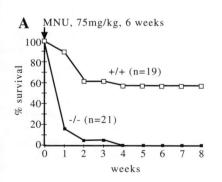 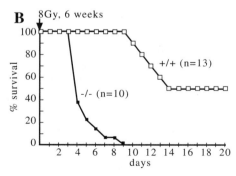

FIGURE 9.4 Survival of PARP wild-type and PARP KO mice after (A) i.p. injection of MNU at 75 mg/kg body weight at 6 weeks of age, (B) γ radiation with 8 Gy at 6 to 8 weeks of age. The percentage of mice alive at the end of a week is plotted against age. (From Ménissier-de Murcia, J. et al., *Proc. Natl. Acad. Sci. U.S.A.*, 94, 7303–7307, ©1997. National Academy of Sciences, U.S.A. With permission.)

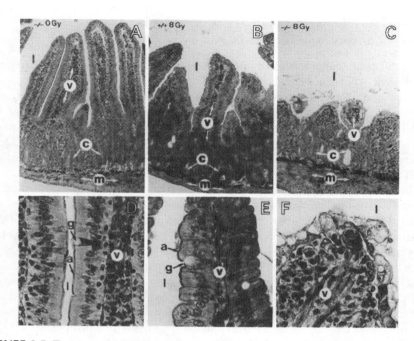

FIGURE 9.5 Transverse histological sections through the duodenum of an untreated PARP−/− mouse (A and D) and an 8 Gy irradiated PARP+/+ mouse (B and E) or an irradiated PARP KO mouse (C and F). (A to C) show the full thickness of the duodenal wall and (D to F) the details of the epithelium near the tips of the villi. Note that the untreated PARP KO duodenum (A and D) is histologically indistinguishable from its wild-type counterpart. *Abbreviations:* a, absorptive cell; c, crypt; g, goblet cell; l, lumen of the small intestine; m, muscularis; v, villi. Original magnifications: ×170 (A to C) and ×750 (D to F). (From Ménissier-de Murcia, J. et al., *Proc. Natl. Acad. Sci. U.S.A.*, 94, 7303–7307, ©1997. National Academy of Sciences, U.S.A. With permission.)

in PARP KO mice, death was caused by dehydration and endotoxicosis secondary to acute radiation toxicity.

Interestingly, *Atm*-deficient mice,[59] as well as SCID mice[60] are as sensitive to ionizing radiation as PARP null mice and exhibit the same phenotype following γ irradiation. In addition in *Atm*-deficient mice PARP activity is not affected and its induction by oxidative damage or γ irradiation is not altered.[61] The selective acute toxicity of specific tissues of these mice to γ irradiation and to CPT-11, a topoisomerase I inhibitor (J. Ménissier-de Murcia, unpublished results), underline the cardinal function of DNA damage–screening programs in rapidly proliferating tissues such as the small intestine epithelium.

9.3.3 PARP Knockout Cells are Severely Affected in the Base Excision Repair Pathway

To analyze further the basis for the altered survival of animals to DNA-damaging agents at the cellular level, primary fibroblasts were derived from mutant and wild-type 13.5-day embryos (mouse embryonic fibroblasts, MEFs). KO cells, totally devoid of PARP or truncated functional domains of it as judged by Western and activity blot, were used to evaluate cell growth and cell cycle progression.[62]

The growth rate of PARP KO primary fibroblasts was indistinguishable from that of wild-type mice in the absence of damage. However, upon exposure to a sublethal dose of MMS, PARP KO MEFs were strongly affected in their ability to proliferate, as their doubling time was raised to 169 h as compared with 72 h for wild-type cells.[63] Mutant MEFs exhibited a marked MMS-dose–dependent accumulation in G2/M, indicating that PARP-deficient cells failed to resume their progression through the cell cycle after DNA damage, whereas under the same conditions wild-type primary fibroblasts were only slightly affected in the cell cycle distribution.[63] The same differential effect could be observed upon treatment with MNU.[56]

Hypersensitivity to MMS, also evidenced by cell viability as monitored by (methyl-^{3}H)thymidine incorporation, could be restored by ectopic expression of the wild-type PARP cDNA,[63] thus demonstrating that this sensitivity to monofunctional alkylating agents was effectively due to the absence of PARP. Accordingly, a reduced survival was also noted in the third PARP KO model when PARP$^{-/-}$ ES stem cells where treated with MMS or γ rays.[57]

Numerous studies have demonstrated that inhibition of PARP activity in cultured cells also increases genomic instability following DNA damage.[53,54,64] An *in vivo* study of SCEs and chromosome breakage was performed in PARP wild-type and KO mice cells. As expected, in bone marrow cells of 2-month-old mutant mice, a high frequency in the reciprocal exchange of sister-chromatid segments (Figure 9.6A and C) and in the formation of chromatid (ctb) and chromosome (csb) breaks (Figure 9.6B and D) were observed after treatment with MNU and whole-body γ irradiation, respectively.

The increased chromosomal instability in PARP KO mouse embryonic fibroblasts cells exposed to 0.05 mM MMS was also demonstrated by measuring the frequency of induction of micronuclei, which represent chromatin fragments that are not incorporated into the nucleus during mitosis. They are considered to be a simple indicator of chromosomal damage. PARP KO cells spontaneously exhibited a two-fold increase in the total number of micronuclei, in the absence of DNA damage. At 24 h after MMS treatment there was a three-fold increase in the frequency of micronuclei in PARP KO cells.[63] Similar results have been obtained using the first PARP KO mouse model.[58] Altogether, these data demonstrate that the absence of PARP leads to a considerable increase in genomic instability, whatever the disrupted exon is in the KO mice model.

Cells exposed to 3-aminobenzamide to inhibit PARP activity have a reduced capacity to repair base-damaged DNA as evidenced by the nucleoid technique[65] or by the alkaline elution method.[66] DNA repair induced by MMS treatment was evaluated in MEFs from PARP wild-type and KO mice, by the single-cell gel electrophoresis assay (comet assay), a highly sensitive method for measuring DNA breaks.[67] Mouse embryonic fibroblasts, mock-treated or exposed to MMS, were immobilized onto a glass microscope slide, detergent-lysed, and subjected to electrophoresis under alkali conditions. Cellular DNA containing breaks migrates toward the anode, giving the appearance of a comet. The product of the percent of DNA in

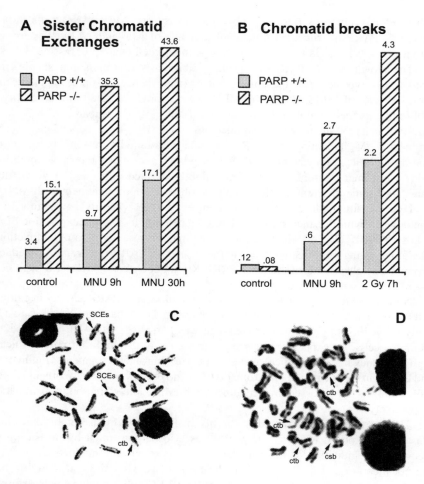

FIGURE 9.6 Chromosome instability in bone marrow cells of mice mock-treated or injected with MNU at 80 mg/kg body weight, or irradiated with 2 Gy. (A) Mean number of SCEs per cell in PARP$^{-/-}$ and PARP$^{+/+}$ mice, before and after exposure to MNU during 9 or 30 h. (B) Mean number of chromatid breaks before and after exposure to MNU during 9 h or after exposure to γ rays 7 h before harvesting. Microphotographs of PARP$^{-/-}$ bone marrow cells isolated (C) 9 h after MNU injection showing a high frequency of SCEs, and (D) after irradiation with 2 Gy showing chromosome breakage (ctb: chromatid breaks, csb: chromosome breaks).

the tail and the mean distance of migration in the tail is taken as a measure of the extent of DNA breakage and is termed *tail moment*. This parameter was found to vary in a linear manner with increasing doses of MMS in the range of 0 to 0.30 mM MMS for each genotype. For cells exposed to 0.15 mM MMS, the repair kinetics showed that, while at 24 h virtually all the DNA breaks resulting from exposure to MMS were resealed in the two cell lines, PARP KO cells display considerably slower rejoining kinetics compared with PARP wild-type cells. Half-lives of MMS-induced DNA strand-breaks of about 1 and 5 h were measured for PARP wild-type and PARP KO cells, respectively.[63] These results are in full agreement with those obtained with the antisense RNA expression[43] and the *trans*-dominant negative[50-53] approaches, but contradict those obtained *in vitro* using purified PARP in DNA repair assays,[68,69] which leads one to conclude an incidental and negative role for PARP in DNA excision repair.

Although the slower rate of repair, as measured by the comet assay, reflects an apparent ligation defect, DNA repair synthesis that might have been affected by the loss of PARP was recently evaluated comparatively in both genotypes. Two distinct, but in some ways overlapping, pathways for completion of BER have been reported in mammalian cells: the major DNA polymerase β–dependent short patch repair (SPR) pathway (replacement of one nucleotide) and the minor long patch repair (LPR) pathway (replacement of 7 to 14 nucleotides).[70] To evaluate the relative contribution of PARP in the two BER pathways, the ability of PARP wild-type and PARP-deficient MEF extracts to repair a single abasic site derived from uracil or 8-oxoguanine located on a circular duplex plasmid molecule was tested by a standard *in vitro* repair assay.[70] Whole-cell extracts from PARP-deficient cells were about half as efficient as wild-type cells to perform repair synthesis involving one nucleotide (SPR), but were completely inefficient in performing the polymerization step of the long patch repair.[61]

In summary, the PARP-deficient murine model recapitulates most of the phenotypes already observed using chemical inhibitors and the various genetic and molecular approaches. These data demonstrate the crucial role of PARP in processing γ irradiation and alkylating agent–induced DNA lesions, which both trigger the base excision repair pathway. Similarly, a delay in the kinetics of DNA resealing[71] and resynthesis in the LPR pathway[72] were also observed with rodent cells bearing a point mutation in XRCC1, a partner of PARP that plays an essential role in the BER pathway (see below).

9.3.4 SOME PARTNERS OF PARP ARE INVOLVED IN DNA REPAIR AND REPLICATION

As mentionned in Section 9.2, PARP interacts with proteins mainly by its DBD and by its automodification domain containing a BRCT motif, involved in specific protein–protein contacts.[26,27,73] The two-hybrid system has been used to identify proteins putatively interacting *in vivo* with PARP and being involved in its biological function. A cDNA expression library from HeLa cells, in fusion with the activating domain of Gal4, was screened using PARP fused to the LexA DBD as a bait. Among 15 independent clones obtained, one had complete identity with a portion of the cDNA of the DNA repair factor XRCC1. XRCC1 complements the rodent cell lines EM9 and EMC11, which are hypersensitive to monofunctional alkylating agents and γ irradiation.[71,74] Its interaction with PARP was further confirmed in mammalian cells and the respective interacting interfaces have been identified.[14] XRCC1 binds to PARP by its first BRCT module (aa 314 to 403) whereas PARP contacts XRCC1 both by its zinc-finger domain and its central region containing the automodification sites included in a BRCT module. This interaction, which is resistant to 1M salt, negatively regulates PARP activity *in vivo* as evidenced by the preferential association of oligo ADP-ribosylated PARP with XRCC1. A conformational change, induced by limited auto-poly(ADP-ribosyl)ation, is believed to take place, increasing the accessibility and/or the affinity of PARP for XRCC1. Similar observations have been made in the study of PARP–histone interactions.[12] Interestingly, XRCC1 has been demonstrated to serve as an adaptor during the BER reaction, through its interaction with DNA polymerase β[75] and DNA ligase III.[28,76] Moreover, PARP interacts with the N-terminal part of DNA ligase III via the automodification domain and its own zinc-finger region (C. Niedergang et al., unpublished results). An interaction was also found between the N-terminal region of DNA polymerase β encompassing the desoxyribose phosphodiesterase activity and PARP (F. Dantzer et al., unpublished results). Altogether, these results provide strong evidence that PARP is a member of a BER multiprotein complex comprising (1) nick sensors (PARP, DNA ligase III), (2) an adaptor factor (XRCC1), and (3) DNA repair effectors (DNA polymerase β, DNA ligase III). These factors and enzymes involved in BER behave as multimodular polypeptides capable of various combinations through protein–protein contacts mediated by small specific motifs such as the BRCT motif present in PARP,[14] XRCC1,[28] and DNA ligase III.[77]

The human homologue of the yeast ubiquitin–conjugating enzyme Ubc9 (hUbc9) was identified by the two-hybrid method as another partner of PARP. The interaction appeared to take place at the automodification domain of PARP.[30] The ubiquitin-conjugating enzyme is part of a nonlysosomal proteolytic pathway in which ubiquitination targets proteins for degradation to a 26S proteasome complex.[78] PARP has been implicated in apoptosis after DNA damage.[53] Ubiquitin-mediated proteolysis has also been suggested to be important in DNA damage–induced apoptosis.[79,80] Specific inhibitors of proteasome function blocked cell death and prevented apoptosis-associated cleavage of PARP induced in thymocytes by ionizing radiation[81] and in neurons by nerve growth factor deprivation.[82] Recently, the proteasome activated by poly(ADP-ribosyl)ation has been shown to remove, more efficiently, oxidatively damaged histones.[83] The PARP–hUbc9 interaction might then be involved, through proteolysis of key proteins, in the commitment of the cell to apoptosis. The ubiquitination–proteolysis pathway may also represent a mechanism for cell cycle regulation, particularly for sister chromatid separation at anaphase in yeast.[84] The interaction between hUbc9 and Rad51 is supposed to mediate the specific breakdown of proteins required for interactions of homologous chromosomes in the synaptonemal complex in meiotic cells.[85] In addition, the pachytene stage of spermatocyte differentiation coincides with a particular abundance of PARP[86] as well as XRCC1,[87] DNA ligase III,[77] DNA polymerase β,[88] and several other DNA repair proteins, suggesting the necessity of protective mechanisms for genome integrity during the regulation of mitosis and meiosis.

PARP interacts *in vivo* and *in vitro* with DNA polymerase α.[13,89,90] The interaction has been demonstrated to occur between the N-terminal zinc-finger domain of PARP and the catalytic subunit of the DNA polymerase α-primase tetramer in HeLa cell extracts. This interaction is maximal during the S phase of the cell cycle suggesting that PARP participates in a survey mechanism implicating its nick-sensor function when DNA breaks are present in the template, as part of the control of replication fork progression. PARP stimulates DNA polymerase activity in the absence of damage, suggesting that it is a component of the replication complex.[89,90] Conversely, compared with wild-type cells, cells lacking PARP exhibit decreased DNA polymerase activity and reduce their progression through the S phase, when damaged by sublethal doses of MMS.[13]

Despite the fact that PARP has not been found in two-hybrid screening, PARP appears to dimerize to form an active molecular complex.[91] Moreover, it was shown that two molecules of PARP, i.e., a catalytic dimer, are required for the automodification reaction and function according to an intermolecular reaction.[92]

9.4 POLY(ADP-RIBOSE) POLYMERASE AND CELL DEATH

Apoptosis or programmed cell death is a fundamental biological process that plays an important role in early development, cell homeostasis, and diseases such as neurodegenerative disorders.[93,95,96] Programmed cell death can occur in response to a number of stimuli, such as genotoxic damage when DNA repair is saturated, withdrawal of growth factors, and activation of specific receptors, like CD95 antigen and TNF receptor. Morphologically, it is characterized by the appearance of membrane blebbing, cell shrinkage, chromatin condensation, DNA cleavage, and finally cell fragmentation into membrane-bound apoptotic bodies. At the biochemical level, there is increasing evidence for a central role for the family of cystein proteases, the caspases, in the pathway that mediates the highly ordered process leading to cell death. Caspases have been identified as the enzymes that, through proteolytic cleavage, disable critical homeostatic and repair enzymes, as well as key structural components during apoptosis.[97] All caspases cleave after an aspartic acid for which they have an absolute specificity. Caspases exist in the cytoplasm as inactive proenzymes that are processed to a large and a small subunit to form the active enzyme.[98]

9.4.1 CLEAVAGE OF PARP DURING APOPTOSIS

PARP is one of the first substrates that was found to be cleaved by caspases during apoptosis.[99] PARP is cleaved at the conserved sequence $_{211}DEVD_{214}$ to an 89-kDa fragment containing the active site and the automodification domain, and to a 24-kDa fragment containing the zinc fingers responsible for its DNA binding activity (see Figure 9.2B). The outcome is the loss of enzymatic activity since the DBD is physically separated from the catalytic domain. PARP cleavage has been shown in almost all forms of apoptosis, including apoptosis induced by irradiation, by chemotherapeutic agents, and upon activation of the death receptors. In all instances where apoptosis is inhibited (by Bcl-2 overexpression, use of caspase-specific inhibitors, etc.), PARP cleavage is also inhibited.[100] In our laboratory we have tried to understand why the cell needs to get rid of this protein to bring about apoptosis. We addressed this question by expressing a noncleavable mutant of PARP (D214A-PARP) in PARP KO cells. The cells expressing the PARP mutant were exposed to anti-CD95 to induce apoptosis. This resulted in a significant delay of cell death.[101] Morphological analysis showed that the kinetics of cell shrinkage and nuclear condensation were significantly retarded. This suggests that the cleavage of PARP during apoptosis facilitates cellular disassembly (particularly of the nucleus) to ensure the completion and irreversibility of this process. Inactivation by cleavage of PARP by apoptotic caspases might serve, therefore, to a double finality: first, disable a key component of genomic surveillance (together with DNA-PK[94]) to avoid unnecessary DNA repair during chromatin degradation and, second, facilitate the accessibility of endonucleases to chromatin and nuclear disintegration. A recent study showed that the 24-kDa product of PARP cleavage by caspase-3 irreversibly binds internucleosomal DNA in apoptotic cells. This may contribute to the irreversibility of apoptosis by preventing the access of DNA to DNA repair enzymes.[102] The preferential localization of PARP in the vicinity of the nuclear envelope[13] also suggests that its cleavage during apoptosis participates in nuclear disassembly, and facilitates downstream events, which are temporarily retarded upon expression of the uncleavable PARP mutant.

A recent study has also used a caspase-resistant PARP mutant to evaluate the type of cell death arising when PARP is not cleaved during apoptosis induced by TNF-α.[103] They found that fibroblasts stably expressing the caspase-resistant PARP exhibited accelerated cell death (apoptosis and necrosis), which was due to NAD$^+$ depletion, and they concluded that PARP is cleaved during apoptosis to avoid necrotic cell death, which might result in tissue injury. This is a very interesting hypothesis; however, the authors failed to explain the reason for the increased apoptotic cell death seen in cells expressing the caspase-resistant PARP, even with a dramatic ATP drop that rather might increase further necrosis but not apoptosis, which itself depends on the ATP level.[104] Moreover, at the time all the experiments were performed (12 h) the cells already showed about 50 to 70% loss in cell viability for the wild-type and the PARP KO cells expressing the uncleavable PARP, respectively. With such a high rate of cell death, the apoptotic process is at a very final step at which PARP is completely cleaved. In our previous study[101] the consequences of the noncleavage of PARP were studied at a time when the viability of cells expressing the PARP-D214A mutant was about 90%. One can then postulate that preventing PARP cleavage might interfere with cellular disassembly and delay the morphological changes of apoptosis at the initial steps of the process;[101] this could result in increased necrosis as a final outcome.[103]

The consequences of the cleavage of specific substrates of caspases in the morphological changes of cells during apoptosis have been studied by expressing uncleavable mutants of these proteins. A recent and elegant example of the effect on apoptosis of a noncleavable caspase-3 substrate is the inhibition of the caspase-activated deoxyribonuclease (CAD) activity upon expression of mutant noncleavable inhibitor (ICAD). In cells expressing this mutant version, DNA was not degraded, but the cells displayed all the characteristics of apoptosis.[105] Mutant lamin-expressing cells showed no signs of chromatin condensation or nuclear shrinkage during apoptosis.[106] Blocking p21-activated kinase cleavage or activation during CD95-induced apoptosis inhibited membrane cell blebbing, whereas the nuclear modifications were unaffected.[107]

It has been reported that inhibition of caspase activity causes a switch from apoptotic to necrotic cell death.[109,110] This switch can ultimately be modulated by the availability of ATP.[111] Inhibition of caspases thus prevents the cell from completing the apoptotic process, but does not inhibit cell death. This indicates that at least some of these proteolytic cleavages are necessary for the apoptotic process, but not for cell death in general. This distinction could be of extreme importance for the organism to prevent necrosis-induced injuries. PARP is also cleaved during necrosis, although the fragments are not the same as those found in apoptosis.[112] These cleavages seemed specific, although not related to caspase activity, and PARP remained intact for longer than in apoptosis.

9.4.2 Apoptosis in PARP-Deficient Cells

Splenocytes from PARP KO mice undergo apoptosis more rapidly upon treatment with an alkylating agent.[56] In contrast, fibroblasts and primary bone marrow cells derived from these mice are as sensitive as the wild-type cells when the apoptotic inducers were able, independently of DNA lesions, to activate PARP.[101] Why these cells lacking PARP are much more fragile and undergo programmed cell death faster than parental cells could be explained by the accumulation of unrepaired DNA damage. This makes the cell unable to cycle and engage the apoptotic pathway to avoid transmission of damaged DNA to a new generation of cells. This view is supported by the fact that PARP KO cells have prolonged kinetics of DNA rejoining as measured by the comet assay.[61]

The sensitivity of PARP KO cells to apoptosis has also been studied for the knockout mice developed by Wang et al.[58] These cells have been shown to be as sensitive as their wild-type counterpart to apoptotic inducers not acting by creating DNA damage.[110] In contrast, a recent study on the role of poly(ADP-ribosyl)ation during CD95-induced apoptosis found that the absence of PARP made the cells resistant to cell death following CD95 treatment.[108] This conflicting result could have arisen from the use of clones of immortalized or stably transfected PARP KO MEFs in this study, which might have influenced the responsiveness to CD95. They concluded that PARP was necessary for apoptotic signaling and that it was a necessary step to activate effector caspases. However, they could not explain the discrepancy between their results and the findings from other groups. Most of the current data on the implication of PARP in apoptosis suggest that PARP is a passive rather than an active player in the apoptotic process. New approaches should be attempted to study the implication of PARP cleavage in the degradation of caspase substrates in the final phase of apoptosis, including DNA fragmentation. The importance of new PARP homologues (see Section 9.5) in this process should be carefully evaluated.

9.5 FUTURE PROSPECTS

There are at least three reasons knowledge of the biology of PARP has been considerably improved during the last 2 years. First, the generation of animal and cellular models deficient in PARP has provided novel and deep insights into the complexity of poly(ADP-ribosyl)ation reactions in two different physiological conditions having

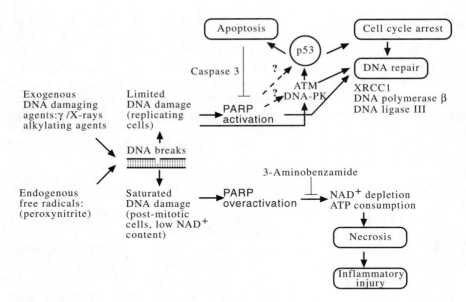

FIGURE 9.7 Model for implication of PARP in the coordination of cell metabolism with DNA BER. PARP activation leads, in the case of limited DNA damage in replicating cells, to cell cycle arrest or apoptosis and in the case of saturated DNA damage, to cell necrosis.

in common, however, the presence of breaks in the genome. It is now evident that PARP plays an essential role as a survival factor in replicating cells injured with limited DNA damage[56] (Figure 9.7). Conversely, in the inflammatory response and in pathologies such as brain ischemia,[113] diabetes,[114,115] and septic shock[116] associated with inflammatory cell damage, the necrotic evolution of the cell death seems to be directly related to energy consumption and overactivation of PARP. During necrosis, PARP apparently is cleaved by different proteases with slower kinetics than apoptosis,[112] which in turn may accentuate the NAD$^+$ depletion. Several mechanisms have been proposed to explain inactivation of PARP (either pharmacologically or in genetically engineered animals lacking PARP) and to improve the clinical outcome of animals in those conditions.[117,118]

Second, the recent discovery of several PARP homologues has made even more complex the two opposed facets (cell survival/cell death) of poly(ADP-ribosyl)ation reactions. Two (out of four) PARP homologues have been cloned and characterized: (1) Tankyrase, a multifunctional protein of 142 kDa with regions of homology to ankyrins and to the PARP catalytic domain binds to and negatively regulates TRF1, a factor involved in telomere maintenance;[119] (2) PARP-2, a novel mammalian DNA-damage–dependent PARP recently cloned in our laboratory.[120] The PARP-2 gene maps to chromosome 14q11.2 in humans. Although it lacks a classical zinc-finger module, the recombinant mouse PARP-2 binds to damaged DNA *in vitro* and catalyzes the formation of poly(ADP-ribose) polymers in a DNA-dependent manner. Two other cDNAs encoding PARP homologues also have been identified recently and need to be further characterized.[121,122] The function of PARP homologues might be considered as possible backup mechanisms in PARP (113-kDa)-deficient cells. The generation of KO mice, deficient in each of these new PARP homologues, will be of importance, even if they account for only 10% of the total PARP activity stimulated by DNA breaks. The eventual lethality of double-KO mice in which PARP activity will be totally abolished may confirm the importance of poly(ADP-ribose) metabolism in development.

Third, over the same period of time, PARP has been shown to interact with a variety of transcription factors other than p53,[123] such as YY1,[32] AP2,[125] OCT1,[31] TEF1,[15] DF4,[125] E47,[127] TFIIF,[128] and RXR-α.[16] A number of experimental evidences have led to the conviction that PARP influences the transcriptional activation of a number of genes by interacting directly with transcription factors, or with enhanceosomes. Further experimental approaches will be needed to elucidate, at the molecular level, the interactions between PARP and the transcription machinery. Interestingly, a connection between PARP and NF-κB has been found very recently by our group[116] and by Hassa and Hottiger[126] showing that the defective activation of NF-κB in PARP KO cells, the subsequent abrogation of NF-κB–dependent TNF-α in the serum, and the downregulation of iNOS render PARP KO mice resistant to endotoxic shock. Therefore, PARP participates not only in the final stages of the inflammatory process by modulating the energy levels, but also early in the regulation of the NF-κB signaling pathway, leading to expression of antiapoptotic genes and to the synthesis of an inflammatory mediator. These findings clearly open the door to new therapeutic strategies for septic shock based on the combined pharmacological inhibition of PARP and NF-κB. However, the genomic surveillance function of PARP may not be compatible with a long-term inhibition.

ABBREVIATIONS

Abbreviations used are PARP: poly(ADP-ribose) polymerase; BER: base excision repair; DBD: DNA-binding domain; MNU: N-methyl-N-nitrosourea; MMS: methylmethanesulfonate; MNNG: N-methyl-N'nitro-N-nitrosoguanidine; NLS: nuclear location signal; XRCC1: X-ray cross-complementing factor-1; MEF: mouse embryonic fibroblast.

ACKNOWLEDGMENTS

We are indebted to our collegues of the PARP group: V. Schreiber, J. C. Amé, F. Dantzer, B. Rinaldi, G. de la Rubia, and A. Huber for their contribution to the work summarized in this chapter. We thank Dr. A. Ruf and G. Schulz for their contribution to the crystal structure and Dr. S. Natarajan for a careful revision of the English text. This work was supported by CNRS, Association pour la Recherche contre le Cancer, Electricité de France, le Commissariat à l'Energie Atomique, and la Fondation pour la Recherche Médicale.

REFERENCES

1. de Murcia, G. and Ménissier-de Murcia, J., Poly(ADP-ribose) polymerase: a molecular nick-sensor, *Trends Biochem. Sci.*, 19, 172, 1994.
2. Oei, S. L., Griesenbeck, J., and Schweiger, M., The role of poly ADP-ribosylation, *Rev. Physiol. Biochem. Pharmacol.*, 131, 4135, 1997.
3. Masson, M., Rolli, V., Dantzer, F., Trucco, C., Schreiber, V., Fribourg, S., Molinete, M., Ruf, A., Alves Miranda, E., Niedergang, C., Hunting, D., Gowans, B., Schulz, G. E., Ménissier-de Murcia, J., and de Murcia, G., Poly(ADP-ribose) polymerase: structure–function relationship, *Biochimie*, 77, 456, 1995.
4. Chambon, P., Weil, J. D., and Mandel, P., Nicotinamide mononucleotide activation of a new DNA-dependent polyadenylic acid synthesizing nuclear enzyme, *Biochem. Biophys. Res. Commun.*, 11, 39, 1963.
5. Kurosaki, T., Ushiro, H., Mitsuuchi, Y., Susuki, S., Matsuda, M., Matsuda, Y., Katunuma, N., Kangawa, K., Matsuo, H., Hirose, T., Inayama, S., and Shizuta, Y., Primary structure of human poly(ADP-ribose) synthetase as deduced from cDNA sequence, *J. Biol. Chem.*, 262, 15990, 1987.
6. Uchida, K., Morita, T., Sato, T., Ogura, T., Yamashita, R., Nogushi, S., Susuki, H., Nyunoya, H., Miwa, M., and Sugimura, T., Nucleotide sequence of a full-length cDNA for human fibroblast poly(ADP-ribose) polymerase, *Biochem. Biophys. Res. Commun.*, 148, 617, 1987.
7. Cherney, B. W., McBride, O. W., Chen, D., Alkhatib, H., Bathia, K., Hensley, P., and Smulson, M. E., cDNA sequence, protein structure, and chromosomal location of the human gene for poly(ADP-ribose) polymerase, *Proc. Natl. Acad. Sci. U.S.A.*, 84, 8370, 1987.
8. Schneider, R., Auer, B., Kühne, C., Herzog, H., Klocker, H., Burtscher, H., Hirsch-Kauffmann, M., Wintersberger, U., and Schweiger, M., Isolation of a cDNA clone for human NAD$^+$: protein ADP-ribosyltransferase, *Eur. J. Cell Biol.*, 44, 302, 1987.
9. Ménissier-de Murcia, J., Molinete, M., Gradwohl, G., Simonin, F., and de Murcia, G., Zinc-binding domain of poly(ADP-ribose) polymerase participates in the recognition of single strand breaks in DNA, *J. Mol. Biol.*, 210, 229, 1989.
10. Le Cam, E., Fack, F., Ménissier-de Murcia, J., Cognet, J. A. H., Barbin, A., Sarantoglou, V., Révet, B., Delain, E., and de Murcia, G., Conformal analysis of a 139 base-pair DNA fragment containing a single-stranded break and its interaction with human poly(ADP-ribose) polymerase, *J. Mol. Biol.*, 235, 1062, 1994.
11. Sawaya, M., Prasad, R., Wilson, S., Kraut, J., and Pelletier, H., Crystal structures of human DNA polymerase β complexed with gapped and nicked DNA: evidence for an induced fit mechanism, *Biochemistry*, 36, 11205, 1997.
12. Griesenbeck, J., Oei, S. L., Mayer-Kuckuk, P., Ziegler, M., Buchlow, G., and Schweiger, M., Protein–protein interaction of the human poly(ADP-ribose)polymerase depends on the functional state of the enzyme, *Biochemistry*, 36, 7297, 1997.
13. Dantzer, F., Nasheuer, H. P., Vonesch, J. L., de Murcia G., and Ménissier-de Murcia, J., Functional association of poly(ADP-ribose) polymerase with DNA polymerase α-primase complex: a link between DNA strand break detection and DNA replication, *Nucl. Acids Res.*, 26, 1891, 1998.
14. Masson, M., Niedergang, C., Schreiber, V., Muller, S., Ménissier-de Murcia, J., and de Murcia, G., XRCC1 is specifically associated with poly(ADP-ribose) polymerase and negatively regulates its activity following DNA damage, *Mol. Cell. Biol.*, 18, 3563, 1998.

15. Butler, A. and Ordahl, C., Poly(ADP-ribose) polymerase binds with transcription enhancer factor 1 to MCAT1 elements to regulate muscle-specific transcription, *Mol. Cell. Biol.*, 19, 296, 1999.
16. Miyamoto, T., Kakizawa, T., and Hashizume, K., Inhibition of nuclear receptor signalling by poly(ADP-ribose) polymerase, *Mol. Cell. Biol.*, 19, 2644, 1999.
17. Ikejima, M., Noguchi, S., Yamashita, R., Ogura, T., Sugimura, T., Gill, D. M., and Miwa, M., The zinc fingers of human poly(ADP-ribose) polymerase are differentially required for the recognition of DNA breaks and nicks and the consequent enzyme activation. Other structures recognize intact DNA, *J. Biol. Chem.*, 265, 21907, 1990.
18. Gradwohl, G., Ménissier-de Murcia, J., Molinete, M., Simonin, F., Koken, M., Hoeijmakers, J. H., and de Murcia, G., The second zinc-finger domain of poly(ADP-ribose) polymerase determines specificity for single-stranded breaks in DNA, *Proc. Natl. Acad. Sci. U.S.A.*, 87, 2990, 1990.
19. Schwabe, J., W. and Klug, A., Zinc mining for protein domains, *Nat. Struct. Biol.*, 1, 345–349, 1994.
20. Schreiber, V., Molinete, M., Boeuf, H., de Murcia, G., and Ménissier-de Murcia, J., The human poly(ADP-ribose) polymerase nuclear localization signal is a bipartite element functionally separate from DNA binding and catalytic activity, *EMBO J.*, 11, 3263, 1992.
21. Kaufmann, S. H., Desnoyers, S., Ottaviano, Y., Davidson, N. E., and Poirier, G. G., Specific proteolytic cleavage of poly(ADP-ribose) polymerase: an early marker of chemotherapy-induced apoptosis, *Cancer Res.*, 53, 3976, 1993.
22. Buki, K., Bauer, P., Hakam, A., and Kun, E., Identification of domains of poly(ADP-ribose) polymerase for protein binding and self-association, *J. Biol. Chem.*, 270, 3370, 1995.
23. Kawaishi, H., Ueda, K., and Hayaishi, O., Multiple autopoly(ADP-ribosyl)ation of rat liver poly(ADP-ribose) synthetase: mode of modification and properties of automodified synthetase, *J. Biol. Chem.*, 256, 9483, 1981.
24. Kreimeyer, A., Wielckens, K., Adamietz, P., and Hiltz, H., DNA repair associated ADP-ribosylation *in vivo*. Modification of histone HI differs from that of the principal acceptor proteins, *Biol. Chem.*, 259, 890, 1984.
25. Zaharadka, P. and Ebisuzaki, K., A shuttle mechanism for DNA–protein interactions. The regulation of poly(ADP-ribose) polymerase, *Eur. J. Biochem.*, 127, 579, 1982.
26. Bork, P., Hofmann, K., Bucher, P., Neuwald, A. F. Altschul, S. F., and Koonin, E. V., A superfamily of conserved domains in DNA damage-responsive cell cycle checkpoint proteins, *FASEB J.*, 11, 68, 1997.
27. Callebaut, I. and Mornon, J. P., From BRCA1 to RAP1: a widespread BRCT module closely associated with DNA repair, *FEBS Lett.*, 400, 25, 1997.
28. Nash, R. A., Caldecott, K., Barnes, D. E., and Lindahl, T., XRCC1 protein interacts with one of two distincts forms of DNA ligase III, *Biochemistry*, 36, 5207, 1997.
29. Critchlow, S., Bowater, R.P., and Jackson, S., Mammalian DNA double-strand break repair protein XRCC4 interacts with DNA ligase IV, *Curr. Biol.*, 7, 588, 1997.
30. Masson, M., Ménissier-de Murcia, J., Mattei, M. G., de Murcia G., and Niedergang, C. P., Poly(ADP-ribose) polymerase interacts with a novel human ubiquitin conjugating enzyme: hUbc9, *Gene*, 190, 287, 1997.
31. Nie, J., Sakamoto, S., Song, D., Qu, Z., Ota, K., and Taniguchi, T., Interaction of Oct-1 and automodification domain of poly(ADP-ribose) synthetase, *FEBS Lett.*, 424, 27, 1998.

32. Oei, S. L., Griesenbeck, J., Schweiger, M., Babich, V., Kropotov, A., and Tomilin, N., Interaction of the transcription factor YY1 with human poly(ADP-ribosyl) transferase, *Biochem. Biophys. Res. Commun.*, 240, 108, 1997.
33. Simonin, F., Höfferer, L., Panzeter, P. L., Muller, S., de Murcia, G., and Althaus, F. R., The carboxyl-terminal domain of human poly(ADP-ribose) polymerase: overproduction in *Escherichia coli*, large-scale purification and characterization, *J. Biol. Chem.*, 268, 13454, 1993.
34. Lepiniec, L., Babiychuk, E., Kushnir, S., Van Montagu, M., and Inzé, D., Characterization of an Arabidopsis thaliana cDNA homologue to animal poly(ADP-ribose) polymerase, *FEBS Lett.*, 364, 103, 1995.
35. Babiychuk, E., Cottrill, P., Storozhenco, S., Fuangthong, M., O'Farrell, M., Van Montagu, M., Inzé, D., and Kushnir, S., Higher plants possess two poly(ADP-ribose) polymerases, *Plant J.*, 15, 635–645, 1998.
36. Giner, H., Simonin, F., de Murcia, G., and Ménissier-de Murcia, J., Overproduction and large-scale purification of the human poly(ADP-ribose) polymerase using a baculovirus expression system, *Gene*, 114, 279, 1992.
37. Jung, S., Alves Miranda, E., Ménissier-de Murcia, J., Niedergang, C., Delarue, M., Schulz, G., and de Murcia, G., Crystallization and X-ray crystallographic analysis of recombinant chicken poly(ADP-ribose) polymerase catalytic domain produced in Sf9 insect cells, *J. Mol. Biol.*, 244, 114, 1994.
38. Ruf, A., Ménissier-de Murcia, J., de Murcia, G., and Schulz, G. E., Structure of the catalytic fragment of poly(ADP-ribose)polymerase from chicken, *Proc. Natl. Acad. Sci. U.S.A.*, 93, 7481, 1996.
39. Ruf, A., de Murcia, G., and Schulz, G. E., Inhibitor and NAD^+ binding to poly(ADP-ribose) polymerase as derived from crystal structures and homology modelling, *Biochemistry*, 37, 3893, 1998.
40. Bell, C. E. and Eisenberg, D., Crystal structure of diphtheria toxin bound to nicotinamide adenine dinucleotide, *Biochemistry*, 35, 1137, 1996.
41. Ruf, A., Rolli, V., de Murcia, G., and Schulz. G. E., The mechanism of the elongation and branching reaction of poly(ADP-ribose) polymerase as derived from crystal structures and mutagenesis, *J. Mol. Biol.*, 278, 57, 1998.
42. Rolli, V., O'Farrell, M., Ménissier-de Murcia, J., and de Murcia, G., Random mutagenesis of the poly(ADP-ribose) polymerase catalytic domain reveals amino acids involved in polymer branching, *Biochemistry*, 36, 12147, 1997.
43. Ding, R., Pommier, Y., Kang, V., and Smulson, M., Depletion of poly(ADP-ribose) polymerase by antisense RNA expression results in a delay in DNA strand breaks rejoining, *J. Biol. Chem.*, 267, 12804, 1992.
44. Ding, R. and Smulson, M., Depletion of nuclear poly(ADP-ribose) polymerase by antisense RNA expression: influences on genomic stability, chromatin organization and carcinogen cytotoxicity, *Cancer Res.*, 54, 4627, 1994.
45. Stevnsner, T., Ding, R., Smulson, M., and Bohr, V. A., Inhibition of gene-specific repair of alkylation damage in cells depleted of poly(ADP-ribose) polymerase, *Nucl. Acids Res.*, 22, 4620, 1994.
46. Chatterjee, S., Petzold, S. J., Berger, S. J., and Berger, N. A., Strategy for selection of cell variants deficient in poly(ADP-ribose) polymerase, *Exp. Cell Res.*, 172, 245, 1987.
47. Chatterjee, S., Hirschler, N. V., Petzold, S. J., Berger, S. J., and Berger, N. A., Mutant cells defective in poly(ADP-ribose) synthesis due to stable alteration in enzyme activity or substrate availability, *Exp. Cell Res.*, 184, 1, 1989.

48. Yoshihara, K., Itaya, A., Hironaka, T., Sakuramoto, S., Tanaka, Y., Tsuyuki, M., Inada, Y., Kamiya, T., Ohnishi, K., Honma, M., Kataoka, E., Mizusawa, H., Uchida, M., Uchida, K., and Miwa, M., Poly(ADP-ribose) polymerase-defective mutant cell clone of mouse L1210 cells, *Exp. Cell Res.*, 200, 126, 1992.
49. Witmer, M., Aboul-Ela, N., Jacobson, M., and Stamato, T., Increased sensitivity to DNA-alkylating agents in CHO mutants with decreased poly(ADP-ribose) polymerase activity, *Mut. Res.*, 314, 249, 1994.
50. Molinete, M., Vermeulen, W., Burkle, A., Ménissier-de Murcia, J., Küpper, J. H., Hoeijmakers, J. H., and de Murcia, G., Overproduction of the poly(ADP-ribose) polymerase DNA-binding domain blocks alkylation-induced DNA repair synthesis in mammalian cells, *EMBO J.*, 12, 2109, 1993.
51. Küpper, J. H., Müller, M., Jacobson, M., Tatsumi-Miyajima, J., Coyle, D., Jacobson, E., and Bürkle, A., *trans*-Dominant inhibition of poly(ADP-ribosylation) sensitizes cells against gamma-irradiation and *N*-methyl-*N'*-nitro-*N*-nitrosoguanidine but does not limit DNA replication of a polyomavirus replicon, *Mol. Cell. Biol.*, 15, 3154, 1995.
52. Küpper, J. H., Müller, M., and Bürkle, A., *trans*-Dominant inhibition of poly ADP-ribosylation potentiates carcinogen-induced gene amplification in SV40-transformed Chinese hamster cells, *Cancer Res.*, 56, 2715, 1996.
53. Schreiber, V., Hunting, D., Trucco, C., Gowans, B., Grunwald, D., de Murcia, G., and Ménissier-de Murcia, J., A dominant-negative mutant of human poly(ADP-ribose) polymerase affects cell recovery and chromosome stability following DNA damage, *Proc. Natl. Acad. Sci. U.S.A.*, 11, 4753, 1995.
54. Shall, S., ADP-ribose in DNA repair: a new component of DNA excision repair, *Adv. Radiat. Biol.*, 11, 1, 1984.
55. Wang, Z. Q., Auer, B., Stingl, L., Berghammer, H., Haidacher, D., Schweiger, M., and Wagner, E. W., Mice lacking ADPRT and poly(ADP-ribosyl)ation develop normally but are susceptible to skin disease, *Genes Dev.*, 9, 509, 1995.
56. Ménissier-de Murcia, J., Niedergang, C., Trucco, C., Ricoul, M., Dutrillaux, B., Mark, M., Olivier, F. J., Masson, M., Dierich, A., LeMeur, M., Walztinger, C., Chambon, P., and de Murcia, G., Requirement of poly(ADP-ribose) polymerase in recovery from DNA damage in mice and in cells, *Proc. Natl. Acad. Sci. U.S.A.*, 94, 7303, 1997.
57. Masutani, M., Nozaki, T., Nishiyama, E., Shimokawa, T., Tachi, Y., Susuki, H., Nakagama, H., Wakabayashi, K., and Sugimura, T., Function of poly(ADP-ribose) polymerase in response to DNA damage: gene-disruption study in mice, *Mol. Cell. Biochem.*, 193, 149, 1999.
58. Wang, Z. Q., Stingl, L., Morrison, C., Jantsch, M., Los, M., Schulze-Osthoff, K., and Wagner, E. W., PARP is important for genomic stability but dispensable in apoptosis, *Genes Dev.*, 11, 2347, 1997.
59. Barlow, C., Hirotsune, S., Paylor, R., Liyanage, M., Eckhaus, M., Collins, F., Shilo, Y., Crawley, J. N., Ried, T., Tagle, D., and Wynshaw-Boris, A., *Atm*-deficient mice: a paradigm of *Ataxia telangiectasia, Cell*, 86, 159, 1996.
60. Biedermann, K. A., Sun, J., Giaccia, A. J., Tosto, L. M., and Brown, J. M., *Scid* mutation in mice confers hypersensitivity to ionizing radiation and a deficiency in DNA double-strand break repair, *Proc. Natl. Acad. Sci. U.S.A.*, 88, 1394, 1991.
61. Dantzer, F., Ménissier-de Murcia, J., Barlow, C., Wynshaw-Boris, A., and de Murcia, G., Poly(ADP-ribose) polymerase activity is not affected in *Ataxia telangiectasia* cells and knockout mice, *Carcinogenesis*, 20, 177, 1999.

62. Trucco, C., Rolli, V., Oliver, F. J., Flatter, E., Masson, M., Dantzer, F., Niedergang, C., Dutrillaux, B., Ménissier-de Murcia, J., and de Murcia, G., A dual approach in the study of poly(ADP-ribose) polymerase: *in vitro* random mutagenesis and generation of deficient mice, *Mol. Cell. Biochem.*, 193, 53, 1999.
63. Trucco, C., Oliver, F. J., de Murcia, G., and Ménissier-de Murcia, J., DNA repair defect in poly(ADP-ribose) polymerase-deficient cell lines, *Nucl. Acids Res.*, 26, 2644, 1998.
64. Oikawa, A., Tohda, H., Kanai, M., Miwa, M., and Sugimura, T., Inhibitors of poly(ADP-ribose) polymerase induce sister chromatid exchanges, *Biochem. Biophys. Res. Commun.*, 97, 1311, 1980.
65. Durkacz, B. W., Irwin, J., and Shall, S., The effect of inhibition of (ADP-ribose)n biosynthesis on DNA repair assayed by the nucleoid technique, *Eur. J. Biochem.*, 121, 65, 1981.
66. Cleaver, J. E., Bodell, W. J., Morgan, W. F., and Zelle, B., Differences in the regulation by poly(ADP-ribose) of repair of DNA damage from alkylating agents and ultraviolet light according to cell type, *J. Biol. Chem.*, 258, 9059, 1983.
67. Olive, P. L., Banath, J. P., and Durand, R. E., Heterogeneity in radiation-induced DNA damage and repair in tumor and normal cells measured using the comet assay, *Radiat. Res.*, 122, 86, 1990.
68. Satoh, M. S. and Lindahl. T., Role of poly(ADP-ribose) formation in DNA repair, *Nature*, 356, 356, 1992.
69. Satoh, M. S., Poirier, G. G., and Lindahl, T., Dual function for poly(ADP-ribose) synthesis in response to DNA strand breakage, *Biochemistry*, 33, 7099, 1994.
70. Frosina, G., Fortini, P., Rossi, O., Carrozzino, F., Raspaglio, G., Cox, L. S., Lane, D. P., Abbondandolo, A., and Dogliotti, E., Two pathways for base excision repair in mammalian cells, *J. Biol. Chem.*, 271, 9573, 1996.
71. Thompson, L. H., Brookman, K. W., Dillehay, L. E., Carrano, A. V., Mazrimas, J. A., Mooney, C. L., and Minkler, J. L., A CHO-cell strain having hypersensitivity to mutagens, a defect in DNA strand-break repair, and an extraordinary baseline frequency of sister-chromatid exchange, *Mutat. Res.*, 95, 427, 1982.
72. Cappelli, E., Taylor, R., Cevasco, M., Abbondandolo, A., Caldecott, K., and Frosina, G., Involvement of XRCC1 and DNA ligase III gene products in DNA base excision, *J. Biol. Chem.*, 272, 23970, 1997.
73. Zhang, X., Moréra, S., Bates, P., Whitehead, P., Coffer, A., Hainbucher, K., Nash, R., Sternberg, M., Lindahl, T., and Freemont, P., Structure of an XRCC1 BRCT domain: a new protein–protein interaction module, *EMBO J.*, 17, 6404, 1998.
74. Thompson, L. H., Brookman, K. W., Jones, N. J., Allen, S. A., and Carrano, A. V., Molecular cloning of the human XRCC1 gene, which corrects defective DNA strand break repair and sister chromatid exchange, *Mol. Cell. Biol.*, 10, 6160, 1990.
75. Kubota, Y., Nash, R. A., Klungland, A., Schär, P., Barnes, D. E., and Lindahl, T., Reconstitution of DNA base excision-repair with purified human proteins: interaction between DNA polymerase beta and the XRCC1 protein, *EMBO J.*, 15, 6662, 1996.
76. Caldecott, K., McKeown, C. K., Tucker, J. D., Ljungquist, S., and Thompson, L. H., An interaction between the mammalian DNA repair protein XRCC1 and DNA ligase III, *Mol. Cell. Biol.*, 14, 68, 1994.
77. Mackey, Z. B., Ramos, W., Levin, D. S., Walter, C. A., McCarrey, J. R., and Tomkinson, A., An alternative splicing event which occurs in mouse pachytene spermatocytes generates a form of DNA ligase III with distinct biochemical properties that may function in meiotic recombination, *Mol. Cell. Biol.*, 17, 989, 1997.

78. Hilt, W. and Wolf, D., Proteasomes: destruction as a program, *Trends Biol. Sci.*, 21, 96, 1996.
79. Delic, J., Morange, M., and Magdelenat, H., Ubiquitin pathway involvement in human lymphocyte gamma-irradiation-induced apoptosis, *Mol. Cell. Biol.*, 13, 4875, 1993.
80. Xu, C., Meikrantz, W., Schlegel, R., and Sager, R., The human papilloma virus 16E6 gene sensitizes human mammary epithelial cells to apoptosis induced by DNA damage, *Proc. Natl. Acad. Sci. U.S.A.*, 92, 7829, 1995.
81. Grimm, L., Goldberg, A., Poirier, G. G., Schwartz, L., and Osborne, B., Proteasomes play an essential role in thymocyte apoptosis, *EMBO J.*, 15, 3835, 1996.
82. Sadoul, R., Fernandez, P. A., Quiquerez, A. L., Martinou, I., Maki, M., Schröter, M., Becherer, D., Irmler, M., Tschopp, J., and Martinou, J. C., Involvement of the proteasome in the programmed cell death of NGF-deprived sympathetic neurons, *EMBO J.*, 15, 3845, 1996.
83. Ullrich, O., Reinheckel, T., Sitte, N., Hass, R., Grune, T., and Davies, K., Poly(ADP-ribose) polymerase activates nuclear proteasome to degrade oxidatively damaged histones, *Proc. Natl. Acad. Sci. U.S.A.*, 96, 6223, 1999.
84. Murray, A., Cyclin unbiquitination: the destructive end of mitosis, *Cell*, 81, 149, 1995.
85. Kovalenko, O., Plug, A., Haaf, T., Gonda, D., Ashley, T., Ward, D., Radding, C., and Golub, E., Mammalian ubiquitin-conjugating enzyme Ubc9 interacts with Rad51 recombination protein and localizes in synaptonemal complexes, *Proc. Natl. Acad. Sci. U.S.A.*, 93, 2958, 1996.
86. Concha, I. I., Figueroa, J., Concha, M. I., Ueda, K., and Burzio, L. O., Intracellular distribution of poly(ADP-ribose) synthetase in rat spermatogenic cells, *Exp. Cell Res.*, 180, 353, 1989.
87. Walter, C. A., Trolian, D., McFarland, M., Street, K., Gurram, G., and McCarrey, J., Xrcc-1 expression during male meiosis in the mouse, *Biol. Reprod.*, 55, 630, 1996.
88. Plug, A. W., Clairmont, C. A., Sapi, E., Ashley, T., and Sweasy, J., Evidence for a role for DNA polymerase β in mammalian meiosis, *Proc. Natl. Acad. Sci. U.S.A.*, 94, 1327, 1997.
89. Simbulan, C. M. G., Suzuki, M., Izuta, S., Sakurai, T., Savoysky, E., Kojima, K., Miyahara, Y., and Yoshida, S., Poly(ADP-ribose) polymerase stimulates DNA polymerase α by physical association, *J. Biol. Chem.*, 268, 93, 1993.
90. Simbulan, C. M. G., Rosenthal, D. S, Hilz, H., Hickey, R., Malkas, L., Applegren, N., Wu, Y., Bers, G., and Smulson, M., The expression of poly(ADP-ribose) polymerase during differentiation-linked DNA replication reveals that it is a component of the multiprotein DNA replication complex, *Biochemistry*, 35, 11622, 1996.
91. Bauer, P., Buki, K., Hakam, A., and Kun, E., Macromolecular association of ADP-ribosyltransferase and its correlation with enzymic activity, *Biochem. J.*, 270, 17, 1990.
92. Mendoza-Alvarez, H. and Alvarez-Gonzalez, R., Poly(ADP-ribose) polymerase is a catalytic dimer and the automodification reaction is intermolecular, *J. Biol. Chem.*, 268, 22575, 1993.
93. Ashkenazi, A. and Dixit, VM., Death receptors: signaling and modulation, *Science*, 281, 1305–1308, 1998.
94. Casciola-Rosen, L., Nicholson, D. W., Chong, T., Rowan, K. R., Thornberry, N. A., Miller, D. K., and Rosen, A., Apopain/CPP32 cleaves proteins that are essential for cellular repair: a fundamental principle of apoptotic death, *J. Exp. Med.*, 183, 1957, 1996.
95. Evan, G. and Littlewood, T. A., matter of life and cell death, *Science*, 281, 1317, 1998.

96. Green, D. R. and Reed, J. C., Mitochondria and apoptosis, *Science,* 281, 1309, 1998.
97. Stroh, C. and Schulze-Osthoff, K., Death by a thousand cuts: an ever increasing list of caspase substrates, *Cell Death Differ.,* 5, 997, 1998.
98. Thornberry, N. A. and Lazebnik, Y., Caspases: enemies within, *Science,* 281, 1312, 1998.
99. Kaufmann, S. H., Desnoyers, S., Ottaviano, Y., Davidson, N. E., and Poirier, G. G., Specific proteolytic cleavage of poly(ADP-ribose) polymerase: an early marker of chemotherapy-induced apoptosis, *Cancer Res.,* 53, 3976, 1993.
100. Duriez, P. J. and Shah, G. M., Cleavage of poly(ADP-ribose) polymerase: a sensitive parameter to study cell death, *Biochem. Cell. Biol.,* 75, 337, 1997.
101. Oliver, F. J., de la Rubia, G., Rolli, V., Ruiz-Ruiz, M. C., de Murcia, G., and Ménissier-de Murcia, J., Role of PARP and PARP cleavage in apoptosis: lesson from an uncleavable mutant, *J. Biol. Chem.* 273, 33533, 1999.
102. Smulson, M. E., Pang, D., Jung, M., Dimtchev, A., Chasovskikh, S., Spoonde, A., and Simbulan-Rosenthal, C., Irreversible binding of poly(ADP)ribose polymerase cleavage product to DNA ends revealed by atomic force microscopy: possible role in apoptosis, *Cancer Res.,* 58, 495, 1998.
103. Herceg, Z. and Wang, Z-Q., Failure of poly(ADP-ribose) polymerase cleavage by caspases leads to induction of necrosis and enhanced apoptosis, *Mol. Cell. Biol.,* 19, 5124, 1999.
104. Leist, M., Single, B., Castoldi, A. F., Kuhnle, S., and Nicotera, P., Intracellular adenosine triphosphate (ATP) concentration: a switch in the decision between apoptosis and necrosis, *J. Exp. Med.,* 185, 1481, 1997.
105. Sakahira, H., Enari, M., and Nagata, S., Cleavage of CAD inhibitor in CAD activation and DNA degradation during apoptosis, *Nature,* 391:96, 1998.
106. Rao, L., Perez, D., and White, E., Lamin proteolysis facilitates nuclear events during apoptosis, *J. Cell Biol.,* 135, 1441, 1996.
107. Rudel, T. B., Membrane and morphological changes in apoptotic cells regulated by caspase-mediated activation of PAK2, *Science,* 276, 1571, 1997.
108. Simbulan-Rosenthal, C. M., Rosenthal, D. S., Iyer, S., Boulares, A. H., and Smulson, M. E., Transient poly(ADP-ribosyl)ation of nuclear proteins and role of poly(ADP-ribose) polymerase in the early stages of apoptosis, *J. Biol. Chem.,* 273, 13703, 1998.
109. Hirsch, T., Marchetti, P., Susin, S. A., Dallaporta, B., Zamzami, N., Marzo, I., and Geuskens, M., The apoptosis-necrosis paradox. Apoptogenic proteases activated after mitochondrial permeability transition determine the mode of cell death, *Oncogene,* 15, 1573, 1997.
110. Lemaire, C., Andreau, K., Souvannavong, V., and Adam, A., Inhibition of caspase activity induces a switch from apoptosis to necrosis, *FEBS Lett.,* 425, 266, 1998
111. Leist, M. S. B., Künstle, G., Volbracht, C., Hentze, H., and Nicotera, P., Apoptosis in the absence of poly(ADP-ribose) polymerase, *Biochem. Biophys. Res. Commun.,* 233, 518, 1997.
112. Guillouf, C., Wang, T. S., Liu, J., Walsh, C. E., Poirier, G. G., Moustacchi, E., and Rosselli, F., Fanconi anemia C protein acts at a switch between apoptosis and necrosis in mitomycin C-induced cell death, *Exp. Cell. Res.,* 246, 384, 1999.
113. Eliasson, M. J., Sampei, K., Mandir, A. S., Hurn, P. D., Traystman, R. J., Bao, J., Pieper, A., Wang, Z. Q., Dawson, T. M., Snyder, S. H., and Dawson, V. L., Poly(ADP-ribose) polymerase gene disruption renders mice resistant to cerebral ischemia, *Nat. Med.,* 3, 1089–1095, 1997.
114. Masutani, M., Suzuki, H., Kamada, N., Watanabe, M., Ueda, O., Nozaki, T., Jishage, K., and Sugimura, T., Poly(ADP-ribose) polymerase gene disruption conferred mice resistant to streptozotocin-induced diabetes, *Proc. Natl. Acad. Sci. U.S.A.,* 96, 2301, 1999.

115. Pieper, A. A., Brat, D. J., Krug, D. K., Watkins, C. C., Gupta, A., Blackshaw, S., Verma, A., Wang, Z. Q., and Snyder, S. H., Poly(ADP-ribose) polymerase-deficient mice are protected from streptozotocin-induced diabetes, *Proc. Natl. Acad. Sci. U.S.A.*, 96, 3059–3064, 1999.
116. Oliver, F. J., Ménissier-de Murcia, J., Nacci, C., Decker, P., Andriantsitohaina, R., Muller, S., de la Rubia, G., Stoclet, J. C., and de Murcia, G., Resistance to endotoxic shock as a consequence of defective NF-κB activation in poly(ADP-ribose) polymerase-1 deficient mice, *EMBO J.*, 18, 4446–4454, 1999.
117. Szabo, C. and Dawson, V. L., Role of poly(ADP-ribose) synthetase in inflammation and ischaemia–reperfusion, *Trends Pharmacol. Sci.*, 19, 287–298, 1998.
118. Pieper, A. A., Verma, A., Zhang, J., and Snyder, S. H., Poly(ADP-ribose) polymerase, nitric oxide and cell death, *Trends Pharmacol. Sci.*, 20, 171, 1999.
119. Smith, S., Giriat, I., Schmitt, A., and de Lange, T., Tankyrase, a poly(ADP-ribose) polymerase at human telomeres, *Science*, 282, 1484–1487, 1998.
120. Ame, J. C., Rolli, V., Schreiber, V., Niedergang, C., Apiou, F., Decker, P., Muller, S., Hoger, T., Ménissier-de Murcia, J. M., and de Murcia G., PARP-2, a novel mammalian DNA damage-dependent poly(ADP-ribose) polymerase, *J. Biol. Chem.*, 274, 17860, 1999.
121. Jean, L., Risler, J. L, Nagase, T., Coulouarn, C., Nomura, N., and Salier, J. P., The nuclear protein PH5P of the inter-alpha-inhibitor superfamily: a missing link between poly(ADP-ribose)polymerase and the inter-alpha-inhibitor family and a novel actor of DNA repair? *FEBS Lett.*, 446, 6, 1999.
122. Johannsson, M. A., Human poly(ADP-ribose) polymerase gene family (ADPRTL): cDNA cloning of two novel poly(ADP-ribose) polymerase homologues, *Genomics*, 57, 442, 1999.
123. Malanga, M., Pleschke, J. M., Kleczkowska, H. E., and Althaus, F. R., Poly(ADP-ribose) binds to specific domains of p53 and alters its DNA binding functions, *J. Biol. Chem.*, 273, 11839, 1998.
124. Kannan, P., Yu, Y., Wankhade, S., and Tainsky, M. A., Poly(ADP-ribose) polymerase is a coactivator for AP-2-mediated transcriptional activation, *Nucl. Acids Res.*, 27, 866, 1999.
125. Plaza, S., Aumercier, M., Bailly, M., Dozier, C., and Saule, S., Involvement of poly(ADP-ribose)-polymerase in the Pax-6 gene regulation in neuroretina, *Oncogene*, 18, 1041, 1999.
126. Hassa, P. O. and Hottiger, M. O., A role of poly(ADP-ribose) polymerase in NF-κB transcriptional activation, *Biol. Chem.*, 380, 953, 1999.
127. Dear, T. N., Hainzl, T., Follo, M., Nehls, M., Wilmore, H., Matena, K., and Boehm, T., Identification of interaction partners for the basic-helix-loop-helix protein E47, *Oncogene*, 14, 891, 1997.
128. Rawling, J. M. and Alvarez-Gonzalez, R., TFIIF a basal transcription factor, that is a substrate for poly(ADP-ribosylation), *Biochem. J.*, 324, 249–253. 1997.
129. Lamarre, D., Talbot, B., de Murcia, G., Laplante, C., Leduc, Y., Mazen, A., and Poirier, G., Structural and functional analysis of poly(ADP-ribose)polymerase: an immunological study, *Biochim. Biophys. Acta*, 950, 147, 1988.

10 Role of Poly(ADP-Ribose) Polymerase in Apoptosis

Marc Germain, A. Ivana Scovassi, and Guy G. Poirier

CONTENTS

10.1 Introduction 209
10.2 General Features of Apoptosis 210
10.3 PARP Activation during Apoptosis 212
       10.3.1 Poly(ADP-Ribose) Synthesis during Apoptosis 212
       10.3.2 DNA Damage–Induced Apoptosis and the p53 Network 213
       10.3.3 Role of PARP Activation during Apoptosis 214
10.4 PARP Cleavage during Apoptosis 215
       10.4.1 Caspase-Mediated Cleavage of PARP 215
       10.4.2 Role of PARP Cleavage during Apoptosis 216
10.5 Conclusions 218
Abbreviations 219
References 220

10.1 INTRODUCTION

Poly(ADP-ribose) polymerase (PARP) is a DNA break sensor enzyme synthesizing a homopolymer of ADP-ribose in response to DNA damage. The enzyme possesses three functional domains. The 43-kDa N-terminal fragment contains the two zinc fingers responsible for the DNA-binding function of the enzyme and a bipartite nuclear localization signal. The central 16-kDa fragment contains the automodification domain as well as a putative leucine zipper, and the catalytic domain is included in the 55-kDa C-terminal portion of the protein.[1,2]

The poly(ADP-ribosyl)ation of PARP and other proteins is likely to modulate their functions during the response to DNA damage. PARP has been involved in the maintenance of genomic integrity and the promotion of cell survival following DNA damage. PARP also has been shown to mediate necrotic cell death in response to excessive DNA damage in many pathological models (see other chapters). On the other hand, PARP is cleaved and inactivated during almost all forms of apoptosis, suggesting an important role for its cleavage. This chapter summarizes the current knowledge on the implication of PARP in apoptosis. Its activation during this cell death process, as well as the potential roles for its cleavage by caspases, will be discussed.

10.2 GENERAL FEATURES OF APOPTOSIS

Apoptosis is a physiological mechanism of cell death controlling the development and homeostasis of multicellular organisms.[3] This cell death process is implicated in organogenesis, the selection of lymphocytes, and the elimination of undesirable cells.[4] A lack of apoptotic cell death is associated with cancer, while its inappropriate occurrence is associated with pathologies like autoimmune and neurodegenerative diseases.[3,5] Apoptotic inducers include DNA damage, chemical agents such as inhibitors of protein or nucleic acid synthesis, deprivation of growth factors, and activation of death receptors.[3,6,7]

The apoptotic sentence follows a precise destruction plan. In particular, a set of structural proteins and enzymes implicated in cellular homeostasis is specifically cleaved by a family of cysteine proteases named caspases. This allows DNA condensation, its degradation into nucleosomal-sized fragments (DNA ladder), and the disruption of the cytoskeleton. Cells are then disassembled into small vesicles and marked for phagocytosis by other cells nearby, preventing the release of the cellular content in the extracellular medium.[3,8] In this process, the death of one cell does not alter surrounding cells. In contrast, necrosis is an uncontrolled mode of cell death characterized by an early disruption of membrane integrity and leakage of the cellular content into the extracellular medium, which is likely to affect surrounding cells. Apoptosis is suggested to prevent some pathologies linked to necrosis, like ischemia and chronic inflammation.[3,8]

Apoptosis is an active mode of cell death requiring energy to proceed. Indeed, cells that are triggered to undergo apoptosis but do not have sufficient energy to perform all the steps of the pathway die by necrosis.[9-12] In particular, the activation of caspase-9 requires the presence of its cofactors APAF-1, cytochrome c, and ATP or dATP.[13,14] Active transport to the nucleus is also likely to be required during the apoptotic process.[15] In addition, the chromatin condensation and the formation of apoptotic bodies are blocked in the absence of ATP.[16]

The apoptotic process is mediated by the caspase family of cysteine protease.[17-19] Caspases are implicated both in the induction and execution of the death sentence. All caspases cleave after an aspartic acid for which they have absolute specificity. Caspases are synthesized as inactive proenzymes, each of which is processed to a large and a small subunit to form the active enzyme. The processing sites within procaspases match the recognition sequences of caspases, indicating that caspases are activated in a proteolytic cascade, initiator caspases processing and activating execution caspases (Figure 10.1). Apoptosis can be induced by activation of death receptors, like Fas, which recruit procaspase-8 via adaptor proteins and promote its autocatalytic activation.[7] On the other hand, DNA-damaging agents and other chemotherapeutic drugs induce cytochrome c release from mitochondria and the activation of caspase-9.[13,20,21] All pathways result in the activation of the effector caspases, caspase-3, -6, and -7, responsible for the proteolytic degradation of most of the proteins cleaved during apoptosis (Table 10.1).[17-19,22]

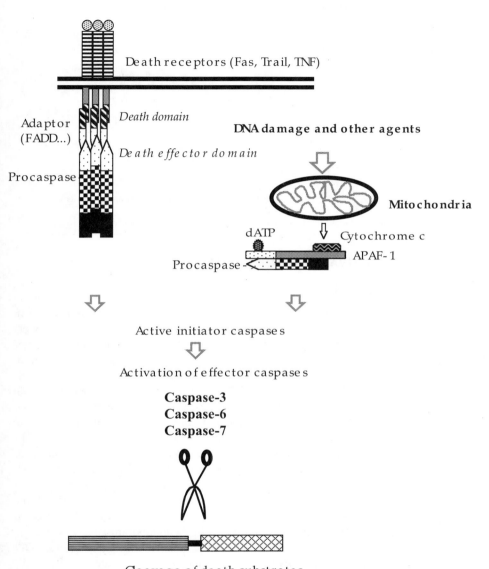

FIGURE 10.1 Activation of the caspase proteases during apoptosis. Caspases are implicated both in the induction and the execution of the apoptotic process. Following the apoptotic stimuli, initiator caspases (caspase-8 or -9) are activated by autocatalysis. The initiator caspases then activate the effector caspases (caspase-3, -6, and -7), which are responsible for most of the protein cleavage during apoptosis.

TABLE 10.1
Caspase-3 and -7 Substrates

Substrate Category	Examples[a]	Potential Role
Cytoskeleton	Fodrin, gelsolin	Apoptotic morphology
Cell adhesion	β-Catenin, plakoglobin	
Cell cycle	Rb, p21	
Replication	RFC 140, MCM3	Disruption of processes involved in cell homeostasis
Transcription	U1-70 kDa sRNP, SREBPs	
DNA repair	PARP, DNA-PK, RAD51	
Signal transduction	PKCδ, ras GAP	Modulation of apoptotic signals

[a] See Stroh and Schulze-Osthoff[22] for an exhaustive list.

10.3 PARP ACTIVATION DURING APOPTOSIS

10.3.1 POLY(ADP-RIBOSE) SYNTHESIS DURING APOPTOSIS

Since PARP is responsible for the synthesis of poly(ADP-ribose) from NAD in response to DNA strand breaks, its activation has been associated with the NAD consumption that has been observed following DNA damage. The drop in cellular NAD occurring during apoptosis also has been associated with PARP activation, since it is partially suppressed by PARP inhibitors like 3-aminobenzamide (3-AB).[23-26] More direct evidence for PARP activation during apoptosis has been reported from studies measuring poly(ADP-ribose) levels. By using permeabilized cells, it was shown that PARP is activated during the course of the apoptotic process.[27-29] Yoon et al.[29] also found, by the same technique, that poly(ADP-ribosyl)ation of histone H1 correlated with the internucleosomal degradation of DNA. More recently, using an antibody against poly(ADP-ribose), it has been demonstrated that apoptotic HeLa and HL60 cells synthesize poly(ADP-ribose).[30,31] Moreover, Smulson and colleagues showed that PARP is activated early in the apoptotic process of osteosarcoma cells grown to confluence,[32] fibroblasts treated by anti-Fas, and HL-60 treated with camptothecin.[33] Our laboratory has also shown, using a new immunological technique to measure endogenous levels of poly(ADP-ribose), that PARP is activated by the occurrence of the DNA ladder in HL60 cells treated with the topoisomerase II inhibitor etoposide, but not before.[34] Many studies have shown a close temporal correlation between the activation of PARP during apoptosis and the occurrence of the DNA ladder.[24,25,27,29,35] On the other hand, as suggested by Nosseri et al.,[24] PARP activation could also occur before the onset of apoptosis induced by DNA-damaging agents. Since PARP is activated by DNA strand breaks, it has been proposed to act as a DNA nick sensor to signal DNA damage and promote subsequent cellular responses like DNA repair and apoptosis. Indeed, some evidence suggests a link between PARP and the p53 pathway of cell arrest and apoptosis.

10.3.2 DNA Damage–Induced Apoptosis and the p53 Network

The p53 protein is normally unstable and its cellular levels are very low. However, following a stress, such as DNA damage, arrest of DNA or RNA synthesis, or nucleotide depletion, the p53 protein is stabilized and transactivated.[36] Several proteins could relay the stress signals to p53. Among them, ATM (*Ataxia telangiectasia* mutated gene product)[37] and DNA-PK (DNA-dependent protein kinase)[38] have been shown to induce p53 activation following different types of DNA damage. In addition, ATM directly associates with p53,[39] and both ATM[39] and DNA-PK[40] can phosphorylate p53. To investigate the possible interaction between PARP and ATM, PARP activity following DNA damage has been analyzed in ATM-deficient cells and mice.[41] The unaltered response of PARP to DNA damage supports the hypothesis that PARP and ATM regulate distinct pathways.

On the other hand, *in vitro* experiments showed that the catalytic activity of DNA-PK toward p53 is stimulated upon binding of its catalytic subunit (DNA-PK$_{cs}$) to PARP and its subsequent poly(ADP-ribosyl)ation.[40] While PARP$^{-/-}$ and SCID mice do not develop lymphomas significantly, the double-mutant mice develop a high frequency of short-latency T-cell lymphoma,[42] suggesting a cooperation between PARP and DNA-PK in promoting the maintenance of genomic integrity. Direct interactions between PARP and p53 have also been reported, either by protein–protein interactions or by poly(ADP-ribosyl)ation. Immunoprecipitation studies have shown that PARP and p53 could associate physically.[43-45] p53 was also shown to be a poly(ADP-ribose) acceptor protein both *in vitro* and *in vivo*[43,44,46] and to possess three conserved poly(ADP-ribose)-binding motifs that could bind free poly(ADP-ribose).[46] Interactions between PARP and p53 could therefore affect the stabilization and transactivation of p53.

A study by Whitacre et al.[47] showed that V79-derived cell lines defective in poly(ADP-ribose) synthesis had lower basal p53 levels than the parental V79 cell line and were unable to activate p53 in response to etoposide. Recent reports on p53 status in PARP$^{-/-}$ cells lend further support to the role of PARP in stabilizing wild-type p53.[48-50] However, this defect in p53 induction was not accompanied by defects in its related functions, such as apoptosis, G_1 arrest, and mitotic spindle checkpoint,[48] even though some defects in p53 transactivation have been reported in PARP$^{-/-}$ cells.[49] Interestingly, an alternatively spliced form of p53, which differs functionally from the regularly spliced form by its constitutive activation in sequence-specific DNA binding, was not affected by the absence of PARP.[49] However, a study by the group of de Murcia[51] has shown that their PARP$^{-/-}$ cells had a faster induction of p53 and subsequent apoptosis in response to the alkylating agent MNU (*N*-methyl-*N*-nitrosourea).[51] In this case, the rapid induction of p53 could be linked to the acute sensitivity of these cells to alkylating agents. A prolongation of the p53 response to DNA damage has also been reported following the depletion of PARP by antisense RNA.[52] Although PARP and p53 are likely to interact *in vivo*, the consequences of these interactions remain unclear. In fact, as PARP$^{-/-}$ mice do not develop spontaneous tumors, PARP is probably dispensable for most of p53 functions.

TABLE 10.2
Effect of the Absence of PARP on the Apoptotic Process

Apoptosis-Inducing Agent	Effect of the Absence of PARP	Ref.
Programmed cell death during development	Normal, mice are healthy and fertile	51, 59, 60
Agents activating PARP		
γ irradiation	No difference	64
Peroxynitrite	No difference at low doses	63, 65
	Causes necrosis in PARP$^{+/+}$ and apoptosis in PARP$^{-/-}$ at high doses	
MNU, MMS	Acceleration of apoptosis in PARP$^{-/-}$ cells	51, 62
Other agents		
Anti-Fas, TNF-α	No difference	62–64, 101
Anti-Fas	Inhibition of apoptosis in the absence of PARP	33
Dexamethasone	No difference	63–65
Etoposide, ionomycin, ceramides, staurosporine, colchicine, MPP	No difference	63
Uncleavable mutants		
Anti-Fas, MMS	Delay of apoptosis	62
TNF-α	Acceleration of apoptosis and induction of necrosis	101

10.3.3 ROLE OF PARP ACTIVATION DURING APOPTOSIS

Studies of the role of PARP activation in the induction of apoptosis mostly have been done using PARP chemical inhibitors. However, these inhibitors have been shown to cause side effects.[53,54] In addition, new ADP-ribose–polymerizing activities which are suppressed by PARP inhibitors have recently been described.[55-58] These activities could be associated with processes that were thought to be mediated by PARP, based on studies using PARP inhibitors. PARP knockouts could therefore give more accurate information on the role of PARP in cell death and other cellular processes.

Three independent PARP knockout mice have been generated.[51,59,60] All are healthy and fertile, indicating that PARP does not play a critical role during development. Two of the knockouts were used to study the role of PARP in apoptosis (Table 10.2; see Le Rhun et al.[61] for a review). de Murcia and colleagues[51,62] found that the absence of PARP sensitizes cells to apoptosis induced by alkylating agents, but not by the topoisomerase I inhibitor CPT-11 or an anti-Fas antibody. The knockout mice developed by Wang et al.[64] and their wild-type counterparts exhibit similar response to apoptosis,[63,64] even in cells treated with DNA-damaging agents that activate PARP (γ radiation in thymocytes,[63] peroxynitrite in neurons[63] and thymocytes[65]). Nevertheless, Simbulan-Rosenthal et al.[33] recently showed that an anti-Fas antibody failed to induce apoptosis in PARP$^{-/-}$ fibroblasts, or fibroblasts depleted of PARP by antisense RNA. These cells regained complete sensitivity when

PARP was reintroduced. They concluded that PARP activation is required for the activation of effector caspases. However, several other studies on Fas-mediated apoptosis in PARP knockout cells have failed to show any difference in their apoptotic response compared with wild-type cells.[62-65] Furthermore, a role for PARP activation in Fas-mediated cell death is difficult to reconcile with the current knowledge on this apoptotic pathway (see Figure 10.1).[7,66] Since the induction of apoptosis by Fas directly stimulates caspase-8 and the effector caspases, it is difficult to ascribe a role for PARP activation, which requires DNA strand breaks, within this pathway. In fact, PARP activation during Fas-induced apoptosis is likely to be caused by the internucleosomal degradation of the DNA. Furthermore, the mechanism of activation of the recently discovered DNase implicated in the apoptotic process (caspase-activated DNase, or CAD)[67,68] presents an argument against a role for PARP activation at this point in the apoptotic process. In normal cells, CAD forms a complex with its inhibitor ICAD. Upon induction of apoptosis, ICAD is cleaved by caspase-3 and CAD is activated. Thus, DNA degradation and subsequent PARP activation occur at a time when the cell is irreversibly engaged in the death process.

As PARP is a DNA nick sensor, its involvement in apoptosis would be more likely following DNA damage. However, the current data on the implication of PARP in DNA damage–induced apoptosis suggest that, despite its DNA nick sensor function and possible interactions with the p53 network, it does not participate in the induction of the apoptotic death sentence.[63,65] Still, PARP$^{-/-}$ mice and cells are highly sensitive to γ irradiation and alkylating agents,[51,60,62,64] suggesting that PARP is implicated in the normal response to such an insult. This acute sensitivity suggests that PARP could be a survival factor rather than an apoptosis-promoting molecule. Interestingly, PARP is cleaved and inactivated during apoptosis, leading to the inhibition of this survival-promoting function.

10.4 PARP CLEAVAGE DURING APOPTOSIS

10.4.1 CASPASE-MEDIATED CLEAVAGE OF PARP

PARP is one of the first proteins that was reported to be cleaved by caspases during apoptosis.[25,69,70] The cleavage of PARP has been observed in almost all apoptotic models. This cleavage occurs early in the execution phase of apoptosis, concomitant with the internucleosomal degradation of DNA. However, the proteolysis of PARP occurs before the appearance of most of the morphological changes observed during apoptosis.[25] The site of caspase cleavage has been mapped to a conserved sequence DEVD (Asp-Glu-Val-Asp-aldehyde)/G, located in the bipartite nuclear localization signal.[69] Often, caspase cleavage sites are located between the catalytic and the regulatory domains of the target protein, leading to its inactivation (PARP, DNA-PK) or its constitutive activation — protein kinase Cδ (PKCδ), gelsolin.[71] PARP cleavage produces a 89-kDa C-terminal fragment containing the active site and the automodification domain, and a 24-kDa fragment containing the zinc fingers responsible for DNA binding activity and the subsequent activation of the enzyme.[25] The 24-kDa fragment binds irreversibly to DNA breaks[72] while the 89-kDa fragment can no longer be stimulated by DNA strand breaks.[25]

Since caspase-3 and -7, two of the effector caspases, have the highest affinity for the synthetic substrate DEVD-pNA (*p*-nitroaniline),[73,74] which mimics the PARP cleavage site, they have been suggested to be responsible for PARP cleavage *in vivo*. Indeed, we recently evidenced that caspase-7 cleaves PARP more efficiently than caspase-3 *in vitro*. Furthermore, caspase-7, but not caspase-3, cleaves poly(ADP-ribosyl)ated PARP more efficiently than the nonmodified enzyme. As PARP is activated by the DNA ladder at the time of its cleavage, these results suggest that caspase-7 would be the real endogenous PARP-cleaving caspase.[34] Furthermore, PARP is cleaved even in the absence of caspase-3.[34,75,76]

10.4.2 Role of PARP Cleavage during Apoptosis

Upon massive DNA damage, PARP is strongly activated, cellular NAD is depleted, and the cell subsequently dies (see other chapters).[77,78] As all of these events can be prevented by PARP inhibitors, Berger[77] and Okamoto and colleagues[78] have proposed that when overstimulated by DNA damage, PARP would cause a suicide response by consuming cellular NAD, which would lead to an ATP depletion. The role of PARP activation in this form of cell death has been recently reinforced by studies on PARP$^{-/-}$ cells and animals.[79-81] However, these events lead to necrotic cell death and not to apoptosis, as the cells rapidly lose their membrane integrity[82] and do not show internucleosomal DNA degradation[65] or typical PARP cleavage.[83] Furthermore, Virág et al.[65] recently reported that the activation of PARP could affect the mode of cell death. Treatment of cells with high doses of peroxynitrite caused PARP activation and necrotic cell death in PARP$^{+/+}$ cells, but apoptosis in the knockout or in the presence of the PARP inhibitor 3-AB. On the other hand, low doses of the oxidant caused apoptosis in both genotypes.

Consequently, as the apoptotic process generates considerable amounts of DNA strand breaks, activation of PARP could switch the type of cell death to necrosis. The cleavage and inactivation of PARP by caspases would then prevent this energy depletion and subsequent switch to necrosis. The inhibition of full-length PARP activity during apoptosis could also be achieved by its transdominant inhibition by the 24-kDa apoptotic fragment. Studies that have been done with the complete 43-kDa DNA-binding domain (DBD) have revealed that it could, in fact, act as a potent transdominant PARP inhibitor.[84-87] Indeed, by inhibiting the activity of the uncleaved PARP, the 24-kDa fragment could prevent a depletion of energy by PARP early in the apoptotic process. This conservation of energy for the later steps of apoptosis may be critical to complete the apoptotic process successfully, since it requires energy to proceed.[9-12] However, although previous studies have failed to detect NAD depletion in PARP$^{-/-}$ cells, a similar lowering of NAD levels in response to DNA-damaging agents has recently been reported for PARP knockout as well as wild-type cells.[55] This suggests that PARP is not the only factor implicated in the NAD consumption following DNA damage. Thus, although it is likely that PARP overactivation following high amounts of DNA damage kills the cell, the exact relation between this activation and NAD and ATP depletion remains unclear.

Another process that requires significant amounts of energy is DNA repair. Attempts to repair the numerous DNA breaks generated during the apoptotic process could also lead to a rapid depletion of energy. The presence of the 24-kDa fragment on these breaks could prevent an attempt from the cell to repair the damages. Overexpression of the complete 43-kDa DBD in cells has been shown to lead to defects in DNA repair, as measured by reduced unscheduled DNA synthesis.[86] Since it is irreversibly linked to DNA breaks, it is likely that the 24-kDa apoptotic fragment blocks DNA repair. Indeed, recent work from G. Poirier's laboratory has shown that the 24-kDa fragment of PARP can inhibit efficiently both PARP activity and DNA repair *in vitro* (unpublished results).

PARP cleavage during apoptosis could also be viewed as a part of a more general mechanism, together with the cleavage of other proteins involved in the maintenance of cellular homeostasis. To maintain the cell integrity following a stressful situation, the cell is equipped with a set of signaling proteins which ensure an appropriate response, among which PARP could be placed.[36,51,62,88] Since apoptosis results in the complete dismantling of the cell, it is likely that such stress-activated proteins would be stimulated by this process, leading to a survival response. This would interfere with the apoptotic cell death either by retarding the onset of death or by causing necrosis. Both events could have deleterious effects on the organism. One way to prevent such events is to inactivate the proteins that would signal, and thus promote, such attempts to survive. Cleavage of proteins implicated in cellular homeostasis by caspases achieves this goal. For example, a set of protein kinases implicated in different survival responses are cleaved during apoptosis resulting in their inactivation.[89] The dying cell could also attempt to repair its DNA following the activation of the apoptotic DNase. As a rapid apoptotic death is likely to be preferable for the organism, inactivation of enzymes implicated in the signaling of DNA damage is essential. It is probably most efficient to target the early events of repair to shut off these mechanisms rapidly. Proteins other than PARP that are implicated in the response to DNA strand breaks, such as p21,[90,91] Rb,[92,93] DNA-PK,[94,95] and RAD51,[96,97] are also cleaved and inactivated during apoptosis. The cleavage p21 would lead to the inhibition of the G_1 cell cycle arrest and disruption of its interaction with proliferating cellular nuclear antigen (PCNA).[90,91] This may have a direct effect on DNA repair.[90,91] In addition, the catalytic subunit of another DNA nick sensor, DNA-PK (DNA-PK$_{CS}$), is inactivated by the caspases.[98] This cleavage could thus result in the clustering of DNA-PK$_{CS}$–Ku complexes on DNA strand breaks, causing interference with DNA repair,[99] as it has been suggested for PARP. It has also been reported that during apoptosis, DNA-PK is activated before its cleavage,[100] and that its catalytically active form is preferentially cleaved by caspase-3.[98] The characteristics of DNA-PK cleavage thus resemble those we recently described for PARP.[34] Results on the cleavage of RAD[51] also suggest a possible differential cleavage depending on the state of activation of the enzyme.[96,97] Altogether, these results suggest that caspase cleavage can selectively inactivate enzymes responsible for the first steps of the cellular stress response. In addition, targeting of active molecules instead of their inactive forms, which may be less dangerous for the cell, would ensure a fast and efficient programmed cell death.

These different possible roles for PARP cleavage are not mutually exclusive and could coexist, ensuring an efficient apoptotic process. Since PARP-deficient mice do not exhibit apparent alterations in apoptosis, some authors suggested that the apoptotic PARP cleavage is an accidental event.[22,63] However, although knockout mice allow the study of a role for PARP activation during apoptosis, they do not permit the elucidation of a role for its cleavage. Cleavage of apoptotic substrates by caspases ensures the dismantling of proteins implicated in cellular homeostasis. PARP is likely to be cleaved to avoid interference with the apoptotic process. In the case of the knockout mice, PARP is absent and therefore one cannot observe the effects of its inactivation by caspases. Experiments using PARP$^{-/-}$ cells transfected with uncleavable PARP would provide answers on the role for PARP cleavage. A first study by Oliver et al.[62] showed that cells transfected with the PARP mutant (DEVD → DEVA) had delayed apoptosis in response to anti-Fas antibody and, to a lesser extent, to the alkylating agent methyl-methane-thiosulfonate (MMS). Similar effects have also been reported for other uncleavable mutants such as p21,[90] RAD51,[97] or Rb.[92] These results support the hypothesis that these proteins are cleaved and inactivated to prevent interference with the apoptotic pathway.

In contrast, Herceg and Wang have reported an acceleration of cell death induced by TNF-α in cells stably transfected with an uncleavable PARP (DEVD → DEVN).[101] This death was associated with both enhanced apoptosis and the induction of necrosis. Their results suggest that PARP is cleaved by caspases to prevent an energy depletion and subsequent necrotic cell death.

Thus, there is at present evidence for both hypotheses (inactivation of survival factor functions and switch to necrosis). The discrepancy between the two studies could come from the apoptotic inducers used, the cell types, or the transfection technique. However, although the exact role for PARP cleavage remains to be elucidated, it is becoming clear that it is not an accidental event during apoptosis.

10.5 CONCLUSIONS

The generation of PARP knockout mice in recent years has allowed a better understanding of the role of PARP in cell death. PARP is likely to act as a survival-promoting factor following DNA damage. Therefore, this enzyme does not participate in the induction of apoptosis, but is rather inactivated during this process. PARP activation during apoptosis is thus probably only an "accidental" event that needs to be controlled to ensure the rapid execution of the death sentence.

PARP is activated early during the response to DNA damage (Figure 10.2). Low amounts of DNA damage are repaired efficiently by the DNA repair machinery. In addition, the activation of PARP by the DNA breaks would promote cell survival. These events lead to cell recovery from the damages. In the case of large amounts of DNA damage, PARP is overactivated, resulting in the necrotic death of the cell. The induction of apoptosis, either by DNA damage or by another stimulus, results in the activation of the apoptotic DNase and subsequent DNA degradation, which in turn stimulates PARP. The cleavage and inactivation of PARP by caspases would thus prevent both the survival signals and PARP overactivation. In addition, the 24-kDa fragment of PARP could inhibit DNA repair. However, much work remains to be done to understand how the catalytic activity of PARP can interfere with the apoptotic process. The relation between PARP activation, energy depletion, and cell

Role of PARP in Apoptosis

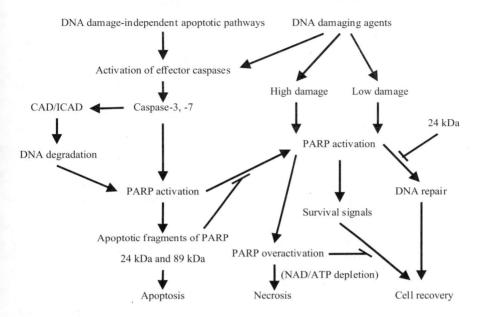

FIGURE 10.2 Schematic model showing the involvement of PARP in the life and death of the cell. PARP is activated following DNA damage. Depending on the severity of the damage, this activation will promote the survival or the death of the cell. The induction of apoptosis results in the internucleosomal degradation of the DNA, which activates PARP. Its concomitant inactivation by caspases results in the inhibition of the survival signals and prevents PARP overactivation. In addition, the 24-kDA apoptotic fragment of PARP binds irreversibly to the damaged DNA and could inhibit DNA repair. These events would ensure a fast and efficient apoptotic cell death.

death should also be studied thoroughly at the molecular level, as it is not clear if PARP is directly or indirectly involved in the depletion of NAD.[55] Much work also remains to be done to understand the molecular basis for the survival-promoting function of PARP. These, and studies using uncleavable mutants of PARP, should permit the elucidation of the role of its cleavage by caspases. Comparison of the phenotypes generated with different uncleavable proteins will also certainly help to understand the biological significance of protein cleavage by caspases.

ABBREVIATIONS

The abbreviations used are 3-AB, 3-aminobenzamide; ATM, *Ataxia telangiectasia* mutated gene product; CAD, caspase-activated DNase; DBD, DNA binding domain; DEVD, Asp-Glu-Val-Asp-aldehyde; DNA-PK, DNA-dependent protein kinase; DNA-PK$_{CS}$, DNA-PK catalytic subunit; ICAD, inhibitor of CAD; MNU, *N*-methyl-*N*-nitrosourea; PARP, poly(ADP-ribose) polymerase; PKCδ, protein kinase C δ; pNA, *p*-nitroaniline; SCID, severe combined immunodeficiency.

REFERENCES

1. Lautier, D., Lagueux, J., Thibodeau, J., Menard, L., and Poirier, G. G., Molecular and biochemical features of poly(ADP-ribose) metabolism, *Mol. Cell. Biochem.*, 122, 171, 1993.
2. D'Amours, D., Desnoyers, S., D'Silva, I., and Poirier, G. G., Poly(ADP-ribosyl)ation reactions in the regulation of nuclear functions, *Biochem. J.*, 342, 249, 1999.
3. Steller, H., Mechanisms and genes of cellular suicide, *Science*, 267, 1445, 1995.
4. Jacobson, M. D., Weil, M., and Raff, M. C., Programmed cell death in animal development, *Cell*, 88, 347, 1997.
5. Leist, M. and Nicotera, P., Apoptosis, excitotoxicity, and neuropathology, *Exp. Cell Res.*, 239, 183, 1998.
6. Green, D. R. and Reed, J. C., Mitochondria and apoptosis, *Science*, 281, 1309, 1998.
7. Ashkenazi, A. and Dixit, V. M., Death receptors: signaling and modulation, *Science*, 281, 1305, 1998.
8. Savill, J., Apoptosis in resolution of inflammation, *J. Leukoc. Biol.*, 61, 375, 1997.
9. Leist, M., Single, B., Castoldi, A. F., Kuhnle, S., and Nicotera, P., Intracellular adenosine triphosphate (ATP) concentration: a switch in the decision between apoptosis and necrosis, *J. Exp. Med.*, 185, 1481, 1997.
10. Leist, M. and Nicotera, P., The shape of cell death, *Biochem. Biophys. Res. Commun.*, 236, 1, 1997.
11. Eguchi, Y., Shimizu, S., and Tsujimoto, Y., Intracellular ATP levels determine cell death fate by apoptosis or necrosis, *Cancer Res.*, 57, 1835, 1997.
12. Tsujimoto, Y., Apoptosis and necrosis: intracellular ATP levels as a determinant for cell death modes, *Cell Death Differ.*, 4, 429, 1997.
13. Li, P., Nijhawan, D., Budihardjo, I., Srinivasula, S. M., Ahmad, M., Alnemri, E. S., and Wang, X., Cytochrome c and dATP-dependent formation of Apaf-1/caspase-9 complex initiates an apoptotic protease cascade, *Cell*, 91, 479, 1997.
14. Eguchi, Y., Srinivasan, A., Tomaselli, K. J., Shimizu, S., and Tsujimoto, Y., ATP-dependent steps in apoptotic signal transduction, *Cancer Res.*, 59, 2174, 1999.
15. Yasuhara, N., Eguchi, Y., Tachibana, T., Imamoto, N., Yoneda, Y., and Tsujimoto, Y., Essential role of active nuclear transport in apoptosis, *Genes Cells*, 2, 55, 1997.
16. Kass, G. E., Eriksson, J. E., Weis, M., Orrenius, S., and Chow, S. C., Chromatin condensation during apoptosis requires ATP, *Biochem. J.*, 318, 749, 1996.
17. Nicholson, D. W. and Thornberry, N. A., Caspases: killer proteases, *Trends Biochem. Sci.*, 22, 299-306, 1997.
18. Salvesen, G. S. and Dixit, V. M., Caspases: intracellular signaling by proteolysis, *Cell*, 91, 443, 1997.
19. Thornberry, N. A. and Lazebnik, Y., Caspases: enemies within, *Science*, 281, 1312, 1998.
20. Kuida, K., Haydar, T. F., Kuan, C. Y., Gu, Y., Taya, C., Karasuyama, H., Su, M. S., Rakic, P., and Flavell, R. A., Reduced apoptosis and cytochrome c-mediated caspase activation in mice lacking caspase 9, *Cell*, 94, 325, 1998.
21. Yoshida, H., Kong, Y. Y., Yoshida, R., Elia, A. J., Hakem, A., Hakem, R., Penninger, J. M., and Mak, T. W., Apaf1 is required for mitochondrial pathways of apoptosis and brain development, *Cell*, 94, 739, 1998.
22. Stroh, C. and Schulze-Osthoff, K., Death by a thousand cuts: an ever increasing list of caspase substrates, *Cell Death Differ.*, 5, 997, 1998.
23. Sugimoto, K., Yamada, K., Egashira, M., Yazaki, Y., Hirai, H., Kikuchi, A., and Oshimi, K., Temporal and spatial distribution of DNA topoisomerase II alters during proliferation, differentiation, and apoptosis in HL-60 cells, *Blood*, 91, 1407–1417, 1998.

24. Nosseri, C., Coppola, S., and Ghibelli, L., Possible involvement of poly(ADP-ribosyl) polymerase in triggering stress-induced apoptosis, *Exp. Cell Res.*, 212, 367, 1994.
25. Kaufmann, S. H., Desnoyers, S., Ottaviano, Y., Davidson, N. E., and Poirier, G. G., Specific proteolytic cleavage of poly(ADP-ribose) polymerase: an early marker of chemotherapy-induced apoptosis, *Cancer Res.*, 53, 3976, 1993.
26. Coppola, S., Nosseri, C., Maresca, V., and Ghibelli, L., Different basal NAD levels determine opposite effects of poly(ADP-ribosyl)polymerase inhibitors on H_2O_2-induced apoptosis, *Exp. Cell Res.*, 221, 462, 1995.
27. Shimizu, T., Kubota, M., Tanizawa, A., Sano, H., Kasai, Y., Hashimoto, H., Akiyama, Y., and Mikawa, H., Inhibition of both etoposide-induced DNA fragmentation and activation of poly(ADP-ribose) synthesis by zinc ion, *Biochem. Biophys. Res. Commun.*, 169, 1172, 1990.
28. Tanizawa, A., Kubota, M., Hashimoto, H., Shimizu, T., Takimoto, T., Kitoh, T., Akiyama, Y., and Mikawa, H., VP-16-induced nucleotide pool changes and poly(ADP-ribose) synthesis: the role of VP-16 in interphase death, *Exp. Cell Res.*, 185, 237, 1989.
29. Yoon, Y. S., Kim, J. W., Kang, K. W., Kim, Y. S., Choi, K. H. and Joe, C. O., Poly(ADP-ribosyl)ation of histone H1 correlates with internucleosomal DNA fragmentation during apoptosis, *J. Biol. Chem.*, 271, 9129, 1996.
30. Negri, C., Donzelli, M., Bernardi, R., Rossi, L., Bürkle, A., and Scovassi, A. I., Multiparametric staining to identify apoptotic human cells, *Exp. Cell Res.*, 234, 174, 1997.
31. Donzelli, M., Negri, C., Mandarino, A., Rossi, L., Prosperi, E., Frouin, I., Bernardi, R., Bürkle, A., and Scovassi, A. I., Poly(ADP-ribose) synthesis: a useful parameter for identifying apoptotic cells, *Histochem. J.*, 29, 831, 1997.
32. Rosenthal, D. S., Ding, R., Simbulan-Rosenthal, C. M., Vaillancourt, J. P., Nicholson, D. W., and Smulson, M., Intact cell evidence for the early synthesis, and subsequent late apopain-mediated suppression, of poly(ADP-ribose) during apoptosis, *Exp. Cell Res.*, 232, 313, 1997.
33. Simbulan-Rosenthal, C. M., Rosenthal, D. S., Iyer, S., Boulares, A. H., and Smulson, M. E., Transient poly(ADP-ribosyl)ation of nuclear proteins and role of poly(ADP-ribose) polymerase in the early stages of apoptosis, *J. Biol. Chem.*, 273, 13703, 1998.
34. Germain, M., Affar, E. B., D'Amours, D., Dixit, V. M., Salvesen, G. S., and Poirier, G. G., Cleavage of automodified poly(ADP-ribose) polymerase during apoptosis: evidence for involvement of caspase-7, *J. Biol. Chem.*, 274, 28379, 1999.
35. Guano, F., Bernardi, R., Negri, C., Donzelli, M., Prosperi, E., Astaldi Ricotti, G., and Scovassi, A. I., Dose-dependent zinc inhibition of DNA ladder in apoptotic HeLa cells regulates the activity of poly(ADP-ribose) polymerase and does not protect from death induced by VP-16, *Cell Death Differ.*, 1, 101, 1994.
36. Agarwal, M. L., Taylor, W. R., Chernov, M. V., Chernova, O. B., and Stark, G. R., The p53 network, *J. Biol. Chem.*, 273, 1, 1998.
37. Kastan, M. B., Zhan, Q., el-Deiry, W. S., Carrier, F., Jacks, T., Walsh, W. V., Plunkett, B. S., Vogelstein, B., and Fornace, A. J., Jr., A mammalian cell cycle checkpoint pathway utilizing p53 and GADD45 is defective in ataxia-telangiectasia, *Cell*, 71, 587, 1992.
38. Woo, R. A., McLure, K. G., Lees-Miller, S. P., Rancourt, D. E., and Lee, P. W., DNA-dependent protein kinase acts upstream of p53 in response to DNA damage, *Nature*, 394, 700, 1998.
39. Khanna, K. K., Keating, K. E., Kozlov, S., Scott, S., Gatei, M., Hobson, K., Taya, Y., Gabrielli, B., Chan, D., Lees-Miller, S. P., and Lavin, M. F., ATM associates with and phosphorylates p53: mapping the region of interaction, *Nat. Genet.*, 20, 398, 1998.

40. Ruscetti, T., Lehnert, B. E., Halbrook, J., Le Trong, H., Hoekstra, M. F., Chen, D. J., and Peterson, S. R., Stimulation of the DNA-dependent protein kinase by poly(ADP-ribose) polymerase, *J. Biol. Chem.*, 273, 14461, 1998.
41. Dantzer, F., Ménissier-de Murcia, J., Barlow, C., Wynshaw-Boris, A., and de Murcia, G., Poly(ADP-ribose) polymerase activity is not affected in *Ataxia telangiectasia* cells and knockout mice, *Carcinogenesis*, 20, 177, 1999.
42. Morrison, C., Smith, G. C., Stingl, L., Jackson, S. P., Wagner, E. F., and Wang, Z. Q., Genetic interaction between PARP and DNA-PK in V(D)J recombination and tumorigenesis, *Nat. Genet.*, 17, 479, 1997.
43. Wesierska-Gadek, J., Schmid, G., and Cerni, C., ADP-ribosylation of wild-type p53 *in vitro:* binding of p53 protein to specific p53 consensus sequence prevents its modification, *Biochem. Biophys. Res. Commun.*, 224, 96, 1996.
44. Wesierska-Gadek, J., Bugajska-Schretter, A., and Cerni, C., ADP-ribosylation of p53 tumor suppressor protein: mutant but not wild-type p53 is modified, *J. Cell. Biochem.*, 62, 90, 1996.
45. Vaziri, H., West, M. D., Allsopp, R. C., Davison, T. S., Wu, Y. S., Arrowsmith, C. H., Poirier, G. G., and Benchimol, S., ATM-dependent telomere loss in aging human diploid fibroblasts and DNA damage lead to the post-translational activation of p53 protein involving poly(ADP-ribose) polymerase, *EMBO J.*, 16, 6018, 1997.
46. Malanga, M., Pleschke, J. M., Kleczkowska, H. E., and Althaus, F. R., Poly(ADP-ribose) binds to specific domains of p53 and alters its DNA binding functions, *J. Biol. Chem.*, 273, 11839, 1998.
47. Whitacre, C. M., Hashimoto, H., Tsai, M. L., Chatterjee, S., Berger, S. J., and Berger, N. A., Involvement of NAD-poly(ADP-ribose) metabolism in p53 regulation and its consequences, *Cancer Res.*, 55, 3697, 1995.
48. Agarwal, M. L., Agarwal, A., Taylor, W. R., Wang, Z. Q., Wagner, E. F., and Stark, G. R., Defective induction but normal activation and function of p53 in mouse cells lacking poly-ADP-ribose polymerase, *Oncogene*, 15, 1035, 1997.
49. Wesierska-Gadek, J., Wang, Z. Q., and Schmid, G., Reduced stability of regularly spliced but not alternatively spliced p53 protein in PARP-deficient mouse fibroblasts, *Cancer Res.*, 59, 28, 1999.
50. Schmid, G., Wang, Z. Q., and Wesierska-Gadek, J., Compensatory expression of p73 in PARP-deficient mouse fibroblasts as response to a reduced level of regularly spliced wild-type p53 protein, *Biochem. Biophys. Res. Commun.*, 255, 399, 1999.
51. Ménissier-de Murcia, J., Niedergang, C., Trucco, C., Ricoul, M., Dutrillaux, B., Mark, M., Oliver, F. J., Masson, M., Dierich, A., LeMeur, M., Walztinger, C., Chambon, P., and de Murcia, G., Requirement of poly(ADP-ribose) polymerase in recovery from DNA damage in mice and in cells, *Proc. Natl. Acad. Sci. U.S.A.*, 94, 7303, 1997.
52. Simbulan-Rosenthal, C. M., Rosenthal, D. S., Ding, R., Bhatia, K., and Smulson, M. E., Prolongation of the p53 response to DNA strand breaks in cells depleted of PARP by antisense RNA expression, *Biochem. Biophys. Res. Commun.*, 253, 864, 1998.
53. Milam, K. M. and Cleaver, J. E., Inhibitors of poly(adenosine diphosphate-ribose) synthesis: effect on other metabolic processes, *Science*, 223, 589, 1984.
54. Hunting, D. J., Gowans, B. J., and Henderson, J. F., Specificity of inhibitors of poly(ADP-ribose) synthesis. Effects on nucleotide metabolism in cultured cells, *Mol. Pharmacol.*, 28, 200, 1985.
55. Shieh, W. M., Ame, J. C., Wilson, M. V., Wang, Z. Q., Koh, D. W., Jacobson, M. K., and Jacobson, E. L., Poly(ADP-ribose) polymerase null mouse cells synthesize ADP-ribose polymers, *J. Biol. Chem.*, 273, 30069, 1998.

56. Smith, S., Giriat, I., Schmitt, A., and de Lange, T., Tankyrase, a poly(ADP-ribose) polymerase at human telomeres, *Science*, 282, 1484, 1998.
57. Ame, J. C., Rolli, V., Schreiber, V., Niedergang, C., Apiou, F., Decker, P., Muller, S., Hoger, T., Ménissier-de Murcia, J., and de Murcia, G., PARP-2, a novel mammalian DNA damage-dependent poly(ADP-ribose) polymerase, *J. Biol. Chem.*, 274, 17860, 1999.
58. Sallmann, F., Vodenicharov, M. D., Wang, Q. Z., and Poirier, G. G., Characterization of sPARP, an alternative of PARP-/gene with poly(ADP-ribose) polymerase activity independent of DNA strand breaks, *J. Biol. Chem.*, in press, 2000.
59. Wang, Z. Q., Auer, B., Stingl, L., Berghammer, H., Haidacher, D., Schweiger, M., and Wagner, E. F., Mice lacking ADPRT and poly(ADP-ribosyl)ation develop normally but are susceptible to skin disease, *Genes Dev.*, 9, 509, 1995.
60. Masutani, M., Nozaki, T., Nishiyama, E., Shimokawa, T., Tachi, Y., Suzuki, H., Nakagama, H., Wakabayashi, K., and Sugimura, T., Function of poly(ADP-ribose) polymerase in response to DNA damage: gene-disruption study in mice, *Mol. Cell. Biochem.*, 193, 149, 1999.
61. Le Rhun, Y., Kirkland, J. B., and Shah, G. M., Cellular responses to DNA damage in the absence of poly(ADP-ribose) polymerase, *Biochem. Biophys. Res. Commun.*, 245, 1, 1998.
62. Oliver, F. J., de la Rubia, G., Rolli, V., Ruiz-Ruiz, M. C., de Murcia, G., and Ménissier-de Murcia, J., Importance of poly(ADP-ribose) polymerase and its cleavage in apoptosis. Lesson from an uncleavable mutant, *J. Biol. Chem.*, 273, 33533, 1998.
63. Leist, M., Single, B., Kunstle, G., Volbracht, C., Hentze, H., and Nicotera, P., Apoptosis in the absence of poly-(ADP-ribose) polymerase, *Biochem. Biophys. Res. Commun.*, 233, 518, 1997.
64. Wang, Z. Q., Stingl, L., Morrison, C., Jantsch, M., Los, M., Schulze-Osthoff, K., and Wagner, E. F., PARP is important for genomic stability but dispensable in apoptosis, *Genes Dev.*, 11, 2347, 1997.
65. Virág, L., Scott, G. S., Cuzzocrea, S., Marmer, D., Salzman, A. L., and Szabó, C., Peroxynitrite-induced thymocyte apoptosis: the role of caspases and poly(ADP-ribose) synthetase (PARS) activation, *Immunology*, 94, 345, 1998.
66. Varfolomeev, E. E., Schuchmann, M., Luria, V., Chiannilkulchai, N., Beckmann, J. S., Mett, I. L., Rebrikov, D., Brodianski, V. M., Kemper, O. C., Kollet, O., Lapidot, T., Soffer, D., Sobe, T., Avraham, K. B., Goncharov, T., Holtmann, H., Lonai, P., and Wallach, D., Targeted disruption of the mouse caspase 8 gene ablates cell death induction by the TNF receptors, Fas/Apo1, and DR3 and is lethal prenatally, *Immunity*, 9, 267, 1998.
67. Enari, M., Sakahira, H., Yokoyama, H., Okawa, K., Iwamatsu, A., and Nagata, S., A caspase-activated DNase that degrades DNA during apoptosis, and its inhibitor ICAD, *Nature*, 391, 43, 1998.
68. Zhang, J., Liu, X., Scherer, D. C., van Kaer, L., Wang, X., and Xu, M., Resistance to DNA fragmentation and chromatin condensation in mice lacking the DNA fragmentation factor 45, *Proc. Natl. Acad. Sci. U.S.A.*, 95, 12480, 1998.
69. Lazebnik, Y. A., Kaufmann, S. H., Desnoyers, S., Poirier, G. G., and Earnshaw, W. C., Cleavage of poly(ADP-ribose) polymerase by a proteinase with properties like ICE, *Nature*, 371, 346, 1994.
70. Tewari, M., Quan, L. T., O'Rourke, K., Desnoyers, S., Zeng, Z., Beidler, D. R., Poirier, G. G., Salvesen, G. S., and Dixit, V. M., Yama/CPP32 beta, a mammalian homolog of CED-3, is a CrmA-inhibitable protease that cleaves the death substrate poly(ADP-ribose) polymerase, *Cell*, 81, 801, 1995.

71. Cryns, V. and Yuan, J., Proteases to die for, *Genes Dev.*, 12, 1551, 1998.
72. Smulson, M. E., Pang, D., Jung, M., Dimtchev, A., Chasovskikh, S., Spoonde, A., Simbulan-Rosenthal, C., Rosenthal, D., Yakovlev, A., and Dritschilo, A., Irreversible binding of poly(ADP)ribose polymerase cleavage product to DNA ends revealed by atomic force microscopy: possible role in apoptosis, *Cancer Res.*, 58, 3495, 1998.
73. Talanian, R. V., Quinlan, C., Trautz, S., Hackett, M. C., Mankovich, J. A., Banach, D., Ghayur, T., Brady, K. D., and Wong, W. W., Substrate specificities of caspase family proteases, *J. Biol. Chem.*, 272, 9677, 1997.
74. Thornberry, N. A., Rano, T. A., Peterson, E. P., Rasper, D. M., Timkey, T., Garcia-Calvo, M., Houtzager, V. M., Nordstrom, P. A., Roy, S., Vaillancourt, J. P., Chapman, K. T., and Nicholson, D. W., A combinatorial approach defines specificities of members of the caspase family and granzyme B. Functional relationships established for key mediators of apoptosis, *J. Biol. Chem.*, 272, 17907, 1997.
75. Benjamin, C. W., Hiebsch, R. R., and Jones, D. A., Caspase activation in MCF7 cells responding to etoposide treatment, *Mol. Pharmacol.*, 53, 446, 1998.
76. Janicke, R. U., Sprengart, M. L., Wati, M. R., and Porter, A. G., Caspase-3 is required for DNA fragmentation and morphological changes associated with apoptosis, *J. Biol. Chem.*, 273, 9357, 1998.
77. Berger, N. A., Poly(ADP-ribose) in the cellular response to DNA damage, *Radiat. Res.*, 101, 4, 1985.
78. Yamamoto, H., Uchigata, Y., and Okamoto, H., DNA strand breaks in pancreatic islets by *in vivo* administration of alloxan or streptozotocin, *Biochem. Biophys. Res. Commun.*, 103, 1014, 1981.
79. Heller, B., Wang, Z. Q., Wagner, E. F., Radons, J., Bürkle, A., Fehsel, K., Burkart, V., and Kolb, H., Inactivation of the poly(ADP-ribose) polymerase gene affects oxygen radical and nitric oxide toxicity in islet cells, *J. Biol. Chem.*, 270, 11176, 1995.
80. Eliasson, M. J., Sampei, K., Mandir, A. S., Hurn, P. D., Traystman, R. J., Bao, J., Pieper, A., Wang, Z. Q., Dawson, T. M., Snyder, S. H., and Dawson, V. L., Poly(ADP-ribose) polymerase gene disruption renders mice resistant to cerebral ischemia, *Nat. Med.*, 3, 1089, 1997.
81. Masutani, M., Suzuki, H., Kamada, N., Watanabe, M., Ueda, O., Nozaki, T., Jishage, K., Watanabe, T., Sugimoto, T., Nakagama, H., Ochiya, T., and Sugimura, T., Poly(ADP-ribose) polymerase gene disruption conferred mice resistant to streptozotocin-induced diabetes, *Proc. Natl. Acad. Sci. U.S.A.*, 96, 2301, 1999.
82. Endres, M., Scott, G. S., Salzman, A. L., Kun, E., Moskowitz, M. A., and Szabó, C., Protective effects of 5-iodo-6-amino-1,2-benzopyrone, an inhibitor of poly(ADP-ribose) synthetase against peroxynitrite-induced glial damage and stroke development, *Eur. J. Pharmacol.*, 351, 377, 1998.
83. Cookson, M. R., Ince, P. G., and Shaw, P. J., Peroxynitrite and hydrogen peroxide induced cell death in the NSC34 neuroblastoma x spinal cord cell line: role of poly (ADP-ribose) polymerase, *J. Neurochem.*, 70, 501, 1998.
84. Küpper, J. H., de Murcia, G., and Bürkle, A., Inhibition of poly(ADP-ribosyl)ation by overexpressing the poly(ADP-ribose) polymerase DNA-binding domain in mammalian cells, *J. Biol. Chem.*, 265, 18721, 1990.
85. Küpper, J. H., Müller, M., Jacobson, M. K., Tatsumi-Miyajima, J., Coyle, D. L., Jacobson, E. L., and Bürkle, A., *trans*-Dominant inhibition of poly(ADP-ribosyl)ation sensitizes cells against gamma-irradiation and N-methyl-N'-nitro-N-nitrosoguanidine but does not limit DNA replication of a polyomavirus replicon, *Mol. Cell. Biol.*, 15, 3154, 1995.

86. Molinete, M., Vermeulen, W., Bürkle, A., Ménissier-de Murcia, J., Küpper, J. H., Hoeijmakers, J. H., and de Murcia, G., Overproduction of the poly(ADP-ribose) polymerase DNA-binding domain blocks alkylation-induced DNA repair synthesis in mammalian cells, *EMBO J.*, 12, 2109, 1993.
87. Schreiber, V., Hunting, D., Trucco, C., Gowans, B., Grunwald, D., de Murcia, G., and Ménissier-de Murcia, J., A dominant-negative mutant of human poly(ADP-ribose) polymerase affects cell recovery, apoptosis, and sister chromatid exchange following DNA damage, *Proc. Natl. Acad. Sci. U.S.A.*, 92, 4753, 1995.
88. Szumiel, I., Monitoring and signaling of radiation-induced damage in mammalian cells, *Radiat. Res.*, 150, S92, 1998.
89. Widmann, C., Gibson, S., and Johnson, G. L., Caspase-dependent cleavage of signaling proteins during apoptosis. A turn-off mechanism for anti-apoptotic signals, *J. Biol. Chem.*, 273, 7141, 1998.
90. Levkau, B., Koyama, H., Raines, E. W., Clurman, B. E., Herren, B., Orth, K., Roberts, J. M., and Ross, R., Cleavage of p21Cip1/Waf1 and p27Kip1 mediates apoptosis in endothelial cells through activation of Cdk2: role of a caspase cascade, *Mol. Cell*, 1, 553, 1998.
91. Gervais, J. L., Seth, P., and Zhang, H., Cleavage of CDK inhibitor p21(Cip1/Waf1) by caspases is an early event during DNA damage-induced apoptosis, *J. Biol. Chem.*, 273, 19207, 1998.
92. Tan, X., Martin, S. J., Green, D. R., and Wang, J. Y. J., Degradation of retinoblastoma protein in tumor necrosis factor- and CD95-induced cell death, *J. Biol. Chem.*, 272, 9613, 1997.
93. Janicke, R. U., Walker, P. A., Lin, X. Y., and Porter, A. G., Specific cleavage of the retinoblastoma protein by an ICE-like protease in apoptosis, *EMBO J.*, 15, 6969, 1996.
94. Song, Q., Lees-Miller, S. P., Kumar, S., Zhang, Z., Chan, D. W., Smith, G. C., Jackson, S. P., Alnemri, E. S., Litwack, G., Khanna, K. K., and Lavin, M. F., DNA-dependent protein kinase catalytic subunit: a target for an ICE- like protease in apoptosis, *EMBO J.*, 15, 3238, 1996.
95. Casciola-Rosen, L. A., Anhalt, G. J., and Rosen, A., DNA-dependent protein kinase is one of a subset of autoantigens specifically cleaved early during apoptosis, *J. Exp. Med.*, 182, 1625, 1995.
96. Flygare, J., Armstrong, R. C., Wennborg, A., Orsan, S., and Hellgren, D., Proteolytic cleavage of HsRad51 during apoptosis, *FEBS Lett.*, 427, 247, 1998.
97. Huang, Y., Nakada, S., Ishiko, T., Utsugisawa, T., Datta, R., Kharbanda, S., Yoshida, K., Talanian, R. V., Weichselbaum, R., Kufe, D., and Yuan, Z. M., Role for caspase-mediated cleavage of Rad51 in induction of apoptosis by DNA damage, *Mol. Cell. Biol.*, 19, 2986, 1999.
98. Casciola-Rosen, L., Nicholson, D. W., Chong, T., Rowan, K. R., Thornberry, N. A., Miller, D. K., and Rosen, A., Apopain/CPP32 cleaves proteins that are essential for cellular repair: a fundamental principle of apoptotic death, *J. Exp. Med.*, 183, 1957, 1996.
99. McConnell, K. R., Dynan, W. S., and Hardin, J. A., The DNA-dependent protein kinase catalytic subunit (p460) is cleaved during Fas-mediated apoptosis in Jurkat cells, *J. Immunol.*, 158, 2083, 1997.
100. Chakravarthy, B. R., Walker, T., Rasquinha, I., Hill, I. E., and MacManus, J. P., Activation of DNA-dependent protein kinase may play a role in apoptosis of human neuroblastoma cells, *J. Neurochem.*, 72, 933, 1999.
101. Herceg, Z. and Wang, Z. Q., Failure of poly(ADP-ribose) polymerase cleavage by caspases leads to induction of necrosis and enhanced apoptosis, *Mol. Cell. Biol.*, 19, 5124, 1999.

11 Poly(ADP-Ribose) Polymerase is an Active Participant in Programmed Cell Death and Maintenance of Genomic Stability

Dean S. Rosenthal, Cynthia M. Simbulan-Rosenthal, Willliam J. Smith, Betty J. Benton, Radharaman Ray, and Mark E. Smulson

CONTENTS

11.1 Introduction 227
11.2 PARP and Cell Death 228
11.3 Poly(ADP-Ribosyl)ation of p53 235
11.4 PARP and Genomic Stability 239
11.5 Conclusions 240
Abbreviations 242
Acknowledgments 242
References 242

11.1 INTRODUCTION

Poly(ADP-ribose) polymerase (PARP) is a major nuclear protein associated with chromatin that contains zinc fingers and binds to either double- or single-strand DNA breaks. PARP is activated upon binding to DNA and forms covalent homopolymers of poly(ADP-ribose) (PAR) attached to a number of nuclear proteins, including itself and proteins involved in DNA replication, DNA repair, and apoptosis. Nuclear NAD, which comprises 95% of the total cellular NAD, is the substrate for polymer formation. PARP has been implicated in numerous biological functions involved with the breaking and rejoining of DNA.[1-6] In addition, other functions have been ascribed for PARP in which the role of DNA strand breaks is not so clear. For example, PARP has been demonstrated to play a role as a coactivator of gene transcription.[7,8] In addition, the binding of PARP to specific nuclear proteins has been shown to alter their activity, in the absence of DNA breaks or NAD.[9]

11.2 PARP AND CELL DEATH

PARP knockout mice have now been independently generated from the interruption of exon 2,[10] exon 4,[11] and most recently, exon 1[12] of the PARP gene on chromosome 1. PARP knockout mice with a disrupted PARP gene neither express intact PARP nor exhibit significant poly(ADP-ribosyl)ation.[10-12] Because poly(ADP-ribosyl)ation is stimulated by DNA fragmentation, the potential role for PARP in cell death via NAD and ATP depletion had been proposed previously.[13,14] This idea has been supported by recent studies in which both exon 1[12] and exon 2[15,16] PARP$^{-/-}$ animals have been shown to be resistant to steptozotocin-induced pancreatic islet cell death, associated with NAD depletion in PARP$^{+/+}$ animals. We have also collaborated in a study that demonstrated that exon 2 PARP$^{-/-}$ animals are resistant to the neurotoxin MPTP-induced parkinsonism.[17] Exon 2 PARP$^{-/-}$ animals are also more resistant to ischemic injury.[18-21]

To investigate whether PARP might play an active role in programmed cell death, we first used a human osteosarcoma cell line that undergoes a "slow," spontaneous apoptotic death.[22] On reaching confluency, approximately 6 days under our culture conditions, these cells undergo the morphological and biochemical changes characteristic of apoptosis. Internucleosomal DNA cleavage was apparent at day 7 and increased until day 10, at which time virtually all of the cells have undergone apoptosis. Cells from duplicate cultures were incubated for up to 10 days and fixed at daily intervals for examination of nuclear poly(ADP-ribosyl)ation with antiserum to PAR. After 3 days, the nuclei of all attached cells stained intensely for the PAR. The *in vivo* synthesis of PAR was markedly reduced afterward (Figure 11.1A). Our results support the idea that nuclear disruption involving strand breaks may be present in the earliest stages of apoptosis, before morphological changes and the appearance of the characteristic nucleosome ladder. The substantial extent of nuclear poly(ADP-ribosyl)ation apparent early during apoptosis is consistent with the appearance of large (1-Mb) chromatin fragments at this reversible stage,[23] given that the activity of PARP is absolutely dependent on DNA strand breaks. A marked decrease in NAD concentration, indicative of increased PAR synthesis, and a subsequent recovery in NAD levels prior to the appearance of internucleosomal DNA cleavage have also been previously observed.[24]

Kaufmann et al.[25] first demonstrated that PARP undergoes proteolytic cleavage during chemotherapy-induced apoptosis. By immunoblot analysis with epitope-specific antibodies, it was demonstrated that programmed cell death was accompanied by early cleavage of PARP into 85- and 24-kDa fragments that contain the active site and the DNA-binding domain (DBD) of the enzyme, respectively. This latter domain is required for full PARP activity. The purification and characterization of caspase-3, responsible for the cleavage of PARP during apoptosis was performed by Nicholson et al.[26] This enzyme is composed of two subunits of 17 and 12 kDa

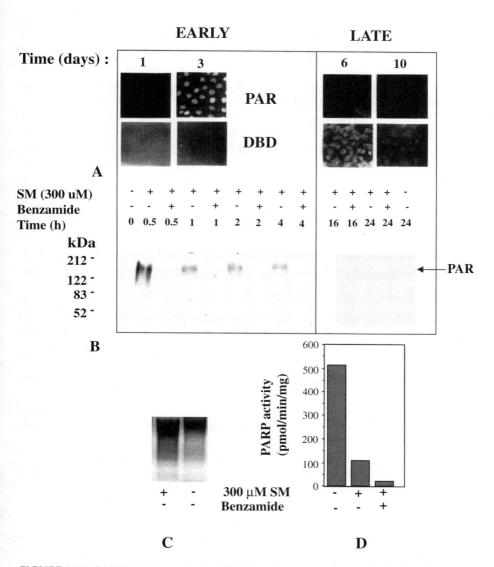

FIGURE 11.1 PARP activity and poly(ADP-ribosyl)ation are characteristic of early stages of apoptosis, while PARP cleavage and inactivation are associated with DNA laddering. (A) Osteosarcoma cells undergoing 10-day spontaneous apoptosis were fixed at the indicated times and subjected to immunofluorescent analysis utilizing antibodies specific for PAR (top), or the caspase-3-cleavage product of PARP (DBD; bottom). (B) Primary human keratinocytes were treated with 300 μM sulfur mustard for the indicated times. Cell extracts were derived and subjected to immunoblot analysis using monoclonal antibody specific for PAR. (C) Keratinocytes were treated with sulfur mustard for 24 h. DNA was isolated, separated by electrophoresis in 1.5% agarose gels, and detected with ethidium bromide. (D) Cell extracts were derived from keratinocytes treated as in (C) and assayed for PARP activity using [^{32}P]NAD as a substrate.

that are derived from a common proenzyme, which is related to interleukin-1β-converting enzyme and to CED-3, the product of a gene required for programmed cell death in *Caenorhabditis elegans*.[27] The identity of this protease was also demonstrated by Tewari et al.[28] To measure PARP cleavage in intact cells, we subjected human osteosarcoma cells to immunofluorescence analysis with antibodies that recognize the DBD but not intact PARP.[22,29] As with the other markers, samples were analyzed each day throughout the total 10-day period; samples from immediate (day 1), early (day 3), mid (day 6), and late (day 10) stages of apoptosis are shown in Figure 11.1A. Immunofluorescence analysis detected the PARP DBD in human osteosarcoma cells only after 6 to 7 days in culture, a time at which the abundance of both PARP and PAR is decreasing, PARP-cleavage activity is increasing, and internucleosomal DNA cleavage is present.[26] The pattern of staining for the DBD differed markedly from that of full-length PARP. Whereas PARP staining was present throughout the nucleus, the DBD showed a more localized punctate pattern in the region of the nucleolus and throughout the nucleus-disrupted cytoplasm.

Therefore, catalytic activation of PARP occurs early in osteosarcoma cell growth, while the cleavage of PARP and the accumulation of a large number of DNA strand breaks occur later in the apoptotic process. The concomitant loss of poly(ADP-ribosyl)ation of target proteins appears to be characteristic of later stages of apoptosis during which cells become irreversibly committed to death. This may conserve NAD and ATP during the later stages of apoptosis. Recently, the requirement for PARP cleavage to prevent necrosis associated with depletion of NAD has been confirmed using PARP[-/-] cells that express a caspase-resistant mutant of PARP.[30]

These results are in contrast to those of Negri et al.,[31] who reported the presence of PAR in cells at the late stages of apoptosis, although, as the authors point out, it is difficult to reconcile the quantitative cleavage of PARP with its activation late in apoptosis (see below). To determine if a minor fraction of uncleaved PARP could be responsible for PAR formation that has been reported late in apoptosis, as well as to determine if transient poly(ADP-ribosyl)ation could be observed in another system, we recently measured PARP activity as well as the total amount of cellular PAR at different stages of apoptosis induced by the alkylating agent sulfur mustard. PAR is strongly induced in the early stages of apoptosis (within 30 min), but not at later stages (Figure 11.1B). Importantly, the appearance of PAR *precedes* the cleavage of DNA fragmentation factor (DFF45; see below), and the appearance of DNA ladders at 24 h (Figure 11.1C). Although PARP is completely cleaved, extracts derived from 24-h apoptotic cells retain approximately 20% of their *in vitro* polymerizing activity (Figure 11.1D), reflecting the low-level DNA-independent activity of the catalytic domain.[32] However, this activity is apparently insufficient to synthesize or sustain detectable steady-state levels of PAR *in vivo* in the presence of PAR-degrading enzymes including poly(ADP-ribose) glycohydrolase. Furthermore, addition of the PARP inhibitor benzamide in this system significantly delays the onset of apoptosis.[33] Clearly, the activation of PARP is associated with DNA breaks, although chromosomal degradation to large fragments precedes the formation of DNA ladders in response to apoptotic signals generated from DNA-damaging agents as well as from stimulation of the Fas receptor (see below).

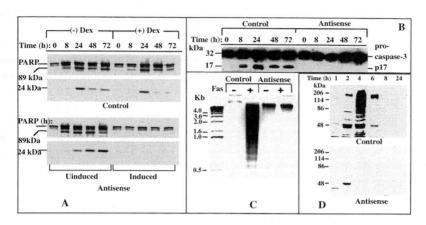

FIGURE 11.2 Effects of PARP depletion by antisense RNA expression on the increase in caspase-3-like activity (A), proteolytic processing of procaspase-3 (B), internucleosomal DNA fragmentation (C), and synthesis of PAR (D) during Fas-mediated apoptosis in 3T3-L1 cells. Mock-transfected (A, top) and PARP-antisense (A, bottom) 3T3-L1 cells were preincubated in the absence or presence of 1 μM Dex for 72 h and then incubated with anti-Fas (100 ng/ml) and cycloheximide (10 μg/ml) for the indicated times. Cytosolic extracts were prepared and assayed for *in vitro* PARP-cleavage activity with [^{35}S]PARP as substrate. (B) 3T3-L1 control and antisense cells were preincubated with Dex for 72 h and then exposed to anti-Fas and cycloheximide for the indicated times as in (A). Cell extracts were subjected to immunoblot analysis with a monoclonal antibody to the p17 subunit of caspase-3 (B) or to PAR (D). The positions of procaspase-3 and p17 are indicated. (C) Total genomic DNA was extracted and internucleosomal DNA ladders characteristic of apoptosis was detected by agarose gel electrophoresis and ethidium bromide staining.

The generality of an early burst of poly(ADP-ribosyl)ation was confirmed with human HL-60 cells, mouse 3T3-L1, and immortalized fibroblasts derived from wild-type mice.[34] The effects of eliminating this early transient modification of nuclear proteins by depletion of PARP protein either by antisense RNA expression or by gene disruption on various morphological and biochemical markers of apoptosis were then examined.

In 3T3-L1 cells stably transfected with a construct expressing dexamethasone (Dex)-inducible PARP antisense RNA, Dex induced a time-dependent depletion of PARP, with only ~5% of the protein remaining after 72 h. A combination of anti-Fas and cycloheximide induced a marked increase in caspase-3-like activity in control 3T3-L1 cells that had been preincubated in the absence or presence of Dex. This effect was maximal 24 h after induction of apoptosis, as indicated by the generation of the 89- and 24-kDa cleavage fragments of PARP in an *in vitro* assay (Figure 11.2A, top). No caspase-3 activity was apparent in PARP-antisense cells that had been depleted of PARP by preincubation with Dex before exposure to anti-Fas and cycloheximide (Figure 11.2A, bottom).

To confirm that procaspase-3 is proteolytically processed to an active p17 subunit during apoptosis in control 3T3-L1 cells, and to determine whether the transient early poly(ADP-ribosyl)ation is necessary for this activation, control and antisense cells were preincubated with Dex, exposed to anti-Fas and cycloheximide for indicated times, and cell extracts were subjected to immunoblot analysis with antibodies to the p17 subunit of caspase-3. Whereas procaspase-3 was proteolytically processed to p17 by 24 h, coinciding with the peak of *in vitro* caspase-3-like PARP-cleavage activity, in control cells, proteolytic processing of procaspase-3 was not apparent in the PARP-depleted antisense cells (Figure 11.2B). Furthermore, using DNA fragmentation analysis as another assay for apoptosis, control 3T3-L1 cells exposed to anti-Fas and cycloheximide for 24 h exhibited marked internucleosomal DNA fragmentation (DNA ladders), but not the PARP-depleted antisense cells exposed to these inducers for the same time (Figure 11.2C). Similar to our previous studies, we noted that the earliest stages of apoptosis were associated with a burst of PAR synthesis. This early synthesis of PAR was eliminated by the expression of PARP antisense RNA (Figure 11.2D).

Cells derived from animals depleted of PARP were also unable to undergo Fas-mediated apoptosis. Anti-Fas and cycloheximide induced a rapid synthesis of PAR in PARP$^{+/+}$ cells, which was not observed in PARP$^{-/-}$ cells (Figure 11.3A). To determine if the activation of PARP and synthesis of PAR was the result of cleavage of the inhibitor of caspase-activated DNase (ICAD/DFF45)[35] and the concomitant oligonucleosomal cleavage of DNA, a time course was first performed for PAR; the same filter was then reprobed with ICAD-specific antibody. Figure 11.3B shows that PAR synthesis occurred at 1 to 2 h, while ICAD cleavage did not occur until 4 to 6 h (Figure 11.3C). In contrast to PARP$^{+/+}$ cells, no processing of ICAD was evident in PARP$^{-/-}$ cells after exposure to anti-Fas and cycloheximide for up to 24 h (not shown).

PARP$^{+/+}$ cells showed substantial nuclear fragmentation and chromatin condensation 24 h after induction of Fas-mediated apoptosis; ~97% of nuclei exhibited apoptotic morphology by this time. In contrast, no substantial changes in nuclear morphology were apparent in the PARP$^{-/-}$ fibroblasts even after exposure to anti-Fas and cycloheximide for 24 or 48 h[34.]

PARP$^{-/-}$ fibroblasts were stably transfected with a plasmid expressing wild-type PARP.[36] Individual as well as pooled clones expressed PARP protein at levels similar to those of PARP$^{+/+}$ cells. These cells were induced to undergo apoptosis by exposure to anti-Fas and cycloheximide for up to 48 h. PARP$^{+/+}$ cells, as well as PARP$^{-/-}$ transfected with PARP, exhibited significant caspase-3-like activity after 48 h. As expected, PARP was not expressed in the PARP$^{-/-}$ fibroblasts nor in PARP$^{-/-}$ cells transfected with vector alone. Consistently, whereas exposure to anti-Fas and cycloheximide induced marked internucleosomal DNA fragmentation in PARP$^{+/+}$ fibroblasts and in PARP$^{-/-}$ cells stably transfected with PARP, no apoptotic DNA ladders were evident in the PARP$^{-/-}$ cells when similarly treated (Figure 11.3D). Furthermore, exposure to anti-Fas plus cycloheximide for 48 h induced apoptotic nuclear morphology in PARP$^{-/-}$ cells transfected with PARP almost to the same extent as the PARP$^{+/+}$ cells.[34]

Thus, depletion of PARP by antisense in 3T3-L1, or by knockout of PARP attenuates Fas plus cycloheximide-mediated apoptosis. In addition, the reintroduction of PARP in independent clones of PARP$^{-/-}$ cells reestablished the response. We interpret these results to indicate that PARP plays an active role early in apoptosis

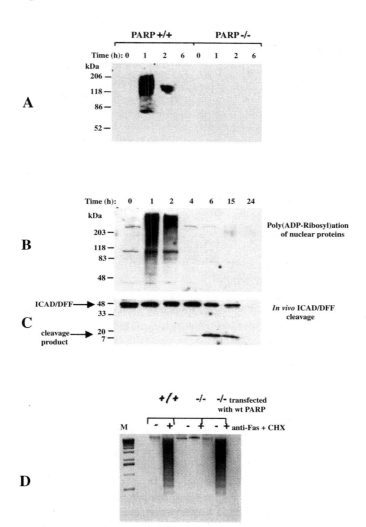

FIGURE 11.3 Effects of anti-Fas and cycloheximide on poly(ADP-ribosyl)ation, ICAD cleavage, and internucleosomal DNA fragmentation in immortalized fibroblasts from wild-type and PARP knockout and mice. (A) PARP$^{+/+}$ and PARP$^{-/-}$ fibroblasts were exposed to anti-Fas (100 ng/ml) and cycloheximide (10 µg/ml) for the indicated times, after which extracts were subjected to immunoblot analysis with antibodies to PAR. (B) PARP$^{+/+}$ fibroblasts were treated as in (A) in an independent experiment, after which extracts were subjected to immunoblot analysis with antibodies to PAR. (C) The identical filter used in (B) was stripped of antibodies by incubation for 30 min at 50°C with a solution containing 100 mM 2-mercaptoethanol, 2% SDS, and 62.5 mM Tris-HCl (pH 6.7), and reprobed using an antibody specific for ICAD. (D) PARP$^{+/+}$ and PARP$^{-/-}$ fibroblasts as well as PARP$^{-/-}$ cells stably transfected with wild-type PARP were exposed to anti-Fas and cycloheximide for 48 h. Apoptosis was monitored by extraction of total genomic DNA and detection of characteristic apoptotic internucleosomal DNA ladders by agarose gel electrophoresis and ethidium bromide staining.

either by depletion of NAD and ATP or via the modification of nuclear proteins involved in apoptosis. Furthermore, these studies are consistent with earlier results using chemical inhibitors (which require a degree of caution in their interpretation), indicating that the activation of PARP is required for apoptosis to occur in some systems.[24,37-39] It should be pointed out that another study indicated that primary PARP$^{-/-}$ fibroblasts underwent similar apoptosis compared to PARP$^{+/+}$ cells.[40] Whether these differing observations are due to the use of different cells or reagents remains to be determined. However, it has been shown recently that expression of caspase-3-resistant PARP in exon 2 PARP$^{-/-}$ cells,[30,41] as well as expression of exogenous wild-type PARP in osteosarcoma cells, results in an earlier onset of the apoptotic response, a finding that is consistent with an active role for PARP and poly(ADP-ribosyl)ation early in apoptosis. Both of these findings are in contrast to a study by Oliver et al.[42] who found a presumptive decrease in apoptotic response, as determined by morphological changes, upon expression of uncleavable PARP in cells derived from different PARP knockout animals. However, as pointed out previously,[30,43] these animals are derived from an interruption in exon 4,[11] and could potentially express a truncated protein encompassing the first zinc finger of the DNA-binding region of PARP. This zinc finger is sufficient to bind both single- and double-stranded DNA breaks,[44] and may therefore act as a dominant-negative mutant.[45,46]

How could PARP play a role in the pathway leading from the Fas/TNF receptor to apoptosis? There is clearly cross talk between the mitochondrial and death receptor-mediated pathways for apoptosis, and in fact "type II" cells do not form a significant amount of death-inducing signaling complex (DISC), comprising Fas, Fas-associating protein with death domain (FADD), and procaspase-8.[47] The characteristics of type II cells include the relatively slow activation of caspase-3, followed by the activation of caspase-8, and direct processing of caspase-3 into a p17 active form without the appearance of the intermediate p20. The immortalized exon 2 PARP$^{-/-}$ fibroblasts, as well as 3T3-L1 cells expressing PARP antisense RNA, utilized in our study, fit both of these criteria. In addition, 3T3 cells demonstrate Bcl-2-inhibitable Fas killing,[48] another hallmark of type II cells.[47]

In type II cells, caspase-3 activation is secondary to the release of proapoptotic factors from the mitochondria, including cytochrome c, procaspase-2, procaspase-9,[49] and apoptosis-inducing factor (AIF).[50] AIF is a noncaspase inducer of apoptosis that can translocate from the mitochondrial intermembrane space to the nucleus and induce DNA cleavage into 50-kb fragments.[51] Cytosols from anti-Fas-treated human lymphoma cells have been shown to accumulate AIF activity.[52]

In this system, the cleavage of ICAD/DFF45 by caspase-3, which allows the translocation of CAD/DFF40 to the nucleus and internucleosomal cleavage, is a relatively late event. However, chromosomal DNA is not directly cleaved into nucleosome ladders during apoptosis. Studies with isolated nuclei[53] and intact cells[54] demonstrate that DNA is first fragmented into high-molecular-weight fragments of 1 to 2 Mb, followed by the appearance of 200- to 800-kb fragments, which may reflect the higher-order chromatin structure of nuclei. Afterward, DNA is further degraded into 50-kb fragments, and then finally into the characteristic nucleosome-sized ladders visible by conventional agarose gel electrophoresis, although certain cell types, such as MCF-7, do not demonstrate this final stage of chromatin degradation. Although elegant studies have demonstrated that ICAD/DFF45 and CAD/DFF40 play important roles in apoptotic DNA fragmentation,[55-57] these studies have not excluded the roles of other nucleases. Hughes et al.[58] have identified a 260-kDa factor that is responsible only for the cleavage of DNA into 30- to 50-kb fragments, while a 25-kDa factor generated both 30- to 50-kb fragments, as well as smaller fragments.[58] Other nucleases that have been shown to play a role in DNA degradation in apoptosis include DNase II,[59] DNase I,[60,61] DNase I-related protein,[62] and cyclophins A, B, and C.[63] Similar to the endonuclease activity induced by AIF, cyclophilin C only cleaves DNA into 50-kb fragments.

Thus, a model may be proposed whereby stimulation of the Fas receptor on immortalized fibroblasts induces the gradual release of mitochondrial factors, such as AIF, that may translocate to the nucleus and induce low levels of caspase-independent cleavage of chromatin into large fragments. These DNA breaks would then stimulate PARP activity, amplifying apoptotic events by either poly(ADP-ribosyl)ating p53 or other factors involved in the upregulation or altered intracellular trafficking of Fas, Bax, or IGFBP3 (see below). Additionally, PARP activation rapidly depletes NAD and ATP, which could contribute to both receptor and mitochondrial pathways of apoptosis. Feldenberg et al.[64] have shown that partial depletion of ATP (approximately 10 to 65% of control) can induce apoptosis of cultured renal epithelial cells including internucleosomal DNA cleavage, morphological changes, and plasma membrane alterations. The ATP-depleted cells display a significant upregulation of Fas, Fas ligand, and FADD, resulting in induction of caspase-8 and caspase-3 activity.[64]

Further depletion of ATP below a threshold level might be expected to inhibit the later events in apoptosis. Eguchi et al.[65] have shown that Fas-induced apoptosis is completely blocked by reducing the intracellular ATP level in both type I and type II cells. In type I cells, ATP-dependent step(s) of Fas-mediated apoptotic signal transduction are only located downstream of caspase-3 activation. However, in type II cells, activation of caspase-3, -8, and -9, as well as cleavage of ICAD/DFF45, was blocked by reduction of intracellular ATP, whereas release of cytochrome c was not affected. This may reflect the requirement for dATP/ATP in the activation of caspase-9.[66-68] Cleavage of PARP at later stages of apoptosis would prevent ATP from falling below this critical level.

11.3 POLY(ADP-RIBOSYL)ATION OF p53

In addition to undergoing automodification, PARP catalyzes the poly(ADP-ribosyl)ation of such nuclear proteins as histones, topoisomerases I and II,[69,70] SV40 large T antigen,[71] DNA polymerase α, proliferating cell nuclear antigen (PCNA), and approximately 15 protein components of the DNA synthesome.[70] The modification of nucleosomal proteins also alters the nucleosomal structure of the DNA containing strand breaks and promotes access of various replicative and repair enzymes to these sites.[72,73] We have obtained some potentially relevant targets for poly(ADP-ribosyl)ation during the burst of PAR synthesis at the early stages of apoptosis, including p53. p53, a tumor suppressor nuclear phosphoprotein, reduces the occurrence of mutations by mediating cell cycle arrest in G_1 or G_2/M or inducing apoptosis in cells that have accumulated substantial DNA damage, thus preventing progression of cells through S phase before DNA repair is complete.[74-76] p53 is induced by a variety of apoptotic stimuli and is required for apoptosis in many cell systems.[77] Overexpression of p53 is sufficient to induce apoptosis in various cell types.[78] Interestingly, p53 can utilize transcription activation of target genes and/or direct protein–protein interaction to initiate p53-dependent apoptosis.

Both PARP activity and p53 accumulation are induced by DNA damage, and both proteins have been implicated in the normal cellular responses to such damage. Whereas PAR synthesis increases within seconds after induction of DNA strand breaks,[79] the amount of wild-type p53, which is usually low because of the short half-life (20 min) of the protein, increases several hours after DNA damage as a result of reduced degradation.[80,81] A functional association of PARP and p53 has recently been suggested by coimmunoprecipitation of each protein *in vitro* by antibodies to the other.[82,83] It was recently shown that p53 is poly(ADP-ribosyl)ated *in vitro* by purified PARP, and that binding of p53 to a specific p53 consensus sequence prevents its covalent modification.[84] We recently showed that modification of p53 by poly(ADP-ribosyl)ation also occurs *in vivo*, and that it represents one of the early acceptors of poly(ADP-ribosyl)ation during apoptosis in human osteosarcoma cells.[85] Given that the *in vivo* half-life of PAR chains on an acceptor has been estimated to be about 1 to 2 min, we additionally explored how this post-translational modification of p53 is altered at the onset of caspase-3-mediated cleavage and inactivation of PARP during the later stages of the death program.

Human osteosarcoma cells were plated under conditions that result in spontaneous apoptosis over a 10-day period.[22,26] Biochemical markers of apoptosis were initially observed at day 5 and maximized around days 7 to 9, including caspase-3-mediated *in vitro* PARP-cleavage activity (Figure 11.4A, top), proteolytic processing of the caspase-3 proenzyme (CPP32) to its active form (p17; Figure 11.4A, middle), and internucleosomal DNA fragmentation.

Consistent with previous studies showing p53 accumulation during early apoptosis in different cell lines, immunoblot analysis with anti-p53 mAbs of extracts of osteosarcoma cells at various stages of spontaneous apoptosis revealed that endogenous levels of p53 protein were significantly increased as early as days 2 to 3, maximized at day 4, and declined thereafter. Immunoblot analysis with antibodies to PARP to monitor *in vivo* PARP cleavage during the same time frame showed that ~50% of endogenous PARP was cleaved to its 89-kDa fragment by day 7 and complete cleavage of PARP was noted by day 9.[85]

When the same extracts were subjected to immunoblot analysis with antibodies to PAR, low levels of polymer were observed at day 2 of apoptosis (Figure 11.4A, bottom), indicating the absence of DNA strand breaks, PARP activity, or both. However, poly(ADP-ribosyl)ation of nuclear proteins was markedly increased at day 3 and was maximal at day 4, a stage at which all the cells were still viable and could be replated, prior to any evidence of internucleosomal DNA fragmentation. Subsequently, a marked decline in poly(ADP-ribosyl)ation of nuclear proteins was observed at later time points (days 7 to 9), concomitant with the onset of substantial DNA fragmentation, proteolytic activation of caspase-3, and caspase-3-mediated *in vitro* and *in vivo* cleavage of PARP.

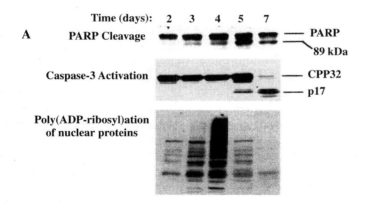

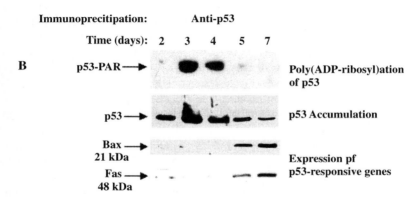

FIGURE 11.4 Time courses of *in vitro* PARP cleavage, activation of caspase-3, and poly(ADP-ribosyl)ation of nuclear proteins (A) vs. poly(ADP-ribosyl)ation of p53 and expression of Bax and Fas (B) during spontaneous apoptosis in human osteosarcoma cells. (A) At the indicated times of confluence-associated spontaneous apoptosis, cell extracts were prepared and caspase-3-like PARP-cleavage activity in cytosolic extracts was assayed with [^{35}S]PARP as substrate (top panel). Extracts were also subjected to immunoblot analysis with mAb to the p17 subunit of caspase-3 (middle panel), or to PAR (bottom panel). (B) At the indicated times during spontaneous apoptosis, cell extracts were prepared and equal amounts of total protein (100 µg) were subjected to immunoprecipitation with an mAb to p53. The immunoprecipitated proteins were then subjected to immunoblot analysis with an mAb to PAR (top panel). The immunoblot shown in A was stripped of antibodies by incubation for 30 min at 50°C with a solution containing 100 mM 2-mercaptoethanol, 2% SDS, and 62.5 mM Tris-HCl (pH 6.7), and reprobed sequentially with antibodies to p53, Bax, and Fas.

To confirm if p53 undergoes poly(ADP-ribosyl)ation *in vivo* during apoptosis in human osteosarcoma cells, cell extracts were derived at various times during spontaneous apoptosis and subjected to immunoprecipitation with an anti-p53 mAb. The immunoprecipitated proteins were then subjected to immunoblot analysis with mAb to PAR. This approach revealed marked poly(ADP-ribosyl)ation of p53 at the early stages of apoptosis (days 3 to 4; Figure 11.4B, top), coincident with the burst of PAR synthesis during this stage. The extent of poly(ADP-ribosyl)ation of p53 declined concomitant with the onset of both *in vitro* and *in vivo* caspase-3-mediated PARP cleavage. Reprobing of the blot with polyclonal antibodies to p53 confirmed that the modified protein was in fact p53 (Figure 11.4B, second panel). The observation that p53 is specifically poly(ADP-ribosyl)ated during the early stages of spontaneous apoptosis in human osteosarcoma cells suggests that this post-translational modification may play a role in regulating its function during the early phases of the cell death cascade.

PARP can modulate the catalytic activity of a number of DNA-binding nuclear enzymes by catalyzing their poly(ADP-ribosyl)ation. In most instances, poly(ADP-ribosyl)ation inhibits the activity of the modified protein, presumably because of a marked decrease in DNA-binding affinity caused by electrostatic repulsion between DNA and PAR. Thus, post-translational modification of p53 may also alter DNA binding to specific DNA sequences in the promoters of target genes associated with the induction of p53-mediated apoptosis, such as those encoding Bax, IGF-BP3,[86] or Fas.[87] The time course of accumulation and poly(ADP-ribosyl)ation of p53 during the early stages of apoptosis was thus correlated with the induction of expression of the p53-responsive genes Bax and Fas. Immunoblot analysis of extracts of cells at various stages of apoptosis in osteosarcoma cells with antibodies to either Bax or Fas revealed that expression of both Bax and Fas (Figure 11.4B, bottom) were negligible before and at the peak of p53 accumulation and poly(ADP-ribosyl)ation (days 3 and 4). Although p53 accumulation was already significantly elevated by day 2, expression of Bax and Fas was markedly induced only at day 5, concomitant with a decline in PAR attached to p53 and the onset of caspase-3-mediated PARP cleavage and inactivation. The coincident decrease in PAR covalently bound to p53 and induction of Bax and Fas expression suggests that poly(ADP-ribosyl)ation may regulate p53 function early in apoptosis; caspase-3-mediated cleavage of PARP may release p53 from poly(ADP-ribosyl)ation-induced inhibition at the later stages of the apoptotic cascade.

Accordingly, p53 may represent a potentially relevant target for poly(ADP-ribosyl)ation during the burst of PAR synthesis at the early periods of apoptosis. Colocalization of PARP and p53 in the vicinity of large DNA breaks and their physical association[82,83] suggest that poly(ADP-ribosyl)ation may regulate the DNA-binding ability and, consequently, the function of p53. The accumulation of p53 may be due to induced expression of the protein by the apoptotic stimuli or stabilization by inhibition of p53 degradation via modification of the protein. These results suggest a negative regulatory role for PARP and/or PAR early in apoptosis, since subsequent degradation of PAR attached to p53 coincided with the increase in caspase-3 (PARP-cleavage) activity as well as the induction of expression of the p53-responsive genes Bax and Fas at a stage when cells are irreversibly committed to death. Although the mechanism(s) of action of Bax/Bcl-2 family of gene products during apoptosis remain to be clarified, induction of Bax expression may influence the decision to commit to apoptosis since homodimerization of Bax promotes cell death and heterodimerization of Bax with Bcl-2 inhibits the antiapoptotic function of Bcl-2.[86] Wild-type p53, but not mutant p53, also upregulates Fas expression during chemotherapy-induced apoptosis, and p53-responsive elements were recently identified within the first intron and the promoter of the Fas gene.[87] Binding of Fas to Fas ligand recruits the adapter molecule FADD via shared protein motifs (death domains), resulting in subsequent activation or amplification of the caspase cascade leading to apoptosis.

Electrophoretic mobility-shift analysis has shown that PAR attached to p53 *in vitro* can block its sequence-specific binding to the palindromic p53 consensus sequence, suggesting that poly(ADP-ribosyl)ation of p53 may regulate p53-mediated transcriptional activation of genes important in the cell cycle and apoptosis.[88] PARP cycles on and off DNA ends in the presence of NAD, and its automodification during DNA repair *in vitro* presumably allows access to DNA-repair enzymes.[4-6] Our results with *in vivo* poly(ADP-ribosyl)ation of p53 suggest that p53 may, similarly, cycle on and off its DNA consensus sequence depending on its level of negative charge based on its poly(ADP-ribosyl)ation state. This may represent a mechanism for regulating transcriptional activation of Bax and Fas by p53 during apoptosis. Alternatively, a polymer-binding site in p53 has been localized near a proteolytic cleavage site,[88] indicating that PAR binding could protect this sequence from proteolysis; similar protection has been noted after binding of monoclonal antibodies adjacent to this region.[89] The significant poly(ADP-ribosyl)ation of p53 early in apoptosis, therefore, suggests that this post-translational modification could also play a role in p53 upregulation by protecting the protein from proteolytic degradation.

11.4 PARP AND GENOMIC STABILITY

Similar to p53, the active role of PARP in cell death may serve to eliminate cells that have accumulated excessive levels of DNA damage, and may therefore function in the maintenance of genomic stability. A number of studies have employed chemical inhibitors,[90-92] dominant negative mutants,[45,46] and PARP antisense RNA[93,94] to examine the function of PARP. These studies have demonstrated that PARP plays a role in reducing the frequency of DNA strand breaks, recombination, gene amplification, micronuclei formation, and sister chromatid exchanges (SCE), all of which are markers of genomic instability, in cells exposed to DNA-damaging agents. PARP-deficient cell lines are hypersensitive to carcinogenic agents and also display increased SCE, implicating PARP as a guardian of the genome that facilitates DNA repair and protects against DNA recombination.[95]

Primary fibroblasts derived from exon 2 PARP knockout mice show an elevated frequency of spontaneous SCE and micronuclei formation in response to treatment with genotoxic agents,[10,40] providing further support for a role of PARP in the maintenance of genomic integrity. Exon 4 PARP knockout mice exhibit extreme sensitivity to γ irradiation and methylnitrosourea and also show increased genomic instability as revealed by a high level of SCE.[11] Immortalized cells derived from these animals are characterized by retarded cell growth, G_2/M block, and chromosomal instability on exposure to DNA-alkylating agents, presumably because of a severe defect in DNA repair.[96]

We recently utilized immortalized fibroblasts derived from exon 2 PARP knockout mice (PARP$^{-/-}$), as well as from control animals of the same strain (PARP$^{+/+}$), to study the role of PARP in genomic stability. FACS analysis initially revealed that these cells exhibit mixed ploidy, including a tetraploid cell population, indicative of genomic instability.[97] The tetraploid population was not observed in PARP[+/+] cells. Further, this tetraploid cell population was no longer apparent in PARP[−/−] cells retransfected with PARP cDNA, suggesting that the reintroduction of PARP into PARP[−/−] cells may have stabilized the genome and resulted in selection against this genomically unstable population (Figure 11.5A).

We characterized the genetic alterations associated with PARP depletion by comparative genomic hybridization (CGH) analysis, a cytogenetic technique that detects chromosomal gains and losses in the test DNA as a measure of genetic instability.[98,99] Although CGH is now commonly used for mapping DNA copy number changes in human tumor genomes, few studies to date have utilized this technique to evaluate genetic instability in transgenic mouse models.[100,101] CGH analysis revealed that PARP[−/−] mice or immortalized PARP[−/−] cells exhibited gains in regions of chromosomes 4, 5, and 14, as well as a deletion in chromosome 14 (Figure 11.5B). We further investigated the effect of stable transfection of immortalized PARP[−/−] fibroblasts with PARP cDNA on the genetic instability of these cells. Reintroduction of PARP cDNA into PARP[−/−] cells appeared to confer stability because these chromosomal gains were no longer detected in these cells, further supporting an essential role for PARP in the maintenance of genomic stability (Figure 11.5B).

In our study, we noted the absence of immunoreactive p53 from these cells as revealed by immunoblot analysis. A previous report indicated that primary fibroblasts from exon 2 PARP[−/−] mice[102] also show reduced basal levels of p53 as well as a defective induction of p53 in response to DNA damage, indicating that PARP-dependent signaling may influence this response. Cells that are unable to synthesize PAR because of unavailability of NAD[103] also show a reduced p53 response. Thus, PARP may regulate genomic stability, at least in part, via p53. Given that the loss of p53 from diploid cells promotes the survival of cells with severe DNA damage and the development of tetraploidy,[104-106] the presence of a tetraploid population among the immortalized PARP[−/−] cells was consistent with the apparent absence of p53. p53 is involved in the maintenance of diploidy as a component of the spindle checkpoint[106] and by regulating centrosome duplication.[107] A functional association of PARP and p53 has been suggested by immunoprecipitation experiments (see above). Thus, both the increased sensitivity of PARP[−/−] mice and cells to DNA-damaging agents[10,11,96] and their genetic instability are consistent with their deficiencies in PARP and p53.

11.5 CONCLUSIONS

PARP has been shown to play active roles in the response to diverse forms of cellular damage resulting from normal metabolic processes, as well as environmental factors, leading to genetic instability. The response may depend upon the level and type of damage, as well as the cell type. In mildly damaged cells, PARP may signal a repair response. In severely damaged cells, PARP activation induces poly(ADP-ribosyl)ation of key nuclear proteins, including p53, and a concomitant lowering of NAD

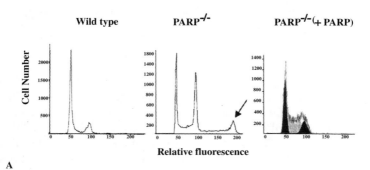

A

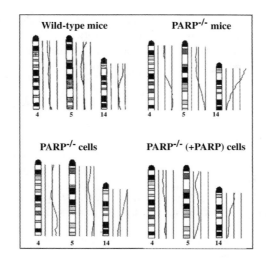

B

FIGURE 11.5 Flow cytometric analysis of immortalized wild-type, PARP$^{-/-}$, and PARP$^{-/-}$ (+PARP) fibroblasts (A), and comparison of the CGH profiles of chromosomes 4, 5, and 14 (B). (A) Cells were harvested 18 h after release from serum deprivation. Nuclei were then prepared and stained with propidium iodide for flow cytometric analysis. In addition to the two major peaks of nuclei at G_0/G_1 and G_2/M apparent in the DNA histograms of wild-type and PARP$^{-/-}$ (+PARP) cells, the DNA histograms of PARP$^{-/-}$ cells exhibit a third peak corresponding to the G_2/M peak of an unstable tetraploid cell population (arrow). (B) Average ratio profiles were computed for all chromosomes and used for the mapping of changes in copy number, with only the results for chromosomes 4, 5, and 14 shown. The three vertical lines to the right of the chromosome ideograms represent values of 0.75, 1, and 1.25 (left to right, respectively) for the fluorescence ratio between the test DNA and the normal control DNA. The ratio profile (curve) was computed as a mean value of at least eight metaphase spreads. A ratio of ≥ 1.25 was regarded as a gain and a ratio of ≤ 0.75 as a loss. PARP$^{-/-}$ (+PARP) fibroblasts did not show the gains at 4C5-ter, 5F-ter, or 14A1-C2 that were apparent in both PARP$^{-/-}$ mice and immortalized PARP$^{-/-}$ cells, although they retained the partial loss at 14D3-ter.

and ATP levels, resulting ultimately in cell death, the form of which (apoptosis vs. necrosis) may depend upon the time of onset of caspase-mediated PARP cleavage.

ABBREVIATIONS

The abbreviations used are AIF, apoptosis-inducing factor; CAD, caspase-activated DNase; CGH, comparative genomic hybridization; DBD, DNA-binding domain; Dex, dexamethasone; DFF, DNA fragmentation factor; DISC, death-inducing signaling complex; FADD, Fas-associating protein with death domain; ICAD, inhibitor of caspase-activated DNase; PAR, poly(ADP-ribose); PARP, poly(ADP-ribose) polymerase; PCNA, proliferating cell nuclear antigen; SCE, sister chromatid exchange.

ACKNOWLEDGMENTS

This work was supported in part by Grants CA25344 and PO1CA74175 from the National Cancer Institute, by the U.S. Air Force Office of Scientific Research (Grant AFOSR-89-0053), and by the U.S. Army Medical Research and Development Command (Contract DAMD17-90-C-0053) (to M.E.S.) and DAMD 17-96-C-6065 (to D.S.R).

REFERENCES

1. Berger, N., A., Petzold, S. J., and Berger, S. J., Association of poly(ADP-ribose) synthesis with cessation of DNA synthesis and DNA fragmentation, *Biochim. Biophys. Acta*, 564, 90, 1979.
2. Stevnsner, T., Ding, R., Smulson, M., and Bohr, V. A., Inhibition of gene-specific repair of alkylation damage in cells depleted of poly(ADP-ribose) polymerase, *Nucl. Acids Res.*, 22, 4620, 1994.
3. Jacobson, M. K. and Jacobson, E. L., *ADP-Ribose Transfer Reactions: Mechanisms and Biological Significance*, Springer-Verlag, New York, 1989, 527.
4. Satoh, M. S. and Lindahl, T., Role of poly(ADP-ribose) formation in DNA repair, *Nature*, 356, 356, 1992.
5. Satoh, M. S., Poirier, G. G., and Lindahl, T., NAD^+-dependent repair of damaged DNA by human cell extracts, *J. Biol. Chem.*, 268, 5480, 1993.
6. Smulson, M., Istock, N., Ding, R., and Cherney, B., Deletion mutants of poly(ADP-ribose) polymerase support a model of cyclic association and dissociation of enzyme from DNA ends during DNA repair, *Biochemistry*, 33, 6186, 1994.
7. Kannan, P., Yu, Y., Wankhade, S., and Tainsky, M. A., PolyADP-ribose polymerase is a coactivator for AP-2-mediated transcriptional activation, *Nucleic Acids Res*, 27, 866, 1999.
8. Meisterernst, M., Stelzer, G., and Roeder, R. G., Poly(ADP-ribose) polymerase enhances activator-dependent transcription in vitro, *Proc. Natl. Acad. Sci. U.S.A.*, 94, 2261, 1997.
9. Simbulan, C., Suzuki, M., Izuta, S., Sakurai, T., Savoysky, E., Kojima, K., Miyahara, K., Shizuta, Y., and Yoshida, S., Poly(ADP-ribose) polymerase stimulates DNA polymerase alpha, *J. Biol. Chem.*, 268, 93, 1993.

10. Wang, Z. Q., Auer, B., Stingl, L., Berghammer, H., Haidacher, D., Schweiger, M., and Wagner, E. F., Mice lacking ADPRT and poly(ADP-ribosyl)ation develop normally but are susceptible to skin disease, *Genes Dev.*, 9, 509, 1995.
11. Ménissier-de Murcia, J., Niedergang, C., Trucco, C., Ricoul, M., Dutrillaux, B., Mark, M., Oliver, F. J., Masson, M., Dierich, A., Le Meur, M., Walztinger, C., Chambon, P., and de Murcia, G., Requirement of poly(ADP-ribose) polymerase in recovery from DNA damage in mice and in cells, *Proc. Natl. Acad. Sci. U.S.A.*, 94, 7303, 1997.
12. Masutani, M., Suzuki, H., Kamada, N., Watanabe, M., Ueda, O., Nozaki, T., Jishage, K., Watanabe, T., Sugimoto, T., Nakagama, H., Ochiya, T., and Sugimura, T., Poly(ADP-ribose) polymerase gene disruption conferred mice resistant to streptozotocin-induced diabetes, *Proc. Natl. Acad. Sci. U.S.A.*, 96, 2301, 1999.
13. Berger, N. A., Sims, J. L., Catino, D. M., and Berger, S. J., Poly(ADP-ribose)polymerase mediates the suicide response to massive DNA damage: studies in normal and DNA-repair defective cells, in *ADP-Ribosylation, DNA Repair and Cancer*, Proceedings of the 13th International Symposium of the Princess Takamatsu Cancer Research Fund, Miwa, M., Hayaishi, O., Shall, S., Smulson, M. E., Sugimura, T., Eds., Japan Scientific Societies Press, Tokyo, 1983, 219.
14. Berger, N. A., Poly(ADP-ribose) in the cellular response to DNA damage, *Radiat. Res.*, 101, 4, 1985.
15. Burkart, V., Wang, Z. Q., Radons, J., Heller, B., Herceg, Z., Stingl, L., Wagner, E. F., and Kolb, H., Mice lacking the poly(ADP-ribose) polymerase gene are resistant to pancreatic beta-cell destruction and diabetes development induced by streptozocin, *Nat. Med.*, 5, 314, 1999.
16. Pieper, A. A., Brat, D. J., Krug, D. K., Watkins, C. C., Gupta, A., Blackshaw, S., Verma, A., Wang, Z. Q., and Snyder, S. H., Poly(ADP-ribose) polymerase-deficient mice are protected from streptozotocin-induced diabetes, *Proc. Natl. Acad. Sci. U.S.A.*, 96, 3059, 1999.
17. Mandir, A. S., Przedborski, S., Jackson-Lewis, V., Wang, Z. Q., Simbulan-Rosenthal, C. M., Smulson, M. E., Hoffman, B. E., Guastella, D. B., Dawson, V. L., and Dawson, T. M., Poly(ADP-ribose) polymerase activation mediates 1-methyl-4-phenyl-1, 2,3,6-tetrahydropyridine (MPTP)-induced parkinsonism, *Proc. Natl. Acad. Sci. U.S.A.*, 96, 5774, 1999.
18. Eliasson, M. J., Sampei, K., Mandir, A. S., Hurn, P. D., Traystman, R. J., Bao, J., Pieper, A., Wang, Z. Q., Dawson, T. M., Snyder, S. H., and Dawson, V. L., Poly(ADP-ribose) polymerase gene disruption renders mice resistant to cerebral ischemia, *Nat. Med.*, 3, 1089, 1997.
19. Endres, M., Wang, Z. Q., Namura, S., Waeber, C., and Moskowitz, M. A., Ischemic brain injury is mediated by the activation of poly(ADP-ribose)polymerase, *J. Cereb. Blood Flow Metab.*, 17, 1143, 1997.
20. Szabó, C., Lim, L. H., Cuzzocrea, S., Getting, S. J., Zingarelli, B., Flower, R. J., Salzman, A. L., and Perretti, M., Inhibition of poly(ADP-ribose) synthetase attenuates neutrophil recruitment and exerts antiinflammatory effects, *J. Exp. Med.*, 186, 1041, 1997.
21. Zingarelli, B., Szabó, C., and Salzman, A. L., Blockade of poly(ADP-ribose) synthetase inhibits neutrophil recruitment, oxidant generation, and mucosal injury in murine colitis, *Gastroenterology*, 116, 335, 1999.
22. Rosenthal, D. S., Ding, R., Simbulan-Rosenthal, C. M. G., Vaillancourt, J. P., Nicholson, D. W., and Smulson, M. E., Intact cell evidence of the early synthesis, and subsequent late apopain-mediated suppression, of poly(ADP-ribose) during apoptosis, *Exp. Cell Res.*, 232, 313, 1997.

23. Neamati, N., Fernandez, A., Wright, S., Kiefer, J., and McConkey, D. J., Degradation of lamin B1 precedes oligonucleosomal DNA fragmentation in apoptotic thymocytes and isolated thymocyte nuclei, *J. Immunol.*, 154, 3788, 1995.
24. Nosseri, C., Coppola, S., and Ghibelli, L., Possible involvement of poly(ADP-ribosyl) polymerase in triggering stress-induced apoptosis, *Exp. Cell Res.*, 212, 367, 1994.
25. Kaufmann, S. H., Desnoyers, S., Ottaviano, Y., Davidson, N. E., and Poirier, G. G., Specific proteolytic cleavage of poly(ADP-ribose)polymerase: an early marker of chemotherapy-induced apoptosis, *Cancer Res.*, 53, 3976, 1993.
26. Nicholson, D. W., Ali, A., Thornberry, N. A., Vaillancourt, J. P., Ding, C. K., Gallant, M., Gareau, Y., Griffin, P. R., Labelle, M., Lazebnik, Y. A., Munday, N. A., Raju, S. M., Smulson, M. E., Yamin, T. T., Yu, V. L., and Miller, D. K., Identification and inhibition of the ICE/CED-3 protease necessary for mammalian apoptosis, *Nature*, 376, 37, 1995.
27. Yuan, J., Shaham, S., Ledoux, S., Ellis, H. M., and Horvitz, H. R., The *C. elegans* death gene *ced*-3 encodes a protein similar to mammalian interleukin-1-b-converting enzyme, *Cell*, 75, 641, 1993.
28. Tewari, M., Quan, L. T., O'Rourke, K., Desnoyers, S., Zeng, Z., Beidler, D. R., Poirier, G. G., Salvesen, G. S., and Dixit, V. M., Yama/CPP32b, a mammalian homolog of CED-3, is a crmA-inhibitable protease that cleaves the death substrate poly(ADP-ribose) polymerase, *Cell*, 81, 801, 1995.
29. Rosenthal, D. S., Ding, R., Simbulan-Rosenthal, C. M. G., Cherney, B., Vanek, P., and Smulson, M. E., Detection of DNA breaks in apoptotic cells utilizing the DNA binding domain of poly(ADP-ribose) polymerase with fluorescence microscopy, *Nucl. Acids Res.*, 25, 1437, 1997.
30. Herceg, Z. and Wang, Z. Q., Failure of poly(ADP-ribose) polymerase cleavage by caspases leads to induction of necrosis and enhanced apoptosis, *Mol. Cell Biol.*, 19, 5124, 1999.
31. Negri, C., Donzelli, M., Bernardi, R., Rossi, L., Burkle, A., and Scovassi, A. I., Multiparametric staining to identify apoptotic human cells, *Exp. Cell Res.*, 234, 174, 1997.
32. Cherney, B. W., Chaudhry, B., Bhatia, K., Butt, T. R., and Smulson, M. E., Expression and mutagenesis of human poly(ADP-ribose) polymerase as a ubiquitin fusion protein from *Escherichia coli, Biochemistry*, 30, 10420, 1991.
33. Rosenthal, D. S., Simbulan-Rosenthal, C. M., Liu, W. F., Smith, W., Benton, B., Ray, R., and Smulson, M. E., Sulfur mustard induces keratinocyte apoptosis via a PARP-dependent pathway, in preparation, 2000.
34. Simbulan-Rosenthal, C. M., Rosenthal, D. S., Iyer, S., Boulares, A. H., and Smulson, M. E., Transient poly(ADP-ribosyl)ation of nuclear proteins and role for poly(ADP-ribose) polymerase in the early stages of apoptosis, *J. Biol. Chem.*, 273, 13703, 1998.
35. Enari, M., Sakahira, H., Yokoyama, H., Okawa, K., Iwamatsu, A., and Nagata, S., A caspase-activated DNase that degrades DNA during apoptosis, and its inhibitor ICAD, *Nature*, 391, 43, 1998.
36. Alkhatib, H. M., Chen, D. F., Cherney, B., Bhatia, K., Notario, V., Giri, C., Stein, G., Slattery, E., Roeder, R. G., and Smulson, M. E., Cloning and expression of cDNA for human poly(ADP-ribose) polymerase, *Proc. Natl. Acad. Sci. U.S.A.*, 84, 1224, 1987.
37. Agarwal, S., Drysdale, B. E., and Shin, H. S., Tumor necrosis factor-mediated cytotoxicity involves ADP-ribosylation, *J. Immunol.*, 140, 4187, 1988.
38. Monti, D., Cossarizza, A., Salvioli, S., Franceschi, C., Rainaldi, G., Straface, E., Rivabene, R., and Malorni, W., Cell death protection by 3-aminobenzamide and other poly(ADP- ribose)polymerase inhibitors: different effects on human natural killer and lymphokine activated killer cell activities, *Biochem. Biophys. Res. Commun.*, 199, 525, 1994.

39. Kuo, M. L., Chau, Y. P., Wang, J. H., and Shiah, S. G., Inhibitors of poly(ADP-ribose) polymerase block nitric oxide-induced apoptosis but not differentiation in human leukemia HL-60 cells, *Biochem. Biophys. Res. Commun.*, 219, 502, 1996.
40. Wang, Z., Stingl, L., Morrison, C., Jantsch, M., Los, M., Schulze-Osthoff, K., and Wagner, E., PARP is important for genomic stability but dispensable in apoptosis, *Genes Dev.*, 11, 2347, 1997.
41. Boulares, A. H., Yakovlev, A. G., Ivanova, V., Stoica, B. A., Wang, G., Iyer, S., and Smulson, M. E., Role of PARP cleavage in apoptosis, *J. Biol. Chem.*, 274, 22932–22940, 1999.
42. Oliver, F. J., de la Rubia, G., Rolli, V., Ruiz-Ruiz, M. C., de Murcia, G., and Murcia, J. M., Importance of poly(ADP-ribose) polymerase and its cleavage in apoptosis. Lesson from an uncleavable mutant, *J. Biol. Chem.*, 273, 33533, 1998.
43. Le Rhun, Y., Kirkland, J. B., and Shah, G. M., Cellular responses to DNA damage in the absence of poly(ADP-ribose) polymerase, *Biochem. Biophys. Res. Commun.*, 245, 1, 1998.
44. Ikejima, M., Noguchi, S., Yamashita, R., Ogura, T., Sugimura, T., Gill, D. M., and Miwa, M., The zinc fingers of human poly(ADP-ribose) polymerase are differentially required for the recognition of DNA breaks and nicks and the consequent enzyme activation. Other structures recognize intact DNA, *J. Biol. Chem.*, 265, 21907, 1990.
45. Schreiber, V., Hunting, D., Trucco, C., Gowans, B., Grunwald, P., de Murcia, G., and Ménissier-de Murcia, J., A dominant negative mutant of human PARP affects cell recovery, apoptosis, and sister chromatid exchange following DNA damage, *Proc. Natl. Acad. Sci. U.S.A.*, 92, 4753, 1995.
46. Kupper, J., Muller, M., and Burkle, A., *trans*-Dominant inhibition of poly(ADP-ribosyl)ation potentiates carcinogen-induced gene amplification in SV40-transformed Chinese hamster cells, *Cancer Res.*, 56, 2715, 1996.
47. Scaffidi, C., Fulda, S., Srinivasan, A., Friesen, C., Li, F., Tomaselli, K. J., Debatin, K. M., Krammer, P. H., and Peter, M. E., Two CD95 (APO-1/Fas) signaling pathways, *EMBO J.*, 17, 1675, 1998.
48. Hueber, A. O., Zornig, M., Lyon, D., Suda, T., Nagata, S., and Evan, G. I., Requirement for the CD95 receptor-ligand pathway in c-Myc-induced apoptosis, *Science*, 278, 1305, 1997.
49. Susin, S. A., Lorenzo, H. K., Zamzami, N., Marzo, I., Brenner, C., Larochette, N., Prevost, M. C., Alzari, P. M., and Kroemer, G., Mitochondrial release of caspase-2 and -9 during the apoptotic process, *J. Exp. Med.*, 189, 381, 1999.
50. Susin, S. A., Lorenzo, H. K., Zamzami, N., Marzo, I., Snow, B. E., Brothers, G. M., Mangion, J., Jacotot, E., Costantini, P., Loeffler, M., Larochette, N., Goodlett, D. R., Aebersold, R., Siderovski, D. P., Penninger, J. M., and Kroemer, G., Molecular characterization of mitochondrial apoptosis-inducing factor [see comments], *Toxicol. Lett.*, 102–103, 121, 1998.
51. Lorenzo, H. K., Susin, S. A., Penninger, J., and Kroemer, G., Apoptosis inducing factor (AIF): a phylogenetically old, caspase-independent effector of cell death, *Cell Death Differ.*, 6, 516, 1999.
52. Susin, S. A., Zamzami, N., Castedo, M., Daugas, E., Wang, H. G., Geley, S., Fassy, F., Reed, J. C., and Kroemer, G., The central executioner of apoptosis: multiple connections between protease activation and mitochondria in Fas/APO-1/CD95- and ceramide-induced apoptosis, *J. Exp. Med.*, 186, 25, 1997.
53. Kokileva, L., Disassembly of genome of higher eukaryotes: pulsed-field gel electrophoretic study of initial stages of chromatin and DNA degradation in rat liver and thymus nuclei by VM-26 and selected proteases, *Comp. Biochem. Physiol. B Biochem. Mol. Biol.*, 121, 145, 1998.

54. Higuchi, Y. and Matsukawa, S., Appearance of 1-2 Mbp giant DNA fragments as an early common response leading to cell death induced by various substances that cause oxidative stress, *Free Radical Biol. Med.*, 23, 90, 1997.
55. Gu, J., Dong, R. P., Zhang, C., McLaughlin, D. F., Wu, M. X., and Schlossman, S. F., Functional interaction of DFF35 and DFF45 with caspase-activated DNA fragmentation nuclease DFF40, *J. Biol. Chem.*, 274, 20759, 1999.
56. Liu, X., Li, P., Widlak, P., Zou, H., Luo, X., Garrard, W. T., and Wang, X., The 40-kDa subunit of DNA fragmentation factor induces DNA fragmentation and chromatin condensation during apoptosis, *Proc. Natl. Acad. Sci. U.S.A.*, 95, 8461, 1998.
57. Liu, X., Zou, H., Slaughter, C., and Wang, X., DFF, a heterodimeric protein that functions downstream of caspase-3 to trigger DNA fragmentation during apoptosis, *Cell*, 89, 175, 1997.
58. Hughes, F. M., Jr., Evans-Storms, R. B., and Cidlowski, J. A., Evidence that non-caspase proteases are required for chromatin degradation during apoptosis, *Cell Death Differ.*, 5, 1017, 1998.
59. Barry, M. A. and Eastman, A., Identification of deoxyribonuclease II as an endonuclease involved in apoptosis, *Arch. Biochem. Biophys.*, 300, 440, 1993.
60. Peitsch, M. C., Polzar, B., Stephan, H., Crompton, T., MacDonald, H. R., Mannherz, H. G., and Tschopp, J., Characterization of the endogenous deoxyribonuclease involved in nuclear DNA degradation during apoptosis (programmed cell death), *EMBO J*, 12, 371, 1993.
61. Peitsch, M. C., Muller, C., and Tschopp, J., DNA fragmentation during apoptosis is caused by frequent single-strand cuts, *Nucl. Acids Res.*, 21, 4206, 1993.
62. Yakovlev, A. G., Wang, G., Stoica, B. A., Simbulan-Rosenthal, C. M., Yoshihara, K., and Smulson, M. E., Role of DNAS1L3 in Ca^{2+}- and Mg^{2+}-dependent cleavage of DNA into oligonucleosomal and high molecular mass fragments, *Nucl. Acids Res.*, 27, 1999, 1999.
63. Montague, J. W., Hughes, F. M., Jr., and Cidlowski, J. A., Native recombinant cyclophilins A, B, and C degrade DNA independently of peptidylprolyl *cis-trans*-isomerase activity. Potential roles of cyclophilins in apoptosis, *J. Biol. Chem.*, 272, 6677, 1997.
64. Feldenberg, L. R., Thevananther, S., del Rio, M., de Leon, M., and Devarajan, P., Partial ATP depletion induces fas- and caspase-mediated apoptosis in MDCK cells, *Am. J. Physiol.*, 276, F837, 1999.
65. Eguchi, Y., Srinivasan, A., Tomaselli, K. J., Shimizu, S., and Tsujimoto, Y., ATP-dependent steps in apoptotic signal transduction, *Cancer Res.*, 59, 2174, 1999.
66. Hu, Y., Benedict, M. A., Ding, L., and Nunez, G., Role of cytochrome c and dATP/ATP hydrolysis in Apaf-1-mediated caspase-9 activation and apoptosis, *EMBO J.*, 18, 3586, 1999.
67. Saleh, A., Srinivasula, S. M., Acharya, S., Fishel, R., and Alnemri, E. S., Cytochrome c and dATP-mediated oligomerization of Apaf-1 is a prerequisite for procaspase-9 activation, *J. Biol. Chem.*, 274, 17941, 1999.
68. Zou, H., Li, Y., Liu, X., and Wang, X., An APAF-1.cytochrome c multimeric complex is a functional apoptosome that activates procaspase-9, *J. Biol. Chem.*, 274, 11549, 1999.
69. Kasid, U. N., Halligan, B., Liu, L. F., Dritschilo, A., and Smulson, M. E., Poly(ADP-ribose)-mediated post-translational modification of chromatin-associated human topoisomerase I. Inhibitory effects on catalytic activity, *J. Biol. Chem.*, 264, 18687, 1989.

70. Simbulan-Rosenthal, C. M. G., Rosenthal, D. S., Hilz, H., Hickey, R., Malkas, L., Applegren, N., Wu, Y., Bers, G., and Smulson, M., The expression of poly(ADP-ribose) polymerase during differentiation-linked DNA replication reveals that this enzyme is a component of the multiprotein DNA replication complex, *Biochemistry*, 35, 11622, 1996.
71. Baksi, K., Alkhatib, H., and Smulson, M. E., *In vivo* characterization of the poly ADP-ribosylation of SV40 chromatin and large T antigen by immunofractionation, *Exp. Cell Res.*, 172, 110, 1987.
72. Poirier, G. G., de Murcia, G., Jongstra-Bilen, J., Niedergang, C., and Mandel, P., Poly(ADP-ribosyl)ation of polynucleosomes causes relaxation of chromatin structure, *Proc. Natl. Acad. Sci. U.S.A.*, 79, 3423, 1982.
73. Butt, T. R., DeCoste, B., Jump, D., Nolan, N., and Smulson, M. E., Characterization of a putative poly adenosine disphosphate ribose chromatin complex, *Biochemistry*, 19, 5243, 1980.
74. Kastan, M. B., Zhan, Q., El-Delry, W. S., Carrier, F., Jacks, T., Walsh, W. V., Plunkett, B. S., Vogelstein, B., and Fornace, A. J. J., A mammalian cell cycle checkpoint pathway utilizing p53 and *GADD45* is defective in ataxia-telangiectasia, *Cell*, 71, 587, 1992.
75. O'Connor, P. M., Jackman, J., Jondle, D., Bhatia, K., Magrath, I., and Kohn, K. W., Role of the p53 tumor suppressor gene in cell cycle arrest and radiosensitivity of Burkitt's lymphoma cell lines, *Cancer Res.*, 53, 4776, 1993.
76. Levine, A., p53, the cellular gatekeeper for growth and division, *Cell*, 88, 323, 1997.
77. Fisher, D., Apoptosis in cancer therapy: crossing the threshold, *Cell*, 78, 539, 1994.
78. Yonish-Rouach, E., Resnitzky, D., Lotem, J., Sachs, L., Kimchi, A., and Oren, M., Wild-type p53 induces apoptosis of myeloid leukaemic cells that is inhibited by interleukin-6, *Nature*, 352, 345, 1991.
79. Berger, N. A. and Petzold, S. J., Identification of the requirements of DNA for activation of poly(ADP-ribose) polymerase, *Biochemistry*, 24, 4352, 1985.
80. Fritsche, M., Haessler, C., and Brandner, G., Induction of nuclear accumulation of the tumor suppressor protein by p53 by DNA damaging agents, *Oncogene*, 8, 307, 1993.
81. Kastan, M. B., Onyekwere, O., Sidransky, D., Vogelstein, B., and Craig, R. W., Participation of p53 protein in the cellular response to DNA damage, *Cancer Res.*, 51, 6304, 1991.
82. Vaziri, H., West, M., Allsop, R., Davison, T., Wu, Y., Arrowsmith, C., Poirier, G., and Benchimol, S., ATM-dependent telomere loss in aging human diploid fibroblasts and DNA damage lead to the post-translational activation of p53 protein involving poly(ADP-ribose) polymerase, *EMBO J.*, 16, 6018, 1997.
83. Wesierska-Gadek, J., Bugajska-Schretter, A., and Cerni, C., ADP-ribosylation of p53 tumor suppressor protein: mutant but not wild-type p53 is modified, *J. Cell. Biochem.*, 62, 90, 1996.
84. Wesierska-Gadek, J., Schmid, G., and Cerni, C., ADP-ribosylation of wild-type p53 in vitro: binding of p53 protein to specific p53 consensus sequence prevents its modification, *Biochem. Biophys. Res. Commun.*, 224, 96, 1996.
85. Simbulan-Rosenthal, C. M., Rosenthal, D. S., and Smulson, M. E., Poly(ADP-ribosyl)ation of p53 during apoptosis in human osteosarcoma cells, *Cancer Res.*, 59, 2190, 1999.
86. Chinnaiyan, A., Orth, K., O'Rourke, K., Duan, H., Poirier, G., and Dixit, V., Molecular ordering of the cell death pathway. Bcl-2 and Bcl-xL function upstream of the CED-3-like apoptotic proteases, *J. Biol. Chem.*, 271, 4573, 1996.

87. Muller, M., Wilder, S., Bannasch, D., Israeli, D., Lelbach, K., Li-Weber, M., Friedman, S., Galle, P., Stremmel, W., Oren, M., and Krammer, P., p53 activates the CD95 (APO-1/Fas) gene in response to DNA damage by anticancer agents, *J. Exp. Med.*, 188, 2033, 1998.
88. Malanga, M., Pleschke, J., Kleczkowska, H., and Althaus, F., Poly(ADP-ribose) binds to specific domains of p53 and alters its DNA binding functions, *J. Biol. Chem.*, 273, 11839, 1998.
89. Li, X. and Coffino, P., Identification of a region of p53 that confers lability, *J. Biol. Chem.*, 271, 4447, 1996.
90. Morgan, W. and Cleaver, J., 3-Aminobenzamide synergistically increases sister-chromatid exchanges in cells exposed to methyl methanesulfonate but not to ultraviolet light, *Mutat. Res.*, 104, 361, 1982.
91. Burkle, A., Heilbronn, R., and zur Hausen, H., Potentiation of carcinogen-induced methotrexate resistance and dihydrofolate reductase gene amplification by inhibitors of poly(adenosinediphosphate-ribose) polymerase, *Cancer Res.*, 50, 5756, 1990.
92. Waldman, A. and Waldman, B., Stimulation of intrachromosomal homologous recombination in mammalian cells by an inhibitor of poly(ADP-ribosyl)ation, *Nucl. Acids Res.*, 19, 5943, 1991.
93. Ding, R., Pommier, Y., Kang, V. H., and Smulson, M. E., Depletion of poly(ADP-ribose) polymerase by antisense RNA expression results in a delay in DNA strand break rejoining, *J. Biol. Chem.*, 267, 12804, 1992.
94. Ding, R. and Smulson, M. E., Depletion of nuclear poly(ADP-ribose) polymerase by antisense RNA expression: influences on genomic stability, chromatin organization and carcinogen cytotoxicity, *Cancer Res.*, 54, 4627, 1994.
95. Chatterjee, S., Berger, S., and Berger, N., Poly(ADP-ribose) polymerase: a guardian of the genome that facilitates DNA repair by protecting against DNA recombination, *Mol. Cell. Biochem.*, 193, 23, 1999.
96. Trucco, C., Oliver, F., de Murcia, G., and Ménissier-de Murcia, J., DNA repair defect in PARP-deficient cell lines, *Nucl. Acids Res.*, 26, 2644, 1998.
97. Andreassen, P., Martineau, S., and Margolis, R., Chemical induction of mitotic checkpoint override in mammalian cells results in aneuploidy following a transient tetraploid state, *Mutat. Res.*, 372, 181, 1996.
98. Kallioniemi, A., Kallioniemi, O.-P., Sudar, D., Rutovitz, D., Gray, J., Waldman, F., and Pinkel, D., Comparative genomic hybridization for molecular cytogenetic analysis of solid tumors, *Science*, 258, 818, 1992.
99. du Manoir, S., Speicher, M., and Jovs, S., Detection of complete and partial chromosome gains and losses by comparative genomic *in situ* hybridization, *Hum. Genet.*, 90, 590, 1993.
100. Shi, Y., Naik, P., Dietrich, W., Gray, J., Hanahan, D., and Pinkel, D., DNA copy number changes associated with characteristic LOH in islet cell carcinomas of transgenic mice, *Genes Chromosomes Cancer*, 2, 104, 1997.
101. Weaver, Z., McCormack, S., Liyanage, M., du Manoir, S., Coleman, A., Schrock, E., Dickson, R., and Ried, T., A recurring pattern of chromosomal aberrations in mammary gland of MMTV-cmyc transgenic mice., *Genes Chromosomes Cancer*, 25, 195, 1999.
102. Agarwal, M., Agarwal, A., Taylor, W., Wang, Z. Q., and Wagner, E., Defective induction but normal activation and function of p53 in mouse cells lacking PARP, *Oncogene*, 15, 1035, 1997.

103. Whitacre, C. M., Hashimoto, H., Tsai, M.-L., Chatterjee, S., Berger, S. J., and Berger, N. A., Involvement of NAD-poly(ADP-ribose) metabolism in p53 regulation and its consequences, *Cancer Res.*, 55, 3697, 1995.
104. Ramel, S., Sanchez, C., Schimke, M., Neshat, K., Cross, S., Raskind, W., and Reid, B., Inactivation of p53 and the development of tetraploidy in the elastase-SV40 Tantigen transgenic mouse pancreas, *Pancreas*, 11, 213, 1995.
105. Yin, X., Grove, L., Datta, N., Long, M., and Prochownik, E., C-myc overexpression and p53 loss cooperate to promote genomic instability, *Oncogene*, 18, 1177, 1999.
106. Cross, S., Sanchez, C., Morgan, C., Schimke, M., Ramel, S., Idzerda, R., Raskind, W., and Reid, B., A p53-dependent mouse spindle ckeckpoint, *Science*, 267, 1353, 1995.
107. Fukasawa, K., Choi, T., Kuriyama, R., Rulong, S., and Vande Woude, G., Abnormal centrosome amplification in the absence of p53, *Science*, 271, 1744, 1996.

12 Pleiotropic Roles of Poly(ADP-Ribosyl)ation of DNA-Binding Proteins

*Cynthia M. Simbulan-Rosenthal,
Dean S. Rosenthal, and Mark E. Smulson*

CONTENTS

12.1 Introduction 252
12.2 Poly(ADP-Ribosyl)ation of Specific DNA-Binding Proteins
during Early Apoptosis 253
    12.2.1 Effects of Preventing Early PARP Activation 253
    12.2.2 Acceptors for Poly(ADP-Ribosyl)ation during Early
         Apoptosis 255
         12.2.2.1 Poly(ADP-Ribosyl)ation of Histone H1 during
               Early Apoptosis 255
         12.2.2.2 Poly(ADP-Ribosyl)ation of p53 during Early
               Apoptosis 255
               12.2.2.2.1 Regulation of p53 Function 256
               12.2.2.2.2 p53 Stabilization and Accumulation 258
         12.2.2.3 Poly(ADP-Ribosyl)ation of Topoisomerases I and IIβ:
               A Role in Protein Stabilization 258
         12.2.2.4 Poly(ADP-Ribosyl)ation of $Ca^{2+}Mg^{2+}$-Dependent
               Endonucleases: Effects on Chromatin 259
12.3 Poly(ADP-Ribosyl)ation of DNA-Binding Component Proteins of the
DNA Sythesome 259
    12.3.1 PARP, A Core Component of the DNA Synthesome 259
    12.3.2 Poly(ADP-Ribosyl)ation of Components of the DNA
         Synthesome 262
         12.3.2.1 Acceptors for Poly(ADP-Ribosyl)ation in the DNA
               Synthesome: PCNA, Topo I, and DNA Pol α 262
         12.3.2.2 Poly(ADP-Ribosyl)ation of DNA Pol α and δ in the
               MRC: Effects on Activity 262
         12.3.2.3 Poly(ADP-Ribosyl)ation of PCNA and Topo I: A Role
               in Recruitment into the DNA Synthesome during Early
               S Phase 264

12.4 Poly(ADP-Ribosyl)ation of DNA-Binding Proteins in Gene
Expression and Transcription 264
  12.4.1 Role of PARP and/or PAR in the Expression of MRC
    Components 264
      12.4.1.1 Expression of DNA Pol α and DNA Primase 264
      12.4.1.2 Expression and Promoter Activity of the Transcription
        Factor E2F-1 264
  12.4.2 Poly(ADP-Ribosyl)ation of DNA-Binding Transcription Factors:
    Effects on Binding to Their Consensus Sequences 265
      12.4.2.1 Physical Association with PARP (TEF-1, AP-2) and
        Poly(ADP-Ribosyl)ation of Specific Transcription
        Factors (TEF-1, TFIIF, YY1, SP-1, TBP, CREB) 265
      12.4.2.2 Poly(ADP-Ribosyl)ation of Transcription Factors p53
        and NFκB *In Vitro*: Effects on Binding to Their
        Respective Consensus Sequences 267
12.5 Conclusions 268
Acknowledgments 270
References 270

12.1 INTRODUCTION

One of the earliest nuclear events that follows DNA strand breakage during DNA repair after exposure to γ irradiation or alkylating agents, as well as during the early stages of DNA replication and apoptosis, is the poly(ADP-ribosyl)ation of nuclear DNA-binding proteins that are localized near DNA strand breaks. The enzyme poly(ADP-ribose) polymerase (PARP, EC 2.4.2.30) catalyzes the poly(ADP-ribosyl)ation of these proteins, including itself, with the respiratory coenzyme NAD as substrate; it is activated only when bound to single- or double-stranded DNA ends via its two zinc fingers, which recognize DNA breaks independent of DNA sequence.[1,2] By undergoing automodification, PARP functions as a "molecular nick sensor" and cycles on and off the DNA ends during DNA repair *in vitro*.[3-6] Inhibition of PARP activity with chemical inhibitors[7] or by dominant negative mutants[8-12] as well as PARP depletion by antisense RNA expression[13-19] have shown that the enzyme plays pleiotropic roles in various nuclear processes, including DNA repair, gene expression, differentiation, DNA replication, and apoptosis. PARP depletion by antisense RNA expression decreases the initial rate of DNA repair in HeLa cells[13] and keratinocytes,[15] reduces cell survival after exposure to mutagenic agents, increases gene amplification,[14] and blocks progression of Fas-mediated apoptosis,[18] as well as differentiation of 3T3-L1 preadipocytes, which is likely attributable to their failure to undergo differentiation-linked DNA replication.[16,17]

The physiological roles of PARP have also been examined increasingly through gene disruption in PARP knockout mice. Several PARP knockout mice, established by disrupting the PARP gene at exon 2,[20] exon 4,[21] and exon 1[22] by homologous recombination, express no immunodetectable PARP protein nor do they exhibit any significant poly(ADP-ribosyl)ation. Although poly(ADP-ribose) (PAR)-synthesizing activity has been detected in PARP[-/-] cells, attributed to a PARP homoloque (PARP-2), this activity is much lower than that in wild-type cells and has not been shown to modify proteins aside from itself; thus, it may not fully compensate for PARP depletion.[23-24] Despite variations in the physiological phenotypes of PARP knockout animals, these mice exhibit increased genomic instability as shown by elevated levels of sister chromatid exchanges.[20,21] Thymocytes from these animals show a delayed recovery after exposure to γ radiation;[20] splenocytes undergo abnormal apoptosis, and primary fibroblasts exhibit proliferation deficiencies and defects in DNA repair.[21,25]

PARP catalyzes the poly(ADP-ribosyl)ation of a limited number of nuclear proteins such as histones,[26,27] topoisomerases I and II,[17,28-30] DNA polymerase α,[19,31] and proliferating cell nuclear antigen (PCNA).[17] The catalytic activity of some DNA-binding enzymes can be modulated by poly(ADP-ribosyl)ation;[19,28-31] and in most instances, this post-translational modification inhibits the activity of the modified protein, presumably because of a marked decrease in DNA-binding affinity caused by electrostatic repulsion between DNA and PAR. Poly(ADP-ribosyl)ation of nucleosomal proteins may also promote access of various replicative and repair enzymes to sites of DNA breaks by altering the nucleosomal structure of the DNA at these sites.[32,33] In this chapter, we focus on the pleiotropic roles of poly(ADP-ribosyl)ation of specific nuclear DNA-binding acceptor proteins by PARP in nuclear processes, such as during the early stages of apoptosis and DNA replication, both of which are characterized by an early transient PARP activation.

12.2 POLY(ADP-RIBOSYL)ATION OF SPECIFIC DNA-BINDING PROTEINS DURING EARLY APOPTOSIS

12.2.1 EFFECTS OF PREVENTING EARLY PARP ACTIVATION

Because PARP is activated by binding to DNA ends or strand breaks, PARP is thought to contribute to cell death via NAD and ATP depletion following DNA damage.[34] PARP is also implicated in the induction of both p53 expression and apoptosis,[35] with the specific proteolysis of the enzyme thought to be a key apoptotic event.[36-38] Caspase-3-mediated cleavage of PARP into 89- and 24-kDa fragments during drug-induced[39] or spontaneous[36,40] apoptosis separates the PARP DNA-binding domain from its catalytic site, thus essentially inactivating the enzyme. Poly(ADP-ribosyl)ation of nuclear proteins occurs early during spontaneous apoptosis in human osteosarcoma cells, prior to commitment to cell death, and is followed by PARP cleavage; only small amounts of PAR remain at the later stages of apoptosis, despite the presence of massive DNA strand breaks.[40] This transient poly(ADP-ribosyl)ation of nuclear proteins during early apoptosis occurs in other cell systems as well, such as during camptothecin-induced apoptosis in HL-60 cells and Fas-mediated apoptosis in Jurkat T-cells, 3T3-L1 cells, and immortalized wild-type mouse fibroblasts (Figure 12.1A) as shown by immunoblot analysis with anti-PAR.

A suppressive effect of PARP inhibitors on DNA fragmentation has earlier suggested a correlation between poly(ADP-ribosyl)ation of nuclear proteins and internucleosomal DNA cleavage[41,42] or nuclear fragmentation[43] during apoptosis. Depletion of PARP from 3T3-L1 by antisense RNA expression prior to the induction

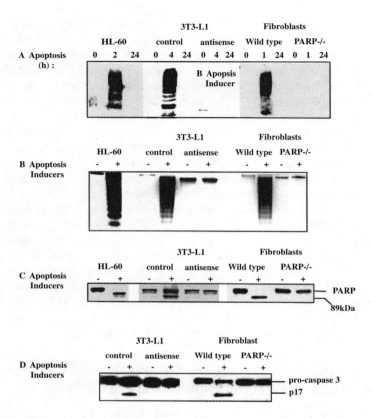

FIGURE 12.1 Time courses of poly(ADP-ribosyl)ation of nuclear proteins (A), DNA fragmentation (B), *in vitro* PARP cleavage (C), and activation of procaspase-3 (D) during apoptosis in HL-60 cells, 3T3-L1 cells, and immortalized wild-type fibroblasts; effects of PARP depletion by antisense RNA expression and gene disruption. Apoptosis was induced in HL-60 cells by incubation with camptothecin (10 μ*M*); PARP-antisense and control cells were preincubated for 72 h in the presence of 1 μ*M* dexamethasone to induce antisense RNA expression and then, together with wild-type and PARP$^{-/-}$ fibroblasts, exposed to anti-Fas (100 ng/ml) and cycloheximide (10 μg/ml) for 24 h. Cell extracts were then prepared and subjected to immunoblot analysis with anti-PAR (A) and with mAb to the p17 subunit of caspase-3 (D). Apoptosis was monitored by detection of characteristic internucleosomal DNA ladders by agarose gel electrophoresis and ethidium bromide staining (B). Caspase-3-mediated PARP-cleavage activity in cytosolic extracts was assayed with [^{35}S]PARP as substrate. The positions of procaspase-3 and p17, as well as PARP and the 89-kDa cleavage fragment, are indicated (D).

of apoptosis, or gene disruption in immortalized fibroblasts derived from PARP$^{-/-}$ mice, also prevents this early PARP activation and blocks subsequent biochemical and morphological changes associated with apoptosis, including internucleosomal DNA fragmentation (Figure 12.1B), *in vitro* PARP cleavage (Figure 12.1C), and proteolytic activation of procaspase-3 into the active caspase-3 (p17; Figure 12.1D), thus correlating the early poly(ADP-ribosyl)ation with later events in the cell death cascade.[18] Stable transfection of PARP$^{-/-}$ fibroblasts with PARP cDNA sensitizes these cells to Fas-mediated apoptosis,[18] suggesting that PARP and/or poly(ADP-ribosyl)ation of relevant nuclear proteins at the early stages of apoptosis may play a role in progression through the death program.

Because PARP is activated only in the presence of DNA strand breaks, the appearance of large chromatin fragments at this early reversible stage of apoptosis[38] is consistent with the substantial nuclear poly(ADP-ribosyl)ation at this time. Our results with pulse field gel electrophoresis of early apoptotic cell extracts also reveal the appearance of large 50-kb DNA fragments at this time during Fas-mediated apoptosis in wild-type mouse fibroblasts (not shown). This early poly(ADP-ribosyl)ation also correlates with a drop in NAD levels, indicative of increased PAR synthesis, and a subsequent recovery prior to internucleosomal DNA cleavage.[44] Similarly, PARP activation occurs during apoptosis induced by DNA-damaging agents, such as alkylating agents, topoisomerase inhibitors, adriamycin, X-rays, ultraviolet radiation, mitomycin C, and cisplatin.[45-49] Rosenthal et al. discusses the role of this transient burst of PARP activity in cell death in further detail in Chapter 11.

12.2.2 Acceptors for Poly(ADP-ribosyl)ation during Early Apoptosis

12.2.2.1 Poly(ADP-Ribosyl)ation of Histone H1 during Early Apoptosis

Poly(ADP-ribosyl)ation of nuclear proteins in response to DNA breaks is transient *in vivo* and is restricted to the pool of potential target proteins located adjacent to DNA breaks.[50] For example, less than 1% of the histone H1 pool is poly(ADP-ribosyl)ated *in vivo*[26] and *in vitro*.[27] Detection of poly(ADP-ribosyl)ated proteins is therefore difficult because only a small proportion of the endogenous protein is expected to be poly(ADP-ribosyl)ated at any one time, and the *in vivo* half-life of PAR chains on an acceptor is ~1 to 2 min as it is rapidly degraded by poly(ADP-ribose) glycohydrolase. Interestingly, poly(ADP-ribosyl)ation of histone H1 early during apoptosis mediated by DNA-damaging agents was shown to facilitate internucleosomal DNA fragmentation by enhancing chromatin susceptibility to cellular endonucleases.[49] Suppression of histone H1 poly(ADP-ribosyl)ation by 3-aminobenzamide decreases micrococcal nuclease susceptibility of chromatin and consequently blocks internucleosomal DNA fragmentation and morphological changes of apoptosis.[49] Identification of relevant acceptors for poly(ADP-ribosyl)ation at the early stages of apoptosis would provide further insights on the apparent requirement for the early transient PARP activation during the death program, at least in the clonal mouse and human cells.

12.2.2.2 Poly(ADP-Ribosyl)ation of p53 during Early Apoptosis

p53 mediates cell cycle arrest in G_1 or G_2/M or induces apoptosis in cells that have accumulated substantial DNA damage.[51,52] Induced by different apoptotic stimuli, p53 is required for apoptosis in many cell lines[53] either via transcriptional activation of p53 target genes and/or direct protein–protein interaction. Purified PARP can poly(ADP-ribosyl)ate p53 *in vitro*, and p53 binding to its consensus sequence prevents its covalent modification.[54] Exposure of cells expressing wild-type p53 to DNA-damaging agents that also stimulate PAR synthesis results in a rapid increase in intracellular p53 levels,[55,56] which may occur through *de novo* protein synthesis[57] or post-translational stabilization of p53.[58] During the course of spontaneous apoptosis in human osteosarcoma cells, biochemical markers of apoptosis are initially observed at day 5 and maximize around days 7 to 9.[36,40] Spontaneous apoptosis in these cells is also associated with a marked increase in the intracellular p53 levels, which were markedly elevated at days 2 to 3, maximized at day 4, and declined thereafter (Figure 12.2E). Immunoblot analysis with anti-PAR showed that poly(ADP-ribosyl)ation of nuclear proteins increased at day 3, peaked at day 4, a stage at which the cells were still viable and could be replated, and subsequently declined (Figure 12.2A), concomitant with caspase-3-mediated *in vivo* PARP cleavage (Figure 12.2B) and proteolytic activation of procaspase-3 (Figure 12.2C). The specificity of the anti-PAR antibody used in these experiments was verified by elimination of the polymer signal after removal of PAR from immunoblots by phosphodiesterase treatment.[17] Immunoprecipitation experiments confirmed that p53 undergoes extensive poly(ADP-ribosyl)ation at days 3 to 4 (Figure 12.2D), coincident with the burst of PAR synthesis at this stage (Figure 12.2A). Subsequent degradation of PAR attached to p53 presumably by poly(ADP-ribose) glycohydrolase occurred at the onset of proteolytic activation of procaspase-3 (Figure 12.2C), caspase-3-mediated PARP cleavage (Figure 12.2B), and internucleosomal DNA fragmentation at the later stages of cell death (Figure 12.3C; days 7 to 9); reprobing of the blot with polyclonal anti-p53 confirmed that the modified protein was p53 (Figure 12.2E). Thus, p53 is one of the early acceptors of poly(ADP-ribosyl)ation during early apoptosis in osteosarcoma cells.[59]

12.2.2.2.1 Regulation of p53 Function

Colocalization of PARP and p53 in the vicinity of large DNA breaks and their physical association,[60,61] as well as the fact that PAR attached to p53 *in vitro* blocks binding to its consensus sequence,[62] together suggest that poly(ADP-ribosyl)ation of p53 may regulate p53-mediated transcriptional activation of genes important in the cell cycle and apoptosis. Given that poly(ADP-ribosyl)ation generally inhibits the activity of acceptor proteins, this modification may also alter p53 binding to DNA sequences in the promoters of p53-responsive genes associated with the induction of p53-mediated apoptosis, such as those encoding Bax, IGF-BP3,[63] or Fas.[64] During spontaneous apoptosis of osteosarcoma cells, expression of the p53-responsive genes Bax and Fas was negligible before and at the peak of p53 accumulation and poly(ADP-ribosyl)ation (days 3 and 4). Although p53 levels were already elevated by day 2, expression of Bax and Fas was markedly induced only at day 5 (Figure 12.2F and G), concomitant with a decline in PAR attached to p53 and caspase-3-mediated PARP cleavage. This coincident decrease in PAR bound to p53

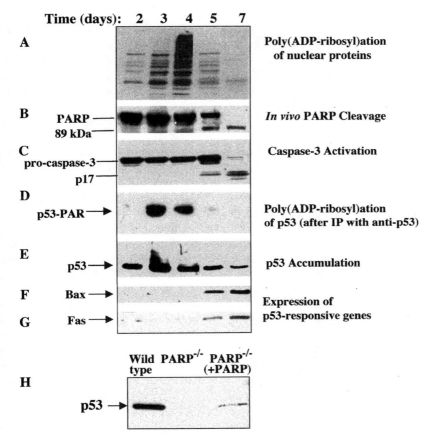

FIGURE 12.2 Time courses of poly(ADP-ribosyl)ation of nuclear proteins (A), *in vivo* PARP cleavage (B), activation of procaspase-3 (C), p53 poly(ADP-ribosyl)ation (D) and accumulation (E), and expression of p53 responsive genes Bax (F) and Fas (G) during spontaneous apoptosis in human osteosarcoma cells; p53 expression in immortalized wild-type, PARP$^{-/-}$, and PARP$^{-/-}$ (+PARP) fibroblasts. Osteosarcoma cells were induced to undergo spontaneous apoptosis for 9 days, and at the indicated times cytosolic extracts were prepared and subjected to immunoblot analyses with antibodies to PAR (A), to PARP (B), to caspase-3 (C), to Bax (F), and to Fas (G). The positions of PARP and of its 89-kDa cleavage product, procaspase-3 and its active form p17, p53, Bax, and Fas are indicated. Equal amounts of protein (100 µg) were also subjected to immunoprecipitation with mAb to p53 and subjected to immunoblot analysis with anti-PAR (D). The immunoblot in D was stripped of antibodies and reprobed with polyclonal antibodies to p53 (E). Cell extracts of wild-type, PARP$^{-/-}$, and PARP$^{-/-}$ (+PARP) fibroblasts (30 µg of protein) were also subjected to immunoblot analysis with anti-p53 (PAb421) (H).

and induction of Bax and Fas expression indicates that poly(ADP-ribosyl)ation may regulate p53 function early in apoptosis; caspase-3-mediated PARP cleavage may then release p53 from PAR-induced inhibition at the later stages of apoptosis when cells are irreversibly committed to death. Alternatively, the role of poly(ADP-ribosyl)ation on the regulation of p53 function may also be indirect since PARP also stimulates DNA-PK activity by poly(ADP-ribosyl)ation,[65] which can then regulate p53 function by phosphorylation.

12.2.2.2.2 p53 Stabilization and Accumulation

Both PARP activity and p53 accumulation are induced by DNA damage, and both proteins have been implicated in the normal cellular responses to such damage. Inhibition of PARP activity by 3-aminobenzamide blocks the rapid p53 accumulation that normally follows radiation-induced DNA damage.[66] Cells that are unable to synthesize PAR because of a deficiency in NAD metabolism show decreased p53 concentration and activity, and they fail to exhibit a p53 response and undergo apoptosis in response to DNA-damaging agents.[35] Furthermore, compared with wild-type cells, primary fibroblasts from PARP-/- mice express lower basal p53 levels and also exhibit a defective induction of p53 in response to DNA damage;[67] thus, PARP-dependent signaling may influence the synthesis or degradation of p53 following DNA damage. Since a PAR-binding site in p53 is localized near a proteolytic cleavage site, it was suggested that PAR binding could protect this sequence from proteolysis;[68] similar protection occurs after binding of monoclonal antibodies adjacent to this region.[69] Thus, the extensive poly(ADP-ribosyl)ation of p53 early in apoptosis suggests that this modification may also play a role in p53 accumulation by protecting the protein from proteolysis. Consistently, immunoblot analysis with anti-p53 detected p53 protein in lysates of wild-type cells, but not in PARP-/- cell extracts, and stable transfection of PARP-/- cells with PARP cDNA partially restored p53 expression in these PARP-/- (+PARP) cells (Figure 12.2H). Indeed, the lack of regularly spliced wild-type p53 in PARP-/- cells recently has been attributed to reduced protein stability, not lower levels of p53 mRNA.[70] Modification of p53 by PARP is therefore also implicated in p53 accumulation and stabilization,[59,68] which may explain the apparent lack of p53 in PARP-/- cells.

12.2.2.3 Poly(ADP-Ribosyl)ation of Topoisomerases I and IIb: A Role in Protein Stabilization

Topoisomerases I and II, nuclear enzymes implicated in the modulation of the topological state of DNA, are poly(ADP-ribosyl)ated *in vitro* and *in vivo* and modification inhibits their catalytic activities.[17,28-30] Topo I activity is also downregulated in response to radiation-induced DNA damage and studies with 3-aminobenzamide have suggested that inhibition of topo I activity is due to poly(ADP-ribosyl)ation of the enzyme.[71] p53 and topo I, which form a complex *in vivo,* are both extensively poly(ADP-ribosyl)ated following γ irradiation, and it has been proposed that rapid poly(ADP-ribosyl)ation of the p53–topo I complex prevents topo I cleavage after DNA damage.[72] This is consistent with a role of poly(ADP-ribosyl)ation in the stabilization of acceptor proteins. Furthermore, PARP$^{-/-}$ cells also exhibit decreased activity and expression of topo IIβ, although levels of topo II mRNA are similar in PARP$^{-/-}$ and wild-type cells.[73] Thus, similar to results with p53, PARP and poly(ADP-ribosyl)ation also appear to play a role in the stabilization of topo I and topo IIβ.

12.2.2.4 Poly(ADP-Ribosyl)ation of Ca^{2+}Mg^{2+}-Dependent Endonucleases: Effects on Chromatin

Ca^{2+}Mg^{2+}-dependent endonucleases (CME), comprising a homologous class of endonucleases, have been implicated in DNA fragmentation during apoptosis.[74,75] They are identical in size and kinetic properties to another CME, DNase γ, which is responsible for DNA fragmentation during thymic apoptosis.[74] CMEs purified from bull seminal plasma as well as rat liver nuclei are inhibited by poly(ADP-ribosyl)ation *in vitro*,[76] suggesting that poly(ADP-ribosyl)ation may be a putative mechanism for the regulation of CME-catalyzed DNA fragmentation during apoptosis. Endonuclease activity assays and immunoblot analysis of poly(ADP-ribosyl)ated CME with anti-PAR confirmed that the activity of purified CME is inhibited by poly(ADP-ribosyl)ation *in vitro* (not shown).

To determine if poly(ADP-ribosyl)ation of CME also correlates with inhibition of DNA fragmentation *in vivo*, antibodies to CME and to PAR were utilized. The CME antibody specifically reacts with nuclear chromatin-bound CMEs from various bovine and human tissues.[77] Although CME levels were essentially the same during the course of spontaneous apoptosis in osteosarcoma cells (Figure 12.3A), immunoblot analysis with anti-PAR revealed marked poly(ADP-ribosyl)ation of the 36-kDa endonuclease during the transient burst of PAR synthesis at the early stages of apoptosis (Figure 12.3B; days 3 to 4). PAR bound to the protein was subsequently degraded and no further modification was detected at the peak of PARP cleavage and DNA fragmentation at the later stages of cell death (Figure 12.3C; days 7 to 9). Thus, at an early reversible stage of apoptosis (days 3 to 4), CME may be in an inactive state, transiently inhibited by PAR chains attached to it. At the onset of caspase-3-mediated PARP cleavage, CME may be activated by loss of inhibitory PAR moieties, thus, at least partially contributing to the massive endonucleolytic DNA cleavage at this time. The generation of nucleosomal DNA ladders increasing in a time-dependent manner during apoptosis may be attributed to increased DNA cleavage resulting from activation of constant levels of endonuclease. Additionally, this may also be due to enhanced susceptibility of chromatin.

12.3 POLY(ADP-RIBOSYL)ATION OF DNA-BINDING COMPONENT PROTEINS OF THE DNA SYNTHESOME

12.3.1 PARP, A Core Component of the DNA Synthesome

PARP exclusively copurifies with core proteins of a 17S multiprotein DNA replication complex (MRC or DNA synthesome) that catalyzes replication of viral DNA

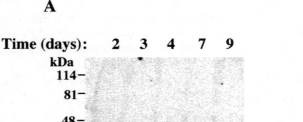
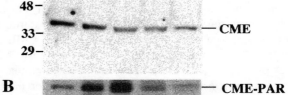
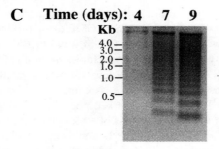

FIGURE 12.3 Time courses of accumulation (A) and poly(ADP-ribosyl)ation CME (B), and internucleosomal DNA fragmentation (C) during spontaneous apoptosis in osteosarcoma cells. Osteosarcoma cells were induced to undergo spontaneous apoptosis for 9 days, and at the indicated times cell extracts were prepared and subjected to immunoblot analysis with anti-CME (A) and with anti-PAR (B). The positions of CME and poly(ADP-ribosyl)ated CME are indicated. Apoptosis was monitored by detection of characteristic internucleosomal DNA ladders by agarose gel electrophoresis and ethidium bromide staining (C).

in vitro and contains DNA polymerases (pol) α, δ, and ε, DNA primase, DNA helicase, DNA ligase, and topo I and II, as well as accessory proteins such as PCNA, RFC, and RPA (Figure 12.4A). The MRC is purified by cell fractionation in a series of centrifugation steps, followed by discontinuous gradient centrifugation on a sucrose cushion, chromatography on a Q-sepharose column, and sucrose gradient centrifugation.[78,79] These purified MRCs migrate as discrete, high-molecular-weight complexes on native polyacrylamide gel electrophoresis.[80] PARP may play a regulatory role within these complexes because it can modulate the catalytic activities of specific replicative enzymes or factors by catalyzing their poly(ADP-ribosyl)ation (DNA pol α, δ, and ε,[17,31,81] topo I and II,[28-30] and RPA[82]).

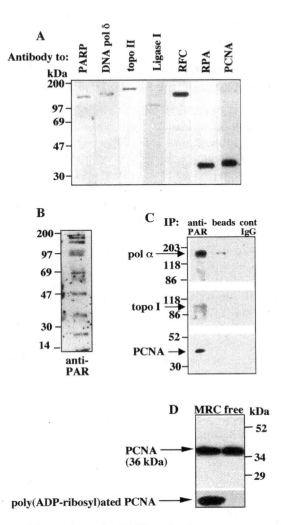

FIGURE 12.4 Immunoblot analyses for PARP and other components of the replication complex (A) and poly(ADP-ribosyl)ated component proteins (B) of purified MRC fractions; identification of modified MRC proteins by immunoprecipitation (C). Purified MRC fractions (SG fraction; 20 μg) were incubated with anti-PARP and with antibodies to the indicated DNA replication proteins (A) as well as with anti-PAR (B) by repeatedly stripping the membranes of antibodies and reprobing with different antibodies. The positions of the molecular weight markers are indicated. MRC fractions were immunoprecipitated with anti-PAR or with protein A-Sepharose (beads) alone or control IgG, and the immunoprecipitates were then separated, transferred, and probed with antibodies to PCNA, topo I and DNA pol α (C). Equal amounts of MRC-enriched and MRC-free fractions (30 μg) were immunoprecipitated with anti-PCNA and then subjected to immunoblot analysis with anti-PCNA; the immunoblot was then stripped of antibodies and reprobed with anti-PAR (D). The positions of PCNA and modified PCNA are indicated.

12.3.2 Poly(ADP-Ribosyl)ation of Components of the DNA Synthesome

12.3.2.1 Acceptors for Poly(ADP-Ribosyl)ation in the DNA Synthesome: PCNA, Topo I, and DNA Pol α

Approximately 15 of the ~40 MRC proteins, including DNA pol α, topo I, and PCNA, are poly(ADP-ribosyl)ated as revealed by immunoblot analysis with anti-PAR (Figure 12.4B).[17] Topo I (100 kDa), DNA pol α (180 kDa), and PCNA (36 kDa) were specifically immunoprecipitated from the MRC by anti-PAR and reacted with their respective antibodies on the immunoblots (Figure 12.4C). Although equal amounts of free and MRC-bound forms of PCNA were detected in cells, only the complexed form was poly(ADP-ribosyl)ated (Figure 12.4D), thus poly(ADP-ribosyl)ation of PCNA in the MRC may play a role in regulating its functions in the MRC or its recruitment into the complex. This modified, complexed form of PCNA corresponds to the ~35% of total cellular PCNA that is associated with replication foci during the peak of S phase.[83]

12.3.2.2 Poly(ADP-Ribosyl)ation of DNA Pol α and δ in the MRC: Effects on Activity

While DNA pol α-primase synthesizes RNA primers and Okazaki fragments and is required for initiation of lagging strand synthesis, DNA pol δ mediates both leading and lagging strand DNA synthesis during the elongation phase of DNA replication,[84] and PCNA is essential for DNA pol δ activity.[85] Consistent with studies using purified enzymes, poly(ADP-ribosyl)ation inhibited the activities of MRC-bound DNA pol α and δ in a manner that was dependent on NAD concentration (Figure 12.5A) and sensitive to 3-aminobenzamide. Thus, PARP may modulate the activities of DNA pol α or δ in the DNA synthesome by catalyzing their poly(ADP-ribosyl)ation. As with other acceptor proteins of PARP, poly(ADP-ribosyl)ation of these enzymes in the presence of increasing NAD presumably confers a large negative charge, which promotes their dissociation from the DNA template primer, thereby inhibiting their activity. Interestingly, physical interaction of PARP with DNA pol α, in the absence of NAD, activates pol α,[86] whereas addition of NAD to the MRC inhibits pol α catalytic activity; thus, PARP appears to play a dual role in the regulation of DNA pol α activity.

To elucidate further the role(s) of PARP and poly(ADP-ribosyl)ation in the MRC, the complex was purified from 3T3-L1 cells at G_0/G_1 and when ~95% of control cells were in S phase and exhibited a transient peak of PARP expression and activity. Under these conditions, quiescent control cells reenter S phase and go through one round of the cell cycle, but not the PARP-depleted antisense cells.[16,17] Although MRC fractions purified from S-phase control cells are competent to support viral DNA replication *in vitro*, only the MRC from S-phase control cells, not those from G_0/G_1 control cells or from PARP-depleted antisense cells, exhibit significant PARP (Figure 12.5B) and DNA pol α and δ activities (Figure 12.5C). Thus, depletion of PARP by antisense RNA expression results in an MRC deficient in DNA pol α and δ activities.

Pleiotropic Roles of Poly(ADP-Ribosyl)ation of DNA-Binding Proteins

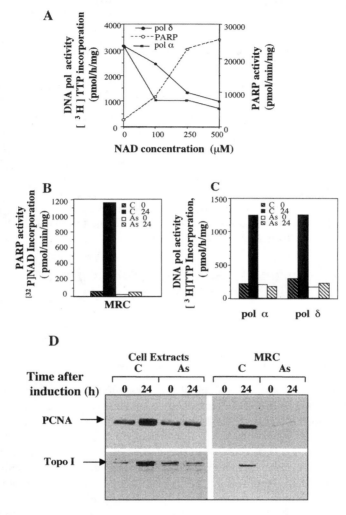

FIGURE 12.5 Effects of poly(ADP-ribosyl)ation (A) and PARP depletion by antisense RNA expression on the activities of DNA pol α (B) and δ (C) in the MRC, and recruitment of PCNA and topo I into the MRC (D). PARP and DNA pol α and δ activities of purified MRC (SG peak fraction) were assayed in the presence of various concentrations of NAD. Data are expressed as pmoles [^{3}H]dTTP incorporated into DNA per hour per milligram (DNA pol) or picomoles of [^{32}P]NAD per minute per milligram of protein (PARP), and are means of triplicate determinations. The MRC (QS peak) fraction was purified from 3T3-L1 control and PARP-antisense cells before and 24 h after induction of differentiation-linked DNA replication, and subjected to PARP (B) and DNA pol α and δ activity assays. Data are means of triplicate determinations, and essentially identical results were obtained in two additional experiments. Equal amounts (20 µg) of total-cell extracts (left panels) and MRC fractions (right panels) from 3T3-L1 control (C) and PARP-antisense (As) cells prepared before and 24 h after induction were subjected to immunoblot analysis with anti-PCNA (upper panels) or to topo I (lower panels) (D). Arrows indicate the positions of PCNA and topo I.

12.3.2.3 Poly(ADP-Ribosyl)ation of PCNA and Topo I: A Role in Recruitment into the DNA Synthesome during Early S Phase

PCNA and topo I were present in the MRC fraction only from S-phase control cells, although they were both detected in total cell extracts from both control and PARP-antisense cells (Figure 12.5D), thus indicating that PARP may play a role in the recruitment of PCNA and topo I into the DNA synthesome during reentry into S phase. The mechanism by which this occurs, however, remains unclear. Immunoprecipitation experiments have shown that PARP physically associates with DNA pol α *in vitro*[86] and *in vivo*,[16] as well as with topo I.[87] This association of PARP and DNA pol α requires the PARP DNA-binding domain and occurs during S and G_2 phases of the cell cycle.[88] Physical association with other proteins may, therefore, represent a mechanism by which PARP can recruit certain proteins into the MRC. A similar mechanism has been proposed for recruitment by PARP-associated polymers of signal proteins, such as p53 and MARCKS protein, to sites of DNA breakage via common polymer binding of 22 amino acids.[89]

12.4 POLY(ADP-RIBOSYL)ATION OF DNA-BINDING PROTEINS IN GENE EXPRESSION AND TRANSCRIPTION

12.4.1 Role of PARP and/or PAR in the Expression of MRC Components

12.4.1.1 Expression of DNA Pol a and DNA Primase

DNA pol α/DNA primase forms a complex of four subunits, the largest of which is the catalytic subunit (~180 kDa),[90] and the two smallest subunits comprise the DNA primase (~58 and ~48 kDa).[91] When quiescent cells are stimulated to proliferate, mRNA levels of all subunits increase simultaneously prior to DNA synthesis; thus, transcription of the genes for DNA pol α and DNA primase are likely regulated by a common mechanism.[92] Immunoblot analysis of cell extracts with antibodies to DNA pol α and DNA primase revealed induced expression of these proteins on S-phase reentry of control cells, but not in PARP-depleted antisense cells (Figure 12.6A). Thus, through mechanisms still unknown, PARP may play a role in the regulation of expression of the corresponding genes when quiescent cells are induced to reenter S phase. The lack of any DNA pol α and DNA primase gene expression could not be attributed to a general inability of the PARP-depleted antisense cells to undergo protein or RNA synthesis as there was no effect on the expression of topo I in these cells.[19]

12.4.1.2 Expression and Promoter Activity of the Transcription Factor E2F-1

How does PARP affect expression of DNA pol α during reentry into S phase? The transcription factor E2F-1 binds to a specific recognition site (5′-TTTCGCGC) and activates the promoters of genes encoding proteins required for DNA replication and execution of the S phase, including DNA pol α, dihydrofolate reductase, thymidine kinase, c-Myc, c-Myb, PCNA, cyclin D, and cyclin E.[93-96] This activity is regulated during the cell cycle by the binding of E2F-1 to the dephosphorylated form of the tumor suppressor Rb; phosphorylation of Rb by cyclin-dependent kinases during entry into S phase releases E2F-1, consequently activating gene expression.[97] Transcription of the E2F-1 gene, in turn, is also regulated during the cell cycle.[98] Interestingly, PARP depletion by antisense RNA expression also prevented the increase in E2F-1 expression during early S phase in 3T3-L1 cells (Figure 12.6B). Control cells, but not PARP-depleted antisense cells, exhibited a marked increase in E2F-1 expression ~1 h after reentry into S phase (Figure 12.6B), consistent with the fact that the E2F-1 gene is an early-response gene,[99] while induction of both DNA pol α and PCNA expression in control cells occurred subsequently, consistent with their being encoded by late-response genes.[94] PARP may therefore regulate the expression of DNA pol α and PCNA genes during reentry into S phase by affecting the expression of the transcriptional factor, E2F-1, which in turn can regulate the transcription of both the DNA pol α and PCNA genes, as well as the E2F-1 gene itself.

A role for PARP in the regulation of the expression of pol α and E2F-1 genes during S-phase reentry in quiescent cells was further investigated with mouse fibroblasts derived from wild-type and PARP$^{-/-}$ mice as well as PARP$^{-/-}$ cells stably transfected with wild-type PARP cDNA — PARP$^{-/-}$ (+PARP). PARP plays an essential role in DNA replication associated with reentry into the cell cycle after release from serum starvation or aphidicolin block.[100] When a construct containing the E2F-1 gene promoter fused to a luciferase reporter gene was transiently transfected into these cells, E2F-1 promoter activity in control and PARP$^{-/-}$ (+PARP) cells increased eightfold 9 h after release from serum deprivation, but not in PARP$^{-/-}$ cells (Figure 12.6C). PARP$^{-/-}$ cells did not show the marked induction of E2F-1 expression during early S phase apparent in control and PARP$^{-/-}$ (+PARP) cells (Figure 12.6D). Reverse transcription polymerase chain reaction (RT-PCR) analysis did not detect the presence of pol α transcripts in PARP$^{-/-}$ cells that was evident in both wild-type and PARP$^{-/-}$ (+PARP) cells after 20 h (Figure 12.6E). These results suggest that PARP is involved in the induction of E2F-1 promoter activity, which then positively regulates both E2F-1 and pol α expression, when quiescent cells are stimulated to reenter the cell cycle. Since PARP depletion by antisense RNA expression also results in significant changes in chromatin structure,[14] effects of the lack of PARP on gene expression may also be attributable, at least in part, to such changes.

12.4.2 Poly(ADP-Ribosyl)ation of DNA-Binding Transcription Factors: Effects on Binding to Their Consensus Sequences

12.4.2.1 Physical Association with PARP (TEF-1, AP-2) and Poly(ADP-Ribosyl)ation of Specific Transcription Factors (TEF-1, TFIIF, YY1, SP-1, TBP, CREB)

The role of PARP in the regulation of transcription of the E2F-1 and pol α genes may be indirect, given that PARP, in the absence of NAD, can enhance activator-

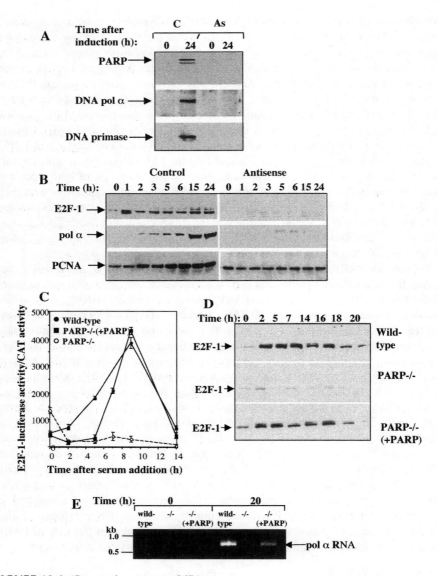

FIGURE 12.6 (See caption on page 267.)

FIGURE 12.6 Effects of PARP depletion by antisense RNA expression and gene knockout on PARP, DNA pol α, DNA primase, E2F-1 expression, E2F-1 promoter activity, and pol α transcription during reentry into S phase. 3T3-L1 control and PARP-antisense cells were exposed to 1 μM dexamethasone and inducers of differentiation-linked DNA replication for indicated times (A,B). Cell extracts were then subjected to immunoblot analyses with antibodies to PARP, DNA pol α, DNA primase (A), as well as PCNA and E2F-1 (B). Arrows indicate the positions of these proteins. Wild-type, PARP$^{-/-}$, and PARP$^{-/-}$ stably transfected with PARP cDNA (PARP$^{-/-}$ (+PARP)) were synchronized by serum deprivation and harvested at indicated times after serum addition (C to E). Prior to serum deprivation, the cells were transiently cotransfected with a construct containing the E2F-1 gene promoter fused to a luciferase reporter gene and pSVCAT; at the indicated times after serum addition, cell extracts were assayed for E2F-1-luciferase and CAT activities (C). The luciferase activity (light units/min) was normalized against transfection efficiencies by CAT activity; data are means of triplicate determinations. Cell extracts (30 μg) were also subjected to immunoblot analysis with anti-E2F-1 (D). Total RNA was purified from cell pellets and subjected to RT-PCR with pol α–specific primers. PCR products were separated on a 1.5% agarose gel and visualized by ethidium bromide staining (E). The positions of the specific pol α product (arrows) and of DNA size standards (in kilobases) are indicated.

---

dependent transcription by interacting with RNA polymerase II–associated factors.[101] Similarly, PARP binds transcription enhancer factor 1 (TEF-1) to enhance muscle-specific gene transcription,[102] as well as the transcription factor AP-2 to coactivate AP-2-mediated transcription.[103] Whether PARP modulates E2F-1-mediated transcription by binding to E2F-1 or to E2F-1 promoter sequences, however, remains to be clarified. PARP appears to also play a dual role in the regulation of transcription; although it acts as a positive cofactor of transcription, the presence of NAD and consequent PARP activation represses RNA pol II–dependent transcription.[101] PARP-dependent silencing of transcription involves poly(ADP-ribosyl)ation of a number of transcription factors, which prevents the formation of active transcription complexes.[104] The basal transcription factor TFIIF and TEF-1 as well as transcription factors TBP, YY1, SP-1, and CREB, but not c-Jun or AP-2, are all highly specific substrates for poly(ADP-ribosyl)ation.[102,104,105] Modification of these proteins prevents binding to their respective DNA consensus sequences.[104]

12.4.2.2 Poly(ADP-Ribosyl)ation of Transcription Factors p53 and NFκB *In Vitro*: Effects on Binding to Their Respective Consensus Sequences

Extensive poly(ADP-ribosyl)ation of p53 early during the apoptotic program in osteosarcoma cells, and coincident degradation of PAR attached to p53 and induction of expression of the p53 responsive genes Bax and Fas, suggests that poly(ADP-ribosyl)ation may play a role in the regulation of p53 function at the later stages of cell death.[59] As with other transcription factors,[104] *in vitro* poly(ADP-ribosyl)ation reactions with purified proteins and subsequent gel shift assays show that the DNA phobicity of modified poly(ADP-ribosyl)ated DNA-binding proteins such as p53 prevents their binding to target DNA consensus sequences (gene promoters). Modification of p53 by poly(ADP-ribosyl)ation by purified recombinant PARP in the presence of NAD markedly inhibited binding to its DNA consensus sequence (Figure 12.7A). Extensive modification of p53 *in vitro* was confirmed by immunoprecipitation and immunoblot analysis of the reactions with anti-p53 and anti-PAR (not shown). To confirm that the inhibitory effect of PAR on p53 binding to DNA is not attributable to the physical interaction between PARP and p53, p53 was also incubated with catalytically inactive mutant PARP and NAD. p53, preincubated with mutant PARP and NAD, remained unmodified and was therefore able to bind tightly to its DNA consensus sequence. Thus, poly(ADP-ribosyl)ation of p53, as with other transcription factors discussed above, can modulate p53 function by altering binding to its DNA consensus sequence *in vitro*.

NFκB, another inducible transcription factor implicated in the regulation of key genes coding for mediators involved in the immune, acute phase, inflammatory, and stress responses,[106,107] can also be modified by poly(ADP-ribosyl)ation by PARP *in vitro* (unpublished data). Activation of NFκB plays a protective role against programmed cell death because blocking the activation of NFκB by a dominant negative form of IκB-a, an inhibitor of NFκB, or by gene knockout induces massive apoptosis. Thus, inactivation of NFκB post-translationally by poly(ADP-ribosyl)ation at the early stages of the death program may contribute to committing cells toward apoptosis. Interestingly, poly(ADP-ribosyl)ation of NFκB by wild-type PARP, similar to p53, blocked binding of the modified protein to its DNA consensus sequence presumably due to electrostatic repulsion between DNA and PAR bound to it (Figure 12.7B). These results suggest that the cycling mechanism earlier proposed for PARP and PAR in DNA repair processes may also be valid for regulation of transcription factor binding to or "cycling on and off" their consensus promoter sequences. PARP cycles on and off DNA ends in the presence of NAD, and its automodification during DNA repair *in vitro* presumably allows access to DNA repair enzymes.[3-5] Thus, p53, NFκB, or other transcription factors may similarly cycle on and off their DNA consensus sequences depending on their level of negative charge based on their poly(ADP-ribosyl)ation state. This may represent a novel mechanism for regulating transcriptional activation of target genes by these transcription factors *in vivo*.

12.5 CONCLUSIONS

PARP catalyzes the poly(ADP-ribosyl)ation of a limited number of nuclear proteins adjacent to DNA breaks in nuclear processes, such as DNA replication, DNA repair, differentiation, gene expression, or apoptosis.

1. This post-translational modification, which is very transient *in vivo*, generally inhibits the activity of the modified protein, presumably because of a decrease in DNA-binding affinity caused by electrostatic repulsion between DNA and PAR covalently bound to the acceptor proteins. Thus, poly(ADP-ribosyl)ation may regulate the functions of replicative enzymes such as DNA pol α and δ in the DNA synthesome or the transactivation of p53-responsive genes such as Bax and Fas by p53 during early apoptosis.
2. Poly(ADP-ribosyl)ation of nucleosomal proteins such as histone H1 may also alter the nucleosomal structure of the DNA to destabilize higher-

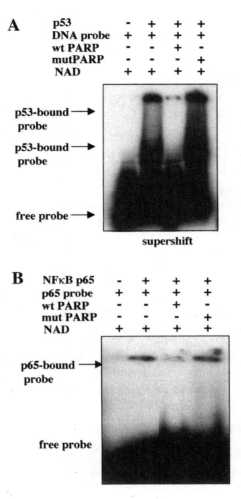

FIGURE 12.7 Effects of poly(ADP-ribosyl)ation of p53 (A) and NFκB (B) by purified wild-type and mutant PARP on binding to their respective DNA consensus sequences *in vitro*. p53 or NFκB p65 (1 μg) were poly(ADP-ribosyl)ated *in vitro* either by purified recombinant wild-type PARP or mutant PARP (0.1 μg) in a reaction mixture (50 μl) containing 50 mM Tris HCl (pH 7.8), 25 mM MgCl$_2$, 1 mM DTT, 4 μg activated DNA, and 100 μM NAD. After incubation for 30 min at 37°C, the reaction was stopped on ice. Electrophoretic mobility supershift assays were then performed according to standard procedures. After preincubation with excess amounts of polydI-dC and BSA, aliquots of the poly(ADP-ribosyl)ation reactions were incubated with anti-p53 or anti-NFκB (Santa Cruz Biotech) for 1 h, and then with [^{32}P]ATP–labeled p53 or NFκB consensus oligonucleotide probes (100,000 cpm/binding reaction) for 15 min at room temperature. The DNA–protein complexes were then resolved by electrophoresis on 5% TBE gels, and the gels were dried and autoradiographed.

order chromatin structures and consequently to promote access of DNA repair and replication enzymes, as well as transcription factors, to these sites or to enhance chromatin susceptibility to endonucleases during apoptosis.

3. PARP and/or poly(ADP-ribosyl)ation play a role in the recruitment of acceptor proteins to certain sites through direct association or through PAR, such as PCNA and topo I to the DNA synthesome during early S phase or signal proteins such as p53 and MARCKS protein to sites of DNA breakage.

4. Extensive poly(ADP-ribosyl)ation of p53, topo I, and topo IIβ *in vivo* is also implicated in the accumulation and stabilization of these acceptor proteins in cells.

5. Finally, PARP and poly(ADP-ribosyl)ation appear to play dual roles in nuclear processes, depending on the levels of the substrate NAD and the presence of PARP-activating DNA breaks. Thus, physical interaction of PARP with DNA pol α, in the absence of NAD, activates pol α, while addition of NAD to the MRC inhibits pol α catalytic activity. Similarly, in the absence of NAD, PARP interacts with different transcription factors to enhance activator-dependent transcription, while the presence of NAD and consequent PARP activation represses transcription presumably by poly(ADP-ribosyl)ation of certain transcription factors. Modification of these DNA binding proteins prevents binding to their respective DNA consensus sequences. Further work clearly is necessary to elucidate the molecular mechanisms by which PARP and poly(ADP-ribosoyl)ation mediate their pleiotropic roles in these various nuclear processes.

ACKNOWLEDGMENTS

The authors thank Drs. Z. Q. Wang (IARC, Lyon, France) for the immortalized wild-type and PARP$^{-/-}$ cells; William G. Kaelin (Dana Farber Cancer Institute, Boston) for the E2F-1 promoter–luciferase construct; Kiochiro Yoshihara (Nara Medical Univeristy, Japan) for the anti-CME; Alexander Spoonde for the purified recombinant wild-type and mutant PARP; Mira Jung for the NFκB cDNA; and Ruibai Luo for expert technical assistance. This work was supported by Grants CA25344 and PO1 CA74175 to M.E.S. from the National Cancer Institute, the U.S. Air Force Office of Scientific Research (Grant AFOSR-89-0053) to M.E.S., and the U.S. Army Medical Research and Development Command Contract DAMD17-90-C-0053 to M.E.S. and Contract DAMD17-96-C-6065 to D.S.R.

REFERENCES

1. Cherney, B. W., McBride, O. W., Chen, D. F., Alkhatib, H., Bhatia, K., Hensley, P., and Smulson, M. E., cDNA sequence, protein structure, and chromosomal location of the human gene for poly(ADP-ribose) polymerase, *Proc. Natl. Acad. Sci. U.S.A.*, 84, 8370, 1987.

2. Ménissier-de Murcia, J., Molinete, M., Gradwohl, G., and de Murcia, G., Zinc-binding domain of poly(ADP-ribose)polymerase participates in the recognition of single strand breaks on DNA, *J. Mol. Biol.*, 210, 229, 1989.
3. Satoh, M. S. and Lindahl, T., Role of poly(ADP-ribose) formation in DNA repair, *Nature*, 356, 356, 1992.
4. Satoh, M. S., Poirier, G. G., and Lindahl, T., NAD^+-dependent repair of damaged DNA by human cell extracts, *J. Biol. Chem.*, 268, 5480, 1993.
5. Smulson, M., Istock, N., Ding, R., and Cherney, B., Deletion mutants of poly(ADP-ribose) polymerase support a model of cyclic association and dissociation of enzyme from DNA ends during DNA repair, *Biochemistry*, 33, 6186, 1994.
6. Smulson, M. E., Pang, D., Jung, M., Dimtchev, A., Chasovskikh, S., Spoonde, A., Simbulan-Rosenthal, C., Rosenthal, D., Yakovlev, A., and Dritschilo, A., Irreversible binding of poly(ADP-ribose) polymerase cleavage product to DNA ends revealed by atomic force microscopy: possible role in apoptosis, *Cancer Res.*, 58, 3495, 1998.
7. Althaus, F. R., Poly-ADP-ribosylation reactions, in *ADP-Ribosylation of Proteins, Enzymology and Biological Significance*, Vol. 37, Althaus, F. R. and Richter, C., Eds., Springer-Verlag, Berlin, 1987, 1.
8. Burkle, A., Inhibition of carcinogen-inducible DNA amplification in a simian virus 40-transformed hamster cell line by ethacridine or ethanol, *Cancer Res.*, 49, 2584, 1989.
9. Kupper, J., Muller, M., Jacobson, M., Tasumi-Miyajima, J., Coyle, D., Jacobson, E., and Burkle, A., *trans*-Dominant inhibition of poly(ADP-ribosyl)ation sensitizes cells against gamma-irradiation and MNNG but does not limit DNA replication of a polyomavirus replicon, *Mol. Cell. Biol.*, 15, 3154, 1995.
10. Kupper, J., Muller, M., and Burkle, A., *trans*-Dominant inhibition of poly(ADP-ribosyl)ation potentiates carcinogen-induced gene amplification in SV40-transformed Chinese hamster cells, *Cancer Res.*, 56, 2715, 1996.
11. Molinete, M., Vermeulen, W., Burkle, A., de Murcia, J. M., Kupper, J. H., Hoeijmakers, J. H. J., and de Murcia, G., Overproduction of the poly(ADP-ribose) polymerase DNA-binding domain blocks alkylation-induced DNA repair synthesis in mammalian cells, *EMBO J.*, 12, 2109, 1993.
12. Schreiber, V., Hunting, D., Trucco, C., Gowans, B., Grunwald, P., de Murcia, G., and Ménissier-de Murcia, J., A dominant negative mutant of human PARP affects cell recovery, apoptosis, and sister chromatid exchange following DNA damage, *Proc. Natl. Acad. Sci. U.S.A.*, 92, 4753, 1995.
13. Ding, R., Pommier, Y., Kang, V. H., and Smulson, M. E., Depletion of poly(ADP-ribose) polymerase by antisense RNA expression results in a delay in DNA strand break rejoining, *J. Biol. Chem.*, 267, 12804, 1992.
14. Ding, R. and Smulson, M. E., Depletion of nuclear poly(ADP-ribose) polymerase by antisense RNA expression: influences on genomic stability, chromatin organization and carcinogen cytotoxicity, *Cancer Res.*, 54, 4627, 1994.
15. Rosenthal, D. S., Shima, T. B., Celli, G., De Luca, L. M., and Smulson, M. E., An engineered human skin model using poly(ADP-ribose) polymerase antisense expression shows a reduced response to DNA damage, *J. Invest. Dermatol.*, 105, 38, 1995.
16. Smulson, M. E., Kang, V. H., Ntambi, J. M., Rosenthal, D. S., Ding, R., and Simbulan-Rosenthal, C. M. G., Requirement for the expression of poly(ADP-ribose) polymerase during the early stages of differentiation of 3T3-L1 preadipocytes, as studied by antisense RNA induction, *J. Biol. Chem.*, 270, 119, 1995.

17. Simbulan-Rosenthal, C. M. G., Rosenthal, D. S., Hilz, H., Hickey, R. J., Malkas, L., Applegren, N., Wu, Y., Bers, G., and Smulson, M. E., The expression of poly(ADP-ribose) polymerase during differentiation-linked DNA replication reveals that this enzyme is a component of the multiprotein DNA replication complex, *Biochemistry*, 35, 11622, 1996.
18. Simbulan-Rosenthal, C. M. G., Rosenthal, D. S., Iyer, S., Boulares, A. H., and Smulson, M. E., Transient poly(ADP-ribosyl)ation of nuclear proteins and role for poly(ADP-ribose) polymerase in the early stages of apoptosis, *J. Biol. Chem.*, 273, 13703, 1998.
19. Simbulan-Rosenthal, C. M. G., Rosenthal, D. S., Boulares, A. H., Hickey, R. J., Malkas, L. H., Coll, J. M., and Smulson, M. E., Regulation of the expression or recruitment of components of the DNA synthesome by poly(ADP-ribose) polymerase, *Biochemistry*, 37, 9363, 1998.
20. Wang, Z. Q., Auer, B., Stingl, L., Berghammer, H., Haidacher, D., Schweiger, M., and Wagner, E. F., Mice lacking ADPRT and poly(ADP-ribosyl)ation develop normally but are susceptible to skin disease, *Genes Dev.*, 9, 509, 1995.
21. Ménissier-de Murcia, J., Niedergang, C., Trucco, C., Ricoul, M., Dutrillaux, B., Mark, M., Oliver, J., Masson, M., Dierich, A., LeMeur, M., Waltzinger, C., Chambon, P., and de Murcia, G., Requirement of poly(ADP-ribose) polymerase in recovery from DNA damage in mice and men, *Proc. Natl. Acad. Sci. U.S.A.*, 94, 7303, 1997.
22. Masutani, M., Suzuki, H., Kamada, N., Watanabe, M., Ueda, O., Nozaki, T., Jishage, K., Watanabe, T., Sugimoto, T., Nakagama, H., Ochiya, T., and Sugimura, T., Poly(ADP-ribose) polymerase gene disruption conferred mice resistant to streptozocin-induced diabetes, *Proc. Natl. Acad. Sci. U.S.A.*, 96, 2301, 1999.
23. Shieh, W. M., Ame, J. C., Wilson, M., Wang, Z. Q., Koh, D., Jacobson, M., and Jacobson, E., Poly(ADP-ribose) polymerase null mouse cells synthesize ADP-ribose polymers, *J. Biol. Chem.*, 273, 30069, 1998.
24. Ame, J., Rolli, V., Schreiber, V., Niedergang, C., Apiou, F., Decker, P., Muller, S., Hoger, T., Ménissier-de Murcia, J., and de Murcia, G., PARP-2, a novel mammalian DNA damage-dependent poly(ADP-ribose) polymerase, *J. Biol. Chem.*, 274, 17860, 1999.
25. Trucco, C., Oliver, F., de Murcia, G., and Ménissier-de Murcia, J., DNA repair defect in PARP-deficient cell lines, *Nucl. Acids Res.*, 26, 2644, 1998.
26. Wong, M., Miwa, M., Sugimura, T., and Smulson, M. E., Relationship between histone H1 poly(adenosine diphosphate ribosylation) and histone H1 phosphorylation using anti-poly(adenosine diphosphate ribose) antibody, *Biochemistry*, 22, 2384, 1983.
27. Malik, N., Miwa, M., Sugimura, T., Thraves, P., and Smulson, M. E., Immunoaffinity fractionation of the poly(ADP-ribosyl)ated domains of chromatin, *Proc. Natl. Acad. Sci. U.S.A.*, 80, 2554, 1983.
28. Kasid, U. N., Halligan, B., Liu, L. F., Dritschilo, A., and Smulson, M. E., Poly(ADP-ribose)-mediated post-translational modification of chromatin-associated human topoisomerase I. Inhibitory effects on catalytic activity, *J. Biol. Chem.*, 264, 18687, 1989.
29. Ferro, A. M. and Olivera, B. M., Poly(ADP-ribosylation) of DNA topoisomerase I from calf thymus, *J. Biol. Chem.*, 259, 547, 1984.
30. Darby, M. K., Schmitt, B., Jongstra, B. J., and Vosberg, H. P., Inhibition of calf thymus type II DNA topoisomerase by poly(ADP-ribosylation), *EMBO J.*, 4, 2129, 1985.
31. Yoshihara, K., Itaya, A., Tanaka, Y., Ohashi, Y., Ito, K., Teraoka, H., Tsukada, K., Matsukage, A., and Kamiya, T., Inhibition of DNA polymerase a, DNA polymerase b, terminal nucleotidyltransferase and DNA ligase II by poly (ADP-ribosyl)ation reaction *in vitro*, *Biochem. Biophys. Res. Commun.*, 128, 61, 1985.

32. Poirier, G. G., de Murcia, G., Jongstra-Bilen, J., Niedergang, C., and Mandel, P., Poly(ADP-ribosyl)ation of polynucleosomes causes relaxation of chromatin structure, *Proc. Natl. Acad. Sci. U.S.A.*, 79, 3423, 1982.
33. Butt, T. R., DeCoste, B., Jump, D., Nolan, N., and Smulson, M. E., Characterization of a putative poly adenosine disphosphate ribose chromatin complex, *Biochemistry*, 19, 5243, 1980.
34. Berger, N., Whitacre, C., Hashimoto, H., Berger, S., and Chaterjee, S., NAD and poly(ADP-ribose) regulation of proteins involved in response to cellular stress and DNA damage, *Biochimie*, 77, 364, 1995.
35. Whitacre, C. M., Hashimoto, H., Tsai, M.-L., Chatterjee, S., Berger, S. J., and Berger, N. A., Involvement of NAD-poly(ADP-ribose) metabolism in p53 regulation and its consequences, *Cancer Res.*, 55, 3697, 1995.
36. Nicholson, D. W., Ali, A., Thornberry, N. A., Vaillancourt, J. P., Ding, C. K., Gallant, M., Gareau, Y., Griffin, P. R., Labelle, M., Lazebnik, Y. A., Munday, N. A., Raju, S. M., Smulson, M. E., Yamin, T. T., Yu, V. L., and Miller, D. K., Identification and inhibition of the ICE/CED-3 protease necessary for mammalian apoptosis, *Nature*, 376, 37, 1995.
37. Tewari, M., Quan, L. T., O'Rourke, K., Desnoyers, S., Zeng, Z., Beidler, D. R., Poirier, G. G., Salvesen, G. S., and Dixit, V. M., Yama/CPP32β, a mammalian homolog of CED-3, is a crmA-inhibitable protease that cleaves the death substrate poly(ADP-ribose) polymerase, *Cell*, 81, 801, 1995.
38. Neamati, N., Fernandez, A., Wright, S., Kiefer, J., and McConkey, D. J., Degradation of lamin B1 precedes oligonucleosomal DNA fragmentation in apoptotic thymocytes and isolated thymocyte nuclei, *J. Immunol.*, 154, 3788, 1995.
39. Kaufmann, S. H., Desnoyers, S., Ottaviano, Y., Davidson, N. E., and Poirier, G. G., Specific proteolytic cleavage of poly(ADP-ribose)polymerase: an early marker of chemotherapy-induced apoptosis, *Cancer Res.*, 53, 3976, 1993.
40. Rosenthal, D. S., Ding, R., Simbulan-Rosenthal, C. M. G., Vaillancourt, J. P., Nicholson, D. W., and Smulson, M. E., Intact cell evidence of the early synthesis, and subsequent late apopain-mediated suppression, of poly(ADP-ribose) during apoptosis, *Exp. Cell Res.*, 232, 313, 1997.
41. Jones, D. P., McConkey, D. J., Nicotera, P., and Orrenius, S., Calcium-activated DNA fragmentation in rat liver nuclei, *J. Biol. Chem.*, 264, 6398, 1989.
42. Bertrand, R., Solary, E., Jenkins, J., and Pommier, Y., Apoptosis and its modulation in human promyelocytic HL-60 cells treated with DNA to poisomerase I and II inhibitors, *Exp. Cell Res.*, 207, 388, 1993.
43. Shiokawa, D., Maruta, H., and Tanuma, S., Inhibitors of PARP suppress nuclear fragmentation and apoptotic body formation during apoptosis in HL-60 cells, *FEBS Lett.*, 413, 99, 1997.
44. Nosseri, C., Coppola, S., and Ghibelli, L., Possible involvement of poly(ADP-ribosyl) polymerase in triggering stress-induced apoptosis, *Exp. Cell Res.*, 212, 367, 1994.
45. Chen, G. and Zeller, W. J., Increased poly(ADP-ribose) formation in cisplatin-resistant rat ovarian tumor cells, *Anticancer Res.*, 14, 1761, 1994.
46. Tanizawa, A., Kubota, M., Hashimoto, H., Shimizu, T., Takimoto, T., Kitoh, T., Akiyama, Y., and Mikawa, H., VP-16-induced nucleotide pool changes and poly(ADP-ribose) synthesis: the role of VP-16 in interphase death, *Exp. Cell Res.*, 185, 237, 1989.
47. Tanizawa, A., Kubota, M., Takimoto, T., Akiyama, Y., Seto, S., Kiriyama, Y., and Mikawa, H., *Biochem. Biophys. Res. Commun.*, 144, 1031, 1987.

48. Manome, Y., Datta, R., Taneja, N., Shafman, T., Bump, E., Hass, R., Weichselbaum, R., and Kufe, D., Coinduction of *c-jun* gene expression and internucleosomal DNA fragmentation by ionizing radiation, *Biochemistry*, 32, 10607, 1993.
49. Yoon, Y., Kim, J., Kang, K., Kim, Y., Choi, K., and Joe, C., Poly(ADP-ribosyl)ation of histone H1 correlates with internucleosomal DNA fragmentation during apoptosis, *J. Biol. Chem.*, 271, 9129, 1996.
50. Thraves, P. J., Kasid, U., and Smulson, M. E., Selective isolation of domains of chromatin proximal to both carcinogen-induced DNA damage and poly-adenosine diphosphate-ribosylation, *Cancer Res.*, 45, 386, 1985.
51. Kastan, M. B., Zhan, Q., El-Delry, W. S., Carrier, F., Jacks, T., Walsh, W. V., Plunkett, B., S., Vogelstein, B., and Fornace, A. J. J., A mammalian cell cycle checkpoint pathway utilizing p53 and *GADD45* is defective in ataxia-telangiectasia, *Cell*, 71, 587, 1992.
52. O'Connor, P. M., Jackman, J., Jondle, D., Bhatia, K., Magrath, I., and Kohn, K. W., Role of the p53 tumor suppressor gene in cell cycle arrest and radiosensitivity of Burkitt's lymphoma cell lines, *Cancer Res.*, 53, 4776, 1993.
53. Fisher, D., Apoptosis in cancer therapy: crossing the threshold, *Cell*, 78, 539, 1994.
54. Wesierska-Gadek, J., Schmid, G., and Cerni, C., ADP-ribosylation of wild-type p53 in vitro: binding of p53 protein to specific p53 consensus sequence prevents its modification, *Biochem. Biophys. Res. Commun.*, 224, 96, 1996.
55. Nelson, W. G. and Kastan, M. B., DNA strand breaks: the DNA template alterations that trigger p53-Dependent DNA damage response pathways, *Mol. Cell. Biol.*, 14, 1815, 1994.
56. Lowe, S. W., Ruley, H. E., Jacks, T., and Housman, D. E., p53-dependent apoptosis modulates the cytotoxicity of anticancer agents, *Cell*, 74, 957, 1993.
57. Kastan, M. B., Onyekwere, O., Sidransky, D., Vogelstein, B., and Craig, R. W., Participation of p53 protein in the cellular response to DNA damage, *Cancer Res.*, 51, 6304, 1991.
58. Blagosklonny, M., Loss of function and p53 protein stabilization, *Oncogene*, 15, 1889, 1997.
59. Simbulan-Rosenthal, C. M., Rosenthal, D. S., and Smulson, M. E., Poly(ADP-ribosyl)ation of p53 during apoptosis in human osteosarcoma cells, *Cancer Res.*, 59, 2190, 1999.
60. Vaziri, H., West, M., Allsop, R., Davison, T., Wu, Y., Arrowsmith, C., Poirier, G., and Benchimol, S., ATM-dependent telomere loss in aging human diploid fibroblasts and DNA damage lead to the post-translational activation of p53 protein involving poly(ADP-ribose) polymerase, *EMBO J.*, 16, 6018, 1997.
61. Wesierska-Gadek, J., Bugajska-Schretter, A., and Cerni, C., ADP-ribosylation of p53 tumor suppressor protein: mutant but not wild-type p53 is modified, *J. Cell Biochem.*, 62, 90, 1996.
62. Malanga, M. and Althaus, F., Poly(ADP-ribose): a negative regulator of p53 functions, in *The 12th International Symposium on ADP-Ribosylation Reactions,* Cancun, Mexico, 1997.
63. Chinnaiyan, A., Orth, K., O'Rourke, K., Duan, H., Poirier, G., and Dixit, V., Molecular ordering of the cell death pathway. Bcl-2 and Bcl-xL function upstream of the CED-3-like apoptotic proteases, *J. Biol. Chem.*, 271, 4573, 1996.
64. Muller, M., Wilder, S., Bannasch, D., Israeli, D., Lelbach, K., Li-Weber, M., Friedman, S., Galle, P., Stremmel, W., Oren, M., and Krammer, P., p53 activates the CD95 (APO-1/Fas) gene in response to DNA damage by anticancer agents, *J. Exp. Med.*, 188, 2033, 1998.

65. Ruscetti, T., Lehnerr, B., Halbrook, J., LeTrong, H., Hoekstra, M., Chen, D., and Peterson, S., Stimulation of DNA PK by poly(ADP-ribose) polymerase, *J. Biol. Chem.*, 273, 14461, 1998.
66. Wang, X., Ohnishi, K., Takahashi, A., and Ohnishi, T., Poly(ADP-ribosyl)ation is required for 53-dependent signal transduction induced by radiation, *Oncogene*, 17, 2819, 1998.
67. Agarwal, M., Agarwal, A., Taylor, W., Wang, Z. Q., and Wagner, E., Defective induction but normal activation and function of p53 in mouse cells lacking PARP, *Oncogene*, 15, 1035, 1997.
68. Malanga, M., Pleschke, J., Kleczkowska, H., and Althaus, F., Poly(ADP-ribose) binds to specific domains of p53 and alters its DNA binding functions, *J. Biol. Chem.*, 273, 11839, 1998.
69. Li, X. and Coffino, P., Identification of a region of p53 that confers lability, *J. Biol. Chem.*, 271, 4447, 1996.
70. Schmid, G., Wang, Z. Q., and Wesierska-Gadek, J., Compensatory expression of p73 in PARP-deficient mouse fibroblasts as response to a reduced level of regularly spliced wild-type p53 protein, *Biochem. Biophys. Res. Commun.*, 255, 399, 1999.
71. Boothman, D., Fukunaga, N., and Wang, M., Down-regulation of topoisomerase I in mammalian cells follwing ionizing radiation, *Cancer Res.*, 54, 4618, 1994.
72. Smith, H. and Grosovsky, A., Poly(ADP-ribosyl)ated regulation of p53 complexed with topoisomerase I following ionizing radiation, *Carcinogenesis*, 20, 1439, 1999.
73. Canitrot, Y., de Murcia, G., and Salles, B., Decreased expression of topoisomerase IIb in poly(ADP-ribose) polymerase deficient cells, *Nucl. Acids Res.*, 26, 5134, 1998.
74. Shiokawa, D., Ohyama, H., Yamada, T., Takahashi, K., and Tanuma, S., Identification of an endonuclease responsible for apoptosis in rat thymocytes, *Eur. J. Biochem.*, 226, 23, 1994.
75. Ribeiro, J. and Carson, D., $Ca^{2+}Mg^{2+}$-dependent endonuclease from human spleen: purification, properties, and role in apoptosis, *Biochemistry*, 32, 9129, 1993.
76. Yoshihara, K., Tanigawa, Y., Burzio, L., and Koide, S. S., Evidence for adenosine diphosphate ribosylation of Ca^{2+}, Mg^{2+}-dependent endonuclease, *Proc. Natl. Acad. Sci. U.S.A.*, 72, 289, 1975.
77. Yoshihara, K., Idel, T., Ota, K., Tanaka, Y., Kameoka, M., Kumamoto, N., and Koide, S., Anti-Ca^{2+},Mg^{2+}-dependent endonuclease antibody detects specifically a class of chromatin-bound endonuclease, *Biochem. Biophys. Res. Commun.*, in press, 2000.
78. Wu, Y., Hickey, R., Lawlor, K., Wills, P., Yu, F., Ozer, H., Starr, R., Quan, J. Y., Lee, M., and Malkas, L., A 17S multiprotein form of murine cell DNA polymerase mediates polyomavirus DNA replication *in vitro*, *J. Cell. Biochem.*, 54, 32, 1994.
79. Applegren, N., Hickey, R., Kleinschmidt, A., Zhou, Q., Coll, J., Bachur, N., Swaby, R., Wu, Y., Quan, J., Lee, M., and Malkas, L., Further characterization of the human cell replication-competent multiprotein form of DNA polymerase, *J. Cell Biochem.*, 59, 91, 1995.
80. Tom, T., Malkas, L., and Hickey, R., Identification of multiprotein complexes containing DNA replication factors by native immunoblotting of HeLa cell protein preparations with T-antigen-dependent SV40 DNA replication activity, *J. Cell Biochem.*, 63, 259, 1996.
81. Eki, T., Poly(ADP-ribose) polymerase inhibits DNA replication by human replicative polymerase alpha, delta, and epsilon *in vitro*, *FEBS Lett.*, 356, 261, 1994.
82. Eki, T. and Hurwitz, J., Influence of poly(ADP-ribose) polymerase on the enzymatic synthesis of SV40 DNA, *J. Biol. Chem.*, 266, 3087, 1991.

83. Morris, G. and Mathews, M., Regulation of proliferating cell nuclear antigen during the cell cycle, *J. Biol. Chem.*, 23, 13856, 1989.
84. Waga, S. and Stillman, B., Anatomy of a DNA replication fork revealed by reconstitution of SV40 DNA replication *in vitro*, *Nature*, 369, 207, 1994.
85. Tsurimoto, T. and Stillman, B., Replication factors required for SV40 DNA replication *in vitro*. I. DNA structure-specific recognition of a primer-template junction by eukaryotic DNA polymerase and their accessory proteins, *J. Biol. Chem.*, 266, 1950, 1991.
86. Simbulan, C., Suzuki, M., Izuta, S., Sakurai, T., Savoysky, E., Kojima, K., Miyahara, K., Shizuta, Y., and Yoshida, S., Poly(ADP-ribose) polymerase stimulates DNA polymerase alpha, *J. Biol. Chem.*, 268, 93, 1993.
87. Ferro, A. M., Higgins, N. P., and Olivera, B. M., Poly (ADP-ribosylation) of a DNA topoisomerase, *J. Biol. Chem.*, 258, 6000, 1983.
88. Dantzer, F., Nasheuer, H., Vonesch, J., de Murcia, G., and Ménissier-de Murcia, J., Functional association of poly(ADP-ribose) polymerase with DNA polymerase a-primase complex: a link between DNA strand-break detection and DNA replication, *Nucl. Acids Res.*, 26, 1891, 1998.
89. Althaus, F., Kleczkowska, H., Malanga, M., Muntener, C., Pleschke, J., Ebner, M., and Auer, B., Poly(ADP-ribosyl)ation: a DNA break signal mechanism, *Mol. Cell. Biochem.*, 193, 5, 1999.
90. Wong, S., Paborsky, L., Fisher, P., Wang, T. S.-F., and Korn, D., Structural and enzymological characterization of immunoaffinity-purified DNA polymerase alpha-DNA primase complex from κB cells, *J. Biol. Chem.*, 261, 7958, 1986.
91. Bambara, R. and Jessee, C., Properties of DNA polymerase delta and epsilon and their roles in eulcaryotic DNA replication, *Biochim. Biophys. Acta*, 1088, 11, 1991.
92. Miyazawa, H., Izumi, M., Tada, S., Takada, R., Masutani, M., Ui, M., and Hanaoka, F., Molecular cloning of the cDNAs for the four subunits of mouse DNA polymerase alpha-primase and their gene expression during cell proliferation and cell cycle, *J. Biol. Chem.*, 268, 8111, 1993.
93. DeGregori, J., Kowalik, T., and Nevins, J., Cellular targets for activation by the E2F1 transcription factor include DNA synthesis and G1/S regulatory genes, *Mol. Cell Biol.*, 15, 4215–4224, 1995.
94. Pearson, B., Nasheuer, H., and Wang, T., Human DNA polymerase a gene: sequences controlling expression in cycling and serum-stimulated cells, *Mol. Cell. Biol.*, 11, 2081, 1991.
95. Blake, M. and Azizkhan, J., Transcription factor E2F is required for efficient expression of hamster dihydrofolate-reductase gene *in vitro* and *in vivo*, *Mol. Cell. Biol.*, 9, 4994, 1989.
96. Nevins, J., EF2: a link between the Rb tumor suppressor protein and viral oncoproteins, *Science*, 258, 424, 1992.
97. Weinberg, R., The retinoblastoma protein and cell cycle control, *Cell*, 81, 323, 1995.
98. Neuman, E., Flemington, E., Sellers, W., and Kaelin, W., Transcription of the E2F-1 gene is rendered cell cycle dependent by E2F DNA-binding sites within its promoter, *Mol. Cell. Biol.*, 14, 6607, 1994.
99. Johnson, D., Ohtani, K., and Nevins, J., Autoregulatory control of E2F1 expression in response to positive and negative regulators of cell cycle progression, *Genes Dev.*, 8, 1514, 1994.
100. Simbulan-Rosenthal, C., Rosenthal, D., Luo, R., and Smulson, M. E., Poly(ADP-ribose) polymerase plays a role in the induction of E2F-1 expression and promoter activity during entry into S-phase, *Oncogene*, 18, 5015, 1999.

101. Meisterernst, M., Stelzer, G., and Roeder, R., Poly(ADP-ribose) polymerase enhances activator-dependent transcription *in vitro, Proc. Natl. Acad. Sci. U.S.A.*, 94, 2261, 1997.
102. Butler, A. and Ordahl, C., Poly(ADP-ribose) polymerase binds with transcription factor 1 to MCAT1 elements to regulate muscle-specific transcription, *Mol. Cell. Biol.*, 19, 296, 1999.
103. Kannan, P., Yu, Y., Wankhade, S., and Tainsky, M., Poly(ADP-ribose) polymerase is a coactivator for AP-2-mediated transcriptional activation, *Nucl. Acids Res.*, 27, 866, 1999.
104. Oei, S., Griesenbeck, J., Schweiger, M., and Ziegler, M., Regulation of RNA polymerase II-dependent transcription by poly(ADP-ribosyl)ation of transcription factors, *J. Biol. Chem.*, 273, 31644, 1998.
105. Rawling, J. and Alvarez-Gonzalez, R., TFIIF, a basal eukaryotic transcription factor, is a substrate for poly(ADP-ribosyl)ation, *Biochem. J.*, 324, 249, 1997.
106. Beg, A. and Baltimore, D., An essential role for NFκB in preventing TNFa-induced cell death, *Science*, 274, 782, 1996.
107. Wang, C., Mayo, M., and Baldwin, A., TNF- and cancer therapy-induced apoptosis: potentiation by inhibition of NFκB, *Science*, 274, 784, 1996.

13 Poly(ADP-Ribose) Polymerase Inhibition by Genetic and Pharmacological Means

Jie Zhang and Jia-He Li

CONTENTS

13.1 Introduction 279
13.2 PARP Primary Structure 280
13.3 The Biochemical Consequence of PARP Inhibition 281
13.4 PARP Inhibition by Gene Manipulation 282
    13.4.1 Lowering PARP Level by PARP Antisense RNA 282
    13.4.2 Dominant-Negative Inhibition by PARP DNA-Binding Domain 283
    13.4.3 Transgenic Mice with PARP Gene Deletion 283
13.5 Chemical Inhibition of PARP 285
    13.5.1 Monocyclic Carboxamides 286
    13.5.2 Bicyclic Lactams and Bicyclic Carboxamides 287
    13.5.3 Tri- and Polycyclic Lactams 288
    13.5.4 Other PARP Inhibitors 292
13.6 Corroborative Results from Genetic and Pharmacological Inactivation of PARP 294
13.7 Conclusion 295
Acknowledgment 295
References 296

13.1 INTRODUCTION

Poly(ADP-ribose) polymerase (PARP, EC 2.4.2.30) is an abundant nuclear protein in most of the eukaryotic tissues.[1] This evolutionarily conserved enzyme appears to be involved in major cellular functions, which include maintaining genomic integrity by limiting sister chromatid exchange, and facilitating chromatin structural changes during DNA repair.[2] PARP has been implicated in participation in, either directly or indirectly, gene expression, DNA replication, DNA rearrangement, differentiation, and mutagenesis. PARP cleavage by caspases during apoptosis may also contribute to the programmed cell death pathway.[3]

Much of the earlier studies of PARP functions rely on the use of PARP inhibitors, e.g., benzamide and 3-aminobenzamide.[4] At high concentrations, however, these PARP inhibitors were found to be associated with side effects that were not mediated by PARP.[5] After the cloning of PARP genes in the later 1980s, PARP research was accelerated by applying molecular biology techniques to study its structure and functions. Site-directed mutagenesis was employed to define key amino acid residues for the catalytic activities of PARP and its binding to DNA, substrate, and inhibitors. Expressing of *trans*-dominant negative fragments or antisense constructs were used to suppress PARP activity. Most recently, the creation of mice with PARP gene deletion offered a direct functional test of PARP in whole animals. In many situations, the results from PARP gene manipulation confirmed studies using PARP inhibitors. Despite limitations in individual genetic or pharmacological methods, together, both approaches greatly increased knowledge of the biological functions of PARP.

Meanwhile, the search for specific, potent, small-molecule PARP inhibitors benefited from an improved understanding of PARP protein structure and became intensified by potential clinical utility of PARP inhibitors. Recent delineation of the structure of PARP catalytic domain enhanced the process of rational design of PARP inhibitors.[6] The potency of the new-generation PARP inhibitors was improved to submicromolar range. Some of the compounds demonstrated remarkable efficacy in reducing tissue damage in animal models of cerebral ischemia, traumatic brain injury, myocardial ischemia, and inflammation.[7-9] Further development of these PARP inhibitors may lead to prospective therapeutic agents for treating the aforementioned diseases.

This chapter is a review of the widely used genetic methods to modulate PARP expression and activity, and a discussion of the structure and activity relationship of PARP inhibition by pharmacological agents. Both approaches have proved to be valuable tools for probing the physiological role of PARP. Studies using the PARP knockout mice have offered compelling evidence that PARP activation is involved in animal models of cerebral ischemia, traumatic brain injury, myocardial ischemia, type 1 diabetes, Parkinson's disease, inflammatory bowel disease, collegen-induced arthritis, and multiple organ failure due to shock.[10,11] PARP inhibitors represent a novel class of promising compounds with the potential to combat these devastating diseases.

13.2 PARP PRIMARY STRUCTURE

Discovered more than three decades ago, PARP is a 113-kDa protein that uses NAD (β-nicotinamide adenine dinucleotide) as its substrate to synthesize poly(ADP-ribose), a branched polymer that can consist of over 200 ADP-ribose units. This major nuclear protein has also been called PARS, ADPRT, or pADPRT. It modifies nuclear proteins with poly(ADP-ribosyl)ation. Many protein acceptors of poly(ADP-ribose) are involved in maintaining DNA integrity. They include histones, topoisomerases, DNA and RNA polymerases, DNA ligases, and Ca^{2+}- and Mg^{2+}-dependent endonuclease. PARP itself serves as a major acceptor through intermolecular auto-ADP-ribosylation.[2] Recently, p53, the tumor suppressor protein, was also identified as a PARP substrate.[12,13]

Molecular cloning of PARP genes from various species reveals that PARP is highly conserved through evolution.[14] Two structures that have been extensively explored for PARP inhibition are the DNA-binding domain and the catalytic domain. The DNA-binding domain resides at the N-terminal region of PARP. Two putative zinc-finger-like motifs in this region are responsible for PARP interaction with DNA. The enzyme binds to DNA with preference to nicked DNA. The binding functions as a sensor of DNA damage, which triggers PARP activation. Truncated protein containing PARP zinc finger is able to compete with the native PARP for DNA binding. This so-called dominant-negative inhibition decreases synthesis of poly(ADP-ribose) by preventing DNA damage–induced PARP activation. Some PARP inhibitors are believed to work by directly antagonizing DNA binding.[15] Most PARP inhibitors appear to antagonize NAD binding at the catalytic site in the C-terminal domain, where polymerization and branching of ADP-ribose polymer take place. Both site-directed and random mutagenesis analyses have identified residues essential for catalytic function and binding of the substrate.[16,17] The binding pocket for a number of representative PARP inhibitors has been defined with X-ray crystallography study of the C-terminal fragment.[18]

13.3 THE BIOCHEMICAL CONSEQUENCE OF PARP INHIBITION

Extensive PARP activation leads to severe depletion of NAD in cells suffering from massive DNA damage.[19] PARP inhibition by chemicals or its inactivation by gene deletion can prevent the drop of NAD.[20,21] When activated by DNA damage, PARP becomes the major consumer of NAD. NAD depletion, which results from overstimulation of PARP, serves as a hallmark of PARP activation. A rapid turnover rate determines the half-life of poly(ADP-ribose) to be less than a minute.[22] Once poly(ADP-ribose) is formed, it is quickly degraded by the constitutively active poly(ADP-ribose) glycohydrolase (PARG), together with phosphodiesterase and (ADP-ribose) protein lyase. PARP and PARG constitute a cycle that converts a large amount of NAD to ADP-ribose. In less than an hour, overstimulation of PARP can cause a drop of NAD and ATP to less than 20% of the normal level.[19] Such a scenario is especially detrimental during ischemia when deprivation of oxygen has already drastically compromised cellular energy output. Calcium overload and subsequent free radical production during reperfusion are assumed to be a major cause of tissue damage. Part of the ATP drop, which is typical during ischemia and reperfusion, could be linked to NAD depletion due to poly(ADP-ribose) turnover.[21] Thus, by maintaining cellular NAD level, PARP inhibitors have therapeutic potential to rescue cells from ischemia and other oxidative stress. Preserving cellular energy level appears to be the main effect that PARP inhibitors exhibit in reducing necrotic cell death.

As a surrogate measurement for PARP inactivation, preventing the drop of cellular NAD level can be a convenient indicator for evaluating the effectiveness of PARP inhibitors. Poly(ADP-ribose) turnover accounts for almost all NAD consumption by PARP in response to DNA damage, since PARP inhibitors can block the drop of NAD. A more direct measurement of PARP inhibition is the decrease of poly(ADP-ribose) accumulation. However, the rapid degradation of poly(ADP-ribose) by PARG often makes the immunohistochemistry determination difficult.

13.4 PARP INHIBITION BY GENE MANIPULATION

PARP functions are mostly studied through inactivating the enzyme in cells or animals to check the "loss of function" consequence, rather than introducing PARP ("knock in") into a system without PARP to see any "gain of function." Such an approach is predetermined by the presence of PARP protein at a high level in many tissues, most notably in the immune system, heart, brain, and germ line cells. The "knock in" is most suitable to reintroduce PARP into cells derived from PARP knockout mice to restore PARP activity. Under normal physiological conditions, there is minimal PARP activity despite its abundance. Once PARP is activated by DNA breaks, no *de novo* protein synthesis is necessary to achieve a high level of poly(ADP-ribose) synthesis.

13.4.1 LOWERING PARP LEVEL BY PARP ANTISENSE RNA

PARP protein has a half-life of longer than 2 days.[23] Depleting PARP by antisense RNA requires stable integration and, typically, inducible expression of antisense transcript for a few days. Smulson and associates[23] designed an antisense construct with the entire human cDNA for PARP in an antisense orientation under the control of mouse mammary tumor virus (MMTV) LTR promoter. Multiple copies of the plasmid DNA were integrated into the genome of HeLa cells, resulting in a cellular system to uncover PARP functions. In 48 to 72 h after dexamethasone induction, antisense PARP transcript caused a decrease of PARP protein by more than 90% and its enzyme activity by 80%. Normal level of PARP was restored in 8 to 16 h after removal of the inducer. Using this reversible PARP inactivation of cell lines, Smulson's group found that diminishing PARP level prolonged the process of DNA repair, even though the damage would eventually be fixed, suggesting that PARP plays an auxiliary role in the repair of DNA strand breaks.[23] After PARP depletion, the cell line became hypersensitive to cytotoxicity of carcinogenic agents.[24] Similarly, Qu and colleagues[25] showed that a metal ion–induced expression of either full-length or a truncated 5′-fragment of PARP cDNA in antisense orientation could decrease PARP activity by more than 90% in permanent transfected human leukemia cells.[25] They found that a decrease of PARP activity correlated with an enhancement of interferon-γ-induced major histocompatibility complex class II gene expression.

In light of recent discovery of multiple PARP homologues, it is not clear if antisense PARP transcripts would also nonspecifically decrease PARP-related proteins.[26-29] Further experiments are required to test the feasibility of adapting the antisense RNA method for gene therapy to inactivate PARP selectively in a tissue-specific fashion and in a reversibly inducible manner to treat chronicle diseases due to PARP activation.

13.4.2 Dominant-Negative Inhibition by PARP DNA-Binding Domain

The *trans*-dominant inhibition takes the advantage of a unique property of PARP, that its activation is dependent on single- or double-strand DNA breaks. The zinc-finger DNA-binding domain serves as a sensor to detect the bending of DNA when it is damaged.[30] The conformation change is transferred to the catalytic domain to activate poly(ADP-ribose) synthesis. Overexpression of the DNA-binding domain competes with endogenous PARP for DNA binding, and thereby blocks PARP activation. In the family of poly(ADP-ribose)-producing enzymes, *trans*-dominant inhibition should be preferentially selective for PARP whose activation is mediated by the zinc-finger DNA-binding domain. In addition, by circumventing interaction with the catalytic domain, this method avoids nonspecific interference associated with some NAD analogue inhibitors. The *trans*-dominant inhibition also takes less time for the DNA-binding domain to accumulate and compete with PARP, in contrast to antisense inhibition that requires longer time to diminish PARP due to its long half-life. It is not clear if the expression of PARP DNA zinc-finger domain would interfere in the functions of other DNA-binding proteins such as transcription factors.

de Murcia and associates[31] developed the inducible *trans*-dominant inhibition system in monkey cells and hamster cells to study PARP function in DNA repair.[31,32] In these cells, PARP activity was inhibited by 90%. Direct injection of recombinant DNA-binding domain into fibroblasts also achieved similar PARP inhibition.[31] Overproduction of PARP DNA-binding domain inhibited unscheduled DNA synthesis induced by *N*-methyl-*N'*-nitro-*N*-nitrosoguanidine (MNNG) without affecting normal cellular proliferation, suggesting PARP may facilitate DNA repair but not directly participate in DNA replication.[31,33] The *trans*-dominant inhibition of PARP also sensitized cells for killing by γ irradiation and alkylating agents.[33] Both spontaneous and MNNG-induced sister chromatid exchange were elevated in cells overexpressing PARP DNA-binding domain.[34]

13.4.3 Transgenic Mice with PARP Gene Deletion

Three strains of mice with PARP gene deletion have been independently generated.[35-37] They provide an invaluable tool to reassess the biological functions of PARP at the whole-animal level. It is perhaps not entirely surprising that mice without PARP can be produced and reproduced, if one considers that PARP activity is largely dependent on DNA damages. Such DNA damage is often elicited by exogenous chemicals, e.g., monoalkylating agents or ionizing γ irradiation, and sometimes due to excessive free radicals produced endogenously in a disease condition. Without exposure to the obnoxious genomic toxins, homozygous PARP knockout (PARP$^{-/-}$) mice are neither embryonic lethal nor grossly abnormal in their life span and in several generations reproduced so far. However, close scrutiny at the cellular level has revealed differences in cell cycle distribution and apoptosis execution between the wild-type and homozygous PARP deletion.

Unlike the use of negative-dominant mutant or antisense expression to decrease PARP, which at most reduces PARP activity by 90%, PARP$^{-/-}$ mice and the derived cells offer a system with no PARP protein at all. Compared with the use of pharmacological agents with potential side effects unrelated to PARP inhibition, PARP$^{-/-}$ mice provide a system in which only PARP is directly eliminated. Such advantages make PARP$^{-/-}$ mice useful for obtaining definitive evidence for whether PARP is involved in certain biological pathways, and for validating whether targeting PARP would be effective in limiting necrotic tissue damage under pathological conditions. On the other hand, lack of PARP throughout the life cycle of the transgenic animals raises the concern whether a compensatory mechanism might cause secondary changes that render the system substantially different from that of a wild type.

Wang and colleagues[35] reported the first PARP knockout mice in 1995. Mice homozygous for PARP deletion were healthy, appeared fertile, and exhibited no macroscopic abnormalities. One striking observation was that cells derived from PARP$^{-/-}$ were not different from wild-type cells in repair of DNA damage. Both the nucleotide excision repair and base excision repair mechanisms could still function in the absence of PARP.[35] After several generations, there has been so far no report for a high rate of tumorogenesis in PARP$^{-/-}$ mice. In contrast, deletion of the p53 tumor suppressor genes resulted in significant increase of tumor growth in multiple organs in 3-month-old mice.[38] Another unexpected observation was the apparent normal execution of apoptosis process in cells from PARP$^{-/-}$ mice, despite the implication that PARP cleavage by caspases plays a major role in mediating the programmed cell death pathway.[39,40] No difference was found among wild-type PARP$^{+/+}$, heterozygous PARP$^{+/-}$, or homozygous PARP$^{-/-}$ cells undergoing apoptosis induced by tumor necrosis factor or anti-Fas antibody. Therefore, the minor role that PARP might play in apoptosis seemed dispensable or compensatable during embryogenesis and postnatal development when apoptosis widely occurred. A major phenotypic abnormality in the first PARP$^{-/-}$ mice was the incidence of epidermal hyperplasia and slight obesity in older mice originated from a mixed genetic background of 129Sv × C57Bl6.[35]

The second PARP knockout mice generated in de Murcia's laboratory displayed some phenotypes different from the first one.[36] Their average litter size was smaller and the adult PARP$^{-/-}$ mice were slightly underweight. Otherwise, they appeared normal and were viable and fertile without skin hyperplasia or obesity. In PARP$^{-/-}$ cells, the cell cycle distribution was disturbed with an elevated G_2/M phase accumulation after DNA damage. Lack of PARP also potentiated the splenocytes to undergo a rapid apoptosis, as revealed by a quick detection of p53 induction in these cells. Exposure to genotoxic dose of alkylating agent or γ irradiation caused an acceleration and increased rate of death in PARP$^{-/-}$ mice, presumably due to acute toxicity on epithelium of small intestine. There is also substantial evidence that base excision repair is compromised in PARP$^{-/-}$ cells.[41]

Most recently, Masutani and colleagues[37,42] reported the third PARP knockout mice. In the embryonic stem cells selected for generating the knockout mice, PARP$^{-/-}$ clones showed enhanced sensitivity to γ irradiation and alkylating agents. The PARP$^{-/-}$ mice were normal in litter size, growth rate, and fertility. Reproduction from heterozygous PARP$^{+/-}$ mice followed a Mendelian pattern in transmitting the PARP deletion mutation, which suggested no disadvantage for embryo development from PARP-deficient zygotes. Like the second knockout, there was no increase of skin lesion or body weight in the third knockout mice.

Although the three PARP knockouts shared many common features, they differed in some other aspects. Part of the difference might be derived from the ways they were generated. The three laboratories targeted different PARP exons for deletions and used different strains of mice in the breeding schemes. In the first knockout mice, Wang and colleagues[35] disrupted the second exon and thus deleted both zinc fingers in the PARP. No DNA-binding domain could be expressed after deleting exon 2. de Murcia and associates targeted the fourth exon, which contained the second but not the first zinc finger. The first zinc finger in PARP was known to be sufficient for DNA binding. However, no expression of PARP or its fragment was detected by multiple antibodies against PARP.[29] The third PARP knockout mice from Masutani's laboratory deleted exon 1 of the mouse PARP gene. These mice were bred from a strain background of 129/SJ × ICR, which was different from 129/Sv × C57Bl6 that were used for breeding in Wang's and de Murcia's laboratories. Furthermore, from the first 129/Sv × C57Bl6 PARP knockout mice, Wang's laboratory also bred a strain of PARP$^{-/-}$ mice with essentially 129/Sv background through several generations of backcrossings. One should bear in mind the differences in strain backgrounds and exon deletions when different PARP knockout mice were subject to similar experiments, as they might contribute, in addition to common PARP inactivation, to variations in experimental results from different groups.[43] Another source of difference might come from the levels of PARP protein in the heterozygotes. In PARP$^{+/-}$ mice, PARP protein levels should always be experimentally determined, rather than assumed to be at the 50% of the wild type. For example, little PARP protein was detected in β-islet cells or cerebral cortical neurons prepared from PARP$^{+/-}$ mice, while a substantial amount of PARP, which could be as high as in the wild type, was present in embryonic stem cells, fibroblast cells, and in a pancreatic preparation from the heterozygotes.[39,44-46]

13.5 CHEMICAL INHIBITION OF PARP

Several classes of PARP inhibitors has been identified and characterized in the past two decades. These compounds have been used as tools to probe the physiological role of PARP and contributed to understanding the biological functions of PARP. Recently, some of the PARP inhibitors have been tested in animal models of diseases, and the results have yielded insights into the pathological role of PARP activation in cell death and tissue damage. The potential therapeutic benefit these compounds may offer has intensified the search for more potent and selective PARP inhibitors. Among the small-molecule PARP inhibitors discovered, most compounds fall into the categories of monoaryl amides, and bi-, tri-, or tetracyclic lactams. A common structural feature for these inhibitors is either a carboxamide attached to an aromatic ring or the carbamoyl group built in a polyaromatic heterocyclic skeleton to form a fused aromatic lactam or imide. Representative compounds like 3-aminobenzamide, nicotinamide, and 1,8-dihydroxyisoquinoline have been widely used as inhibitor in PARP research because of their availability (Figure 13.1).

Most PARP inhibitors exhibit competitive mode, which suggests that they block NAD binding to the catalytic domain of the enzyme.[47-49] Nicotinamide, the coproduct of poly(ADP-ribose) synthesis, was known to exert feedback inhibition on PARP,

nicotinamide   3-aminobenzamide   1,5-dihydroxyisoquinoline

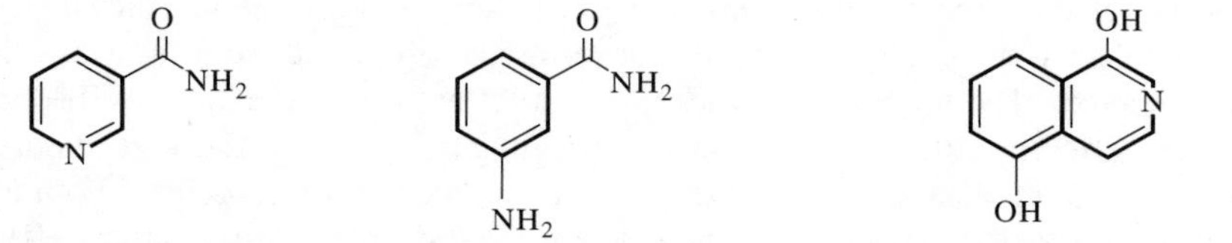

FIGURE 13.1 Examples of the most widely used PARP inhibitors.

albeit only at high concentration with IC_{50} around 30 to 200 µM.[47,50-57] Benzamide, a close analogue to nicotinamide, was soon found to be a more potent compound with IC_{50} from 3 to 20 µM.[48,49,58] Only a few agents interfere in the DNA-binding domain to antagonize PARP activation. For example, 6-nitroso-1,2-benzopyrone was known to oxidize one of the zinc fingers to inhibit PARP activation.[59]

There are widespread variations of inhibition constants for PARP inhibitors.[4,60] Direct comparison for the absolute values of IC_{50} and K_i values among different laboratories was hindered by slight differences in PARP assays due to subtle changes of conditions. The complex reaction catalyzed by PARP includes initiation, elongation, and branching in perhaps a processive motion. Intermolecular auto-ADP-ribosylation is self-inhibitory to the enzyme. PARP activity also seems to be very sensitive to solvents.[61,62] Any small changes in these parameters can affect PARP activity, which may explain the wide range of inhibition values for PARP inhibitors. For example, the IC_{50} values for 3-aminobenzamide inhibition of PARP encompasses from 5.4 to 250 µM.[63,64] Interpretation of potency of PARP inhibitors from different laboratories should be taken cautiously.

13.5.1 Monocyclic Carboxamides

Nicotinamide (niacinamide or vitamin B_3), 3-aminobenzamide, and thiophene-2-carboxamide are prototype inhibitors in this category.[65] As a by-product, nicotinamide became the first PARP inhibitor to be used in pharmacological study. With an IC_{50} value of about 100 µM, nicotinamide is only a weak PARP inhibitor.[57] Sims et al.[66] studied 14 of the pyridine family including different members with carboxamide group at position 2, and 4, N-substituted of pyridine compounds, and pyrazinamide analogues.[66] Nicotinamide proved to be the most potent inhibitor in the group.

In 1980, several investigators found benzamide and substituted benzamide were more potent than nicotinamide.[48,49,58] Apparently, the ring nitrogen in nicotinamide was not necessary for PARP inhibition. Since then, more than 100 benzamide derivatives have been compared for PARP inhibition.[67-69] Substitution studies of structure–activity relationship demonstrated that a single 3-substituted benzamide was more effective than the 2-, 4-, 5-, 6-positions for PARP inhibition. Additional multi-substitution at 2-, 4-, 5-, or 6-positions of 3-substituted benzamides failed to improve potency further, and in most cases, reduced inhibition. It suggests that optimal binding depends on a 3-position electron donor, such as a hydroxyl, methyl, or amino group.[57,67] For example, in one study, 3-aminobenzamide *in vitro* displayed about twofold more potency than benzamide. But, 3-nitrobenzamide (IC_{50} value of

210 μM) lost its inhibition effect, perhaps due to an electron-deficient benzene ring.[70] In general, benzamide derivatives are still weak inhibitors with IC$_{50}$ values above 1 μM. The nonspecific side effects of benzamide and its derivatives at high concentration further limit their use *in vivo*.[64,71-73]

Various monocyclic-ring carboxamides, especially monoaryl-heterocyclic carboxamides, were tested as PARP inhibitors. Cyclohexanecarboxamide, a saturated form of benzamide, was found to be threefold weaker than benzamide.[74] Most of the monoaromatic-heterocyclic carboxamides, such as pyrazine carboxamide analogues[66] and thiophene-3-carboxamide[65,75] turned out to be poor PARP inhibitors. As an exception, substituted thiophene-2-carboxamides could inhibit PARP with a potency similar to compounds with a direct benzene ring analogue of benzamide derivatives.[75] It seems that the heterocyclic thiophene ring can substitute for the benzene ring without significant loss of potency. More heterocyclic carboxamide compounds need to be tested to see if they are effective PARP inhibitors.

13.5.2 BICYCLIC LACTAMS AND BICYCLIC CARBOXAMIDES

Constraining the monoaryl amide compounds by formation of lactam generated bicyclic compounds. Extensive efforts have been devoted to this category of chemicals for the past decade. In general, two-ring PARP inhibitors are superior in potency and specificity over the monoaryl amide series. Suto et al.[69] and Griffin et al.[76-78] have systemically designed constrained 3-aminobenzamide analogues by using nicotinamide or 3-aminobenzamide as a template.

The amide group of nicotinamide or 3-aminobenzamide is free to rotate relative to the plane of the aromatic ring. Only certain orientation of the amide group with respect to the nitrogen of the pyridine ring of nicotinamide or the substitution at the 3-position of benzamide might be accommodated for PARP inhibition. *Ab initio* calculations with nicotinamide indicated that, of the *cis* and *trans* configuration for amide, the C–N bond of the carboxamide *cis* to the 1,2-bond of the ring is preferred (Figure 13.2).

To maintain the *cis*-conformation of the arylamide, planar bicyclics with lactam functionality of isoquinolin-1(2H)-one were synthesized.[69,79] An alternative strategy

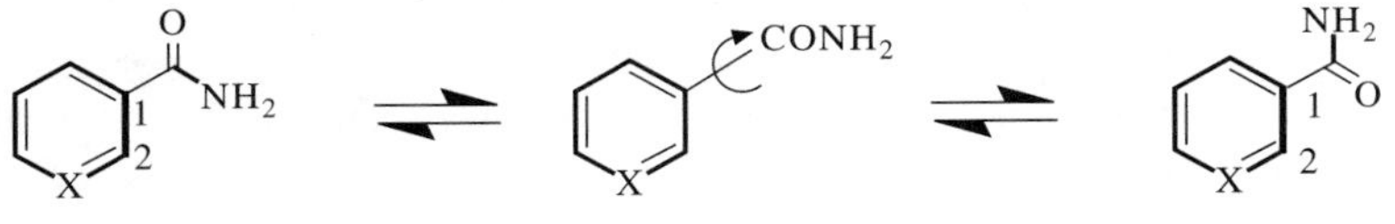

A: *cis*

B: *trans*

X = N: nicontinamide
X = C-NH$_2$: 3-aminobenzamide

FIGURE 13.2 Preferred conformation for optimal bindings between benzamide derivatives and PARP.

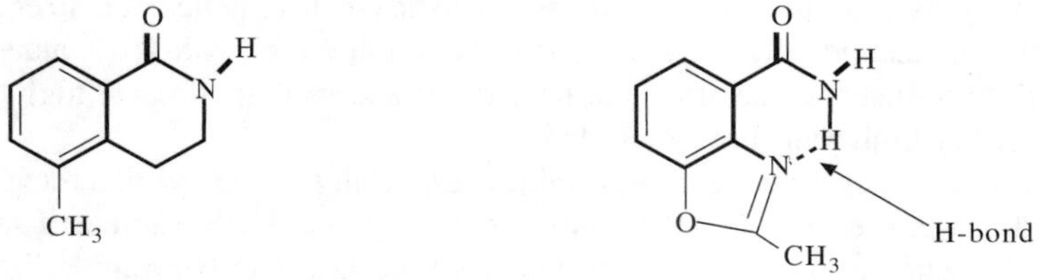

PD 128763

NU 1056

FIGURE 13.3 Constraining amide group into the active conformation by forming lactam or intramolecular hydrogen bond.

for constraining the carboxamide to the active orientation was achieved by an intramolecular hydrogen bond between the carbamoyl NH and an adjacent heterocyclic nitrogen.[80] For example, 3,4-dihydro-5-methyl isoquinolin-1(2H)-one (PD 128763) and benzoxazole-4-carboxamide (NU 1056) were found to be more potent PARP inhibitors than 3-aminobenzamide (Figure 13.3 and Table 13.1). In addition to the preferred *cis* region-arrangement for the C–N and the 1,2-bond as depicted in Figure 13.2, the N–H bond (solid line in Figure 13.3) *cis* to the carbonyl bond is also essential for PARP inhibition.

Incorporation of the carbamoyl substituent into a ring system resulted in compounds such as PD 128763 and NU 1056, which apparently orient the functional groups in the preferred binding conformation. This series of bicyclic lactams is among the most potent of PARP inhibitors. It includes dihydroisoquinolin-1(2H)-nones,[67,69] 1,6-naphthyridine-5(6H)-ones,[79] quinazolin-4(3H)-ones,[77,78] thieno[3,4-c]pyridin-4(5H)ones, and thieno[3,4-d]pyrimidin-4(3H)ones.[65] 1,5-Dihydroxyisoquinoline also belongs to this family.[74] Another member in this category is 2-methylquinazolin-4[3H]-one, which was isolated from the bacillus culture extract.[81] Table 13.1 is a comprehensive list of inhibitors in the bicyclic lactams and bicyclic carboxamides category.

13.5.3 TRI- AND POLYCYCLIC LACTAMS

PARP inhibitors with structures of three or more rings have also been identified. 1,8-Naphthalimide derivatives and (5H)-phenanthridin-6-ones are representative for the tricyclic family. Both types of the compounds have been claimed to be 100-fold more potent than 3-aminobenzamide.[74] The IC_{50} values for 4-amino-1,8-naphthalimide (Figure 13.4, **2**) and (5H)-2-nitrophenanthridin-6-ones (Figure 13.4, **3**) are 0.18 and 0.35 μM, respectively.[61] An inherent disadvantage for these planar heteroaromatic compounds is the poor solubility in water and many organic solvents. Mono- and multisubstituted (5H)-phenanthridin-6-ones have been reported recently.[82] Also disclosed as PARP inhibitors are tetracyclic lactams,[82,83] such as [4H]-cyclopanta[*imn*]phenanthridine-5-ones (Figure 13.4, **4**), [1]benzopyrano[4,3,2,-*de*]phthalazinone (Figure 13.4, **5**) and [1]benzopyrano[4,3,2,-*de*]isoquinolinones (Figure 13.4, **6**). The potency of these PARP inhibitors approaches low nanomolar range, with improved water solubility in some of the compounds. One member of the multiple-

TABLE 13.1
The Structures and Activities of the Bicyclic Lactams and Amides of PARP Inhibitors

Class	Name	X	R1	R2	R3	R4	R5	R6	IC$_{50}$ (mM)	Ref.
A	5-Hydroxy-1(2H)-isoquinolinone	C	OH	H	H				0.14	69
A	5-Methoxy-1(2H)-isoquinolinone	C	OMe	H	H				0.58	69
A	5-Nitro-1(2H)-isoquinolinone	C	NO$_2$	H	H				0.32	69
A	5-Amino-1(2H)-isoquinolinone	C	NH$_2$	H	H				0.24	69
A	1(2H)-Isoquinolinone	C	H	H	H				6.20	69
A	5-Methoxy-3-phenyl-1(2H)-isoquinolinone	C	OMe	H	Ph				2.12	118
A	4-Bromo-1-isoquinolinol	C	H	Br	H				0.36	118
A	Isoquinoline-5-carboxylic acid	C	H	COOH	H				79%@10 µM	67
A	5-Iodoisoquinolin-1-one	C	H	I	H				96%@10 µM	67
A	E-3-(1-Oxoisoquinolin-5-yl)propenoic acid	C	CH=CHCOOH	H	H				81%@10 µM	67
A	1,6-Naphthyridine-5(6H)-one	N	H	H	H				5.00	79
A	7-Methyl-1,6-naphthyridine-5(6H)-one	N	H	H	Me				1.20	79
A	7,8-Dihydro-1,6-naphthyridine-5(6H)-one	N	H	H	H	H	H	H	2.50	79
B	3,4-Dihydro-7-methoxy-1(2H)-isoquinolinone	C	H	H	H	H	OMe	H	120.00	69
B	3,4-Dihydro-6-methoxy-1(2H)-isoquinolinone	C	H	H	H	H	H	OMe	39.00	69
B	3,4-Dihydro-5-hydroxy-1(2H)-isoquinolinone	C	OH	H	H	H	H	H	0.10	69
B	3,4-Dihydro-7-hydroxy-1(2H)-isoquinolinone	C	H	H	H	H	OH	H	9.50	69

TABLE 13.1 (continued)
The Structures and Activities of the Bicyclic Lactams and Amides of PARP Inhibitors

Class	Name	X	R1	R2	R3	R4	R5	R6	IC$_{50}$ (mM)	Ref.
B	3,4-Dihydro-6-hydroxy-1(2H)-isoquinolinone	C	H	H	H	H	H	OH	2.00	69
B	3,4-Dihydro-8-hydroxy-1(2H)-isoquinolinone	C	H	H	H	OH	H	H	11.00	69
B	3,4-Dihydro-5-nitro-1(2H)-isoquinolinone	C	NO$_2$	H	H	H	H	H	3.20	69
B	3,4-Dihydro-7-nitro-1(2H)-isoquinolinone	C	H	H	H	H	NO$_2$	H	13.00	69
B	5-Amino-3,4-dihydro-1(2H)-isoquinolinone	C	NH$_2$	H	H	H	H	H	0.41	69
B	7-Amino-3,4-dihydro-1(2H)-isoquinolinone	C	H	H	H	H	NH$_2$	H	8.00	69
B	3,4-Dihydro-1(2H)-isoquinolinone	C	H	H	H	H	H	H	1.50	69
B	3,4-Dihydro-5-[2-hydroxy-3-(1piperidinyl)propoxy]-1(2H)-isoquinolinone	C	See name	H	H	H	H	H	0.32	118
B	5-(Acetyloxy)-3,4-dihydro-1(2H)-isoquinolinone	C	See name	H	H	H	H	H	2.11	118
B	3,4-Dihydro-5-(phenylmethoxy)-1(2H)-isoquinolinone	C	OBn	H	H	H	H	H	5.20	118
B	3,4-Dihydro-5-[(phenylmethyl)amino]-1(2H)-isoquinolinone	C	NHBn	H	H	H	H	H	2.55	118
B	3,4-Dihydro-5-(acetylamino0-1(2H)-isoquinolinone	C	NHAc	H	H	H	H	H	56.70	118
B	3,4-Dihydro-5-methyl-1(2H)-isoquinolinone	C	Me	H	H	H	H	H	0.16	118
B	5-[Dimethylamino)methoxy]-3,4-dihydro-1(2H)-isoquinolinone	C	OCH$_2$NMe$_2$	H	H	H	H	H	0.97	118
B	5-Ethyl-3,4-dihydro-1(2H)-isoquinolinone	C	Et	H	H	H	H	H	1.00	118

	Compound								
B	5-Chloro-3,4-dihydro-1(2H)-isoquinolinone	C	Cl	H	H	H	H	0.31	118
B	3,4-Dihydro-3,5-dimethyl-1(2H)-isoquinolinone	C	Me	H	Me	H	H	0.74	118
B	4-Bromo-5-methyl-1(2H)-isoquinolinone	C	H	Br	Me	H	H	0.041	118
B	4-Amino-1(2H)-isoquinolinone	C	H	NH$_2$	H	H	H	0.66	118
B	4-Bromo-5-hydroxy-1(2H)-isoquinolinone	C	OH	Br	H	H	H	0.061	118
B	3,4-Dihydro-5-[3-(1-piperidinyl)propoxy]-1(2H)-isoquinolinone	C	See name	H	H	H	H	0.83	118
B	3,4-Dihydro-5-[2-hydroxy-3-(1-pyrrolidinyl)propoxy]-1(2H)-isoquinolinone	C	See name	H	H	H	H	0.77	118
B	3,4-Dihydro-5-[2-hydroxy-3-(4-mopholinyl)propoxy]-1(2H)-isoquinolinone	C	See name	H	H	H	H	1.50	118
B	5-(3-Diethylamino-2-hydroxy)propoxy-3,4-dihydro-1(2H)-isoquinolinone	C	See name	H	H	H	H	1.00	118
B	3,4-Dihydro-5-[2-hydroxy-3-(methylamino)propoxy]-1(2H)-isoquinolinone	C	See name	H	H	H	H	0.20	118
B	3,4-Dihydro-5-[2-(1-piperidinyl)ethoxy]-1(2H)-isoquinolinone	C	See name	H	H	H	H	0.82	118
B	3,4-Dihydro-5-[4-(1-piperidinyl)butoxy]-1(2H)-isoquinolinone	C	See name	H	H	H	H	0.80	118
B	5-Ethoxy-3,4-dihydro-1(2H)-isoquinolinone	C	OEt	H	H	H	H	2.60	118
B	3,4-Dihydro-5-propoxyl-1(2H)-isoquinolinone	C	O-Pr	H	H	H	H	0.11	118
B	4-Butoxyl-3,4-dihydro-1(2H)-isoquinolinone	C	O-Bu	H	H	H	H	6.00	118
B	3,4-Dihydro-5-(2-hydroxy-3-methoxypropoxy)-1(2H)-isoquinolinone	C	See name	H	H	H	H	1.50	118
B	3,4-Dihydro-5-(2-hydroxy-3-pheoxypropoxy)-1(2H)-isoquinolinone	C	See name	H	H	H	H	0.65	118
B	3,4-Dihydro-5-(2-hydroxy-3-phenylpropoxy)-1(2H)-isoquinolinone	C	See name	H	H	H	H	0.34	118
B	3,4-Dihydro-5-(2phenylethoxy)-1(2H)-isoquinolinone	C	See name	H	H	H	H	0.60	118
B	(R)-3,4-Dihydro-5-[2-hydroxy-3-(1-piperidinyl)propoxy]-1(2H)-isoquinolinone	C	See name	H	H	H	H	0.33	118
B	(S)-3,4-Dihydro-5-[2-hydroxy-3-(1-piperidinyl)propoxy]-1(2H)-isoquinolinone	C	See name	H	H	H	H	0.27	118

TABLE 13.1 (continued)
The Structures and Activities of the Bicyclic Lactams and Amides of PARP Inhibitors

Class	Name	X	R1	R2	R3	R4	R5	R6	IC$_{50}$ (mM)	Ref.
C	2-Phenylbenzoxazole-4-carboxamide (NU1051)		Ph						2.10	120
C	2-(4-Methoxyphenyl) benzoxazole-4-carboxamide (NU1054)		PhOMe						1.10	120
C	2-Methylbenzoxazole-4-carboxamide (NU1056)		Me						9.50	120
D	8-Hydroxy-2-methyl-quinazolin-4-[3H]one (NU1025)		OH	Me					0.40	119
D	8-Hydroxyquinazolin-4-[3H]one		OH	H					2.00	119
D	8-Hydroxy-2-(4-nitrophenyl)-quinazolin-4-[3H]one		OH	PhNO$_2$					0.23	119
D	8-Methoxy-2-methylquinazolin-4-[3H]one		OMe	Me					0.78	119
D	8-Methoxy-2-phenylquinazolin-4-[3H]one		OMe	Ph					4.20	119
D	8-Hydroxy-2-phenylquinazolin-4-[3H]one		OH	Ph					5.30	119
D	2,8-Dimethylquinazolin-4-[3H]one		Me	Me					0.20	119
E	6-Methylthieno[3,4-c]pyridin-4(5H)one	C	H	H	Me				90%10 μM	65
E	6-Phenylthieno[3,4-c]pyridin-4(5H)one	C	NO$_2$	H	Ph				93%10 μM	65
E	6-Methyl-1-nitrothieno[3,4-c]pyridin-4(5H)one	C	H	H	Me				54%10 μM	65
E	1-Nitro-6-phenylthieno[3,4-c]pyridin-4(5H)one	C	NO$_2$	H	Ph				86%10 μM	65
E	6-Methyl-7-nitrothieno[3,4-c]pyridin-4(5H)one	C	H	NO$_2$	Me				89%10 μM	65
E	7-Nitro-6-phenylthieno[3,4-c]pyridin-4(5H)one	C	H	NO$_2$	Ph				47%10 μM	65
E	2-Methylthieno[3,4-d]pyrimidin-4(3H)one	N	H	H	Me				81%10 μM	65
E	2-Phenylthieno[3,4-d]pyrimidin-4(3H)one	N	H	H	Ph				42%10 μM	65

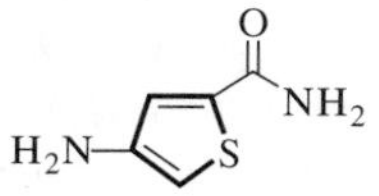
1

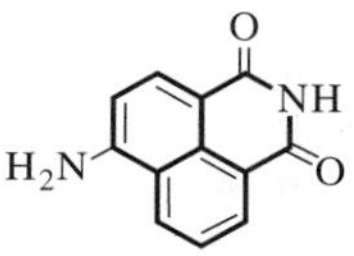
2

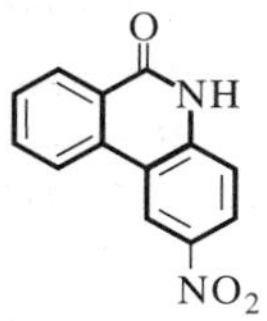
3

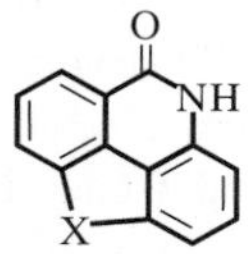
X=C=0, CH2, CHOH, CHNH2

4

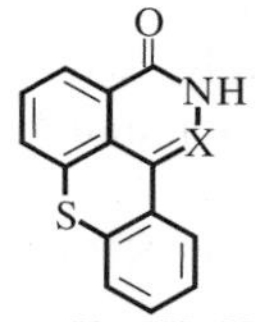
X=CH, N

5

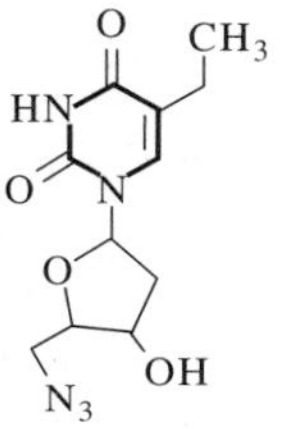
6

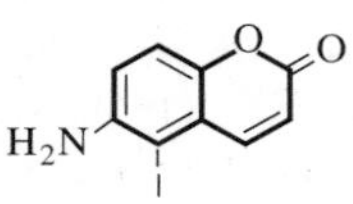
7

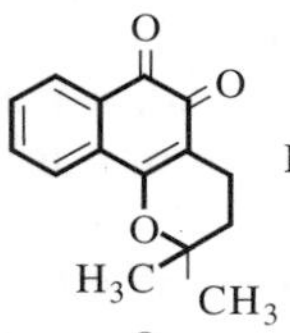
8

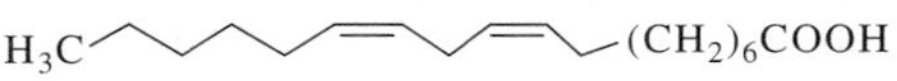
9

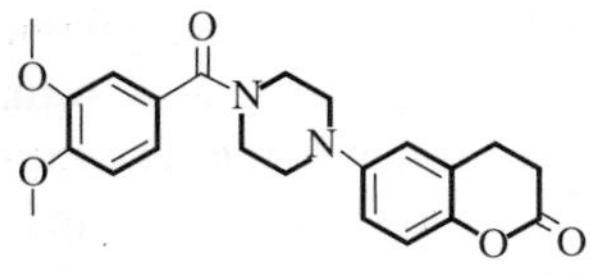
10

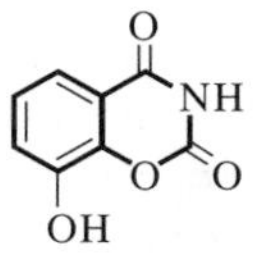
11

FIGURE 13.4 Examples of different classes of PARP inhibitors.

ring PARP inhibitor, GPI 6150, has demonstrated remarkable efficacy in a number of rodent models of cerebral ischemia, head trauma injury, myocardial ischemia, and inflammatory response.[7-9,84]

13.5.4 Other PARP Inhibitors

The similarity of poly(ADP-ribose) structure to that of RNA led to a search for nucleic acid derivatives and nucleoside analogues as PARP inhibitors. A few groups reported that certain nucleic acid analogues can weakly inhibit PARP. They include pyrimidine derivatives such as thymidine, 5-iodouracil, 5-iodouridine, and purine analogues such as 3,7-dimethylxanthine, and theophyline.[51,63,74,85] Deoxyuridine analogues with combined substituents at both the 5-position of the pyrimidine ring and the 3'- or 5'- position of deoxyribose were generally more potent inhibitors than 3-aminobezamide, e.g., 3'-amino-2', 3'-dideoxy-(E)-5-(2'-bromovinyl)uridine ($K_i = 0.7$ μM) and 5'-azido-2',5'-dideoxy-5-ethyluridine (Figure 13.4, **6**) ($K_i = 0.8$ μM).[86] Since these compounds lack aromatic-fused lactam or amide, it is not clear whether and how they bind to the PARP catalytic center.

Another family of PARP inhibitors is the benzopyrone derivatives. The first prototype of this type of PARP inhibitor is 1,2-benzopyrone (IC$_{50}$ value of 47 μM) and it appeared to be a noncompetitive inhibitor.[15] Later, Banasik et al.[74] reported both 1,2-benzopyrone and 1,4-benzopyrone as weak PARP inhibitors with IC$_{50}$ values of 2.8 and 0.4 mM, respectively. Szabó and collaborators[88,89] have used 6-amino-1,2-benzopyrone and 5-iodo-6-amino-1,2-benzopyrone (Figure 13.4, **7**) to demonstrate their protective effects in animal models of cerebral ischemia and inflammation.[87-89] The mechanism of PARP inhibition by this class of compounds remains to be elucidated.

Several chemically distinctive and diverse compounds exhibit PARP inhibition activities. Essential fatty acids, such as linolenic acid (Figure 13.4, **9**), have been reported as moderate PARP inhibitors. The IC$_{50}$ values for these vitamins and their analogues were around 100 μM.[90] PARP inhibition recently has been ascribed to explain the pharmacological activity of β-lapachone (3,4-dihydro-2,2-dimethyl-2H-naphtho[1,2-b]pyran-5,6-dione; Figure 13.4, **8**), a naturally occurring o-naphthoquinone.[91] Inorganic arsenite (NaAsO$_2$) was shown to decrease PARP activity with an IC$_{50}$ value at 10 μM. The arsenic cation might interact with vicinal dithiol groups in the zinc-finger region of PARP.[92] A positive inotropic agent vesnavinone, 1-(3,4-dimethoxybenzoyl)-4-(1,2,3,4-tetrahydro-2-oxo-6-quinolinyl)piperazine (Figure 13.4, **10**) and a natural product benadrostin, 8-hydroxy-2H-1,3-benzoxazine-2,4-dione (Figure 13.4, **11**) were also reported as PARP inhibitors.[93,94]

In summary, a common structural feature of several classes of PARP inhibitors is either the presence of a carboxamide or an imide group built in a polyaromatic heterocyclic skeleton, or a carbamoyl group attached to an aromatic ring. The oxygen atom from this carbonyl group appears to serve as a hydrogen acceptor and the hydrogen atom from the amide or imide group as a proton donor in the hydrogen bond interaction with the enzyme. Consensus structural requirements for PARP inhibitors acting at this nicotinamide-binding site include:

1. Amide or lactam functionality is essential for effective interaction with the binding pocket.
2. An NH proton of this amide or lactam functionality should be conserved for effective bonding.
3. An amide group attached to an aromatic ring or a lactam group fused to an aromatic ring has better inhibition than an amide group attached to a nonaromatic ring or a lactam group fused to a nonaromatic ring.
4. Optimal *cis*-configuration of the amide in the aromatic plane is required for maximal inhibitory activity.
5. Constraining mono-aryl carboxamide into heteropolycyclic lactams usually increases potency.

Although potent inhibitors have an amide group built in a second ring system fused to an aromatic ring, the carbonyl group of the amide, however, is not indispensable for the inhibitory action. For example, weak inhibitors, such as phthalazine, quinazoline, norharman, and isoquinoline, do not contain a carbonyl group, but have a C=N double bond in analogous positions.[74] Thiobenzamide and thionicotinamide with a thiocarbamoyl group in place of a carbamoyl group can also weakly inhibit PARP.[75]

13.6 CORROBORATIVE RESULTS FROM GENETIC AND PHARMACOLOGICAL INACTIVATION OF PARP

The availability of PARP knockout mice and potent PARP inhibitors makes it possible to conduct thorough testing of PARP functions *in vivo*. In most cases, the results from genetic and pharmacological approaches are consistent with and complement each other. For example, maintaining genomic stability is among many functions attributed to PARP. It is based on an early experiment that demonstrated elevated sister chromatid exchange rates when PARP was inhibited.[58] In cells derived from the PARP$^{-/-}$ mice, both the basal and the alkylating agent–enhanced levels of sister chromatid exchange were significantly elevated.[36,39] Early studies also indicated PARP inhibitors could serve as sensitizers to enhance killing of cancer cells and other rapidly growing cells by alkylating agents or γ irradiation.[95,96] Recent experiments confirmed that PARP$^{-/-}$ embryonic stem cells were more susceptible to high doses of γ irradiation and methyl methanesulfonate, and the survival rate of PARP$^{-/-}$ mice dropped drastically after exposure to a genotoxic dose of γ irradiation or *N*-methyl-*N*-nitrosourea.[36,42]

The PARP knockout mice also have been used extensively either to confirm or to explore the role of PARP activation in various animal models of human diseases. One of the early implications for its role in pathogenesis came from the finding that PARP activation was the culprit of streptozotocin-induced diabetes, an animal model to study type 1 diabetes.[97] In β-islet cells, free radicals derived from streptozotocin catabolism damaged DNA and activated PARP. Consequently, depletion of NAD due to PARP activation prohibited the release of insulin and led to cell death, as PARP inhibitors such as nicotinamide and 3-aminobenzamide prevented stretptozotocin toxicity.[98] Three groups have now independently replicated the studies in PARP knockout mice, and all demonstrated that PARP gene deletion renders mice resistant to β-islet cell destruction and hyperglycemia after streptozotocin treatment.[36,45,46] In the nervous system, PARP activation was found to mediate glutamate excitotoxicity and neurotoxicity elicited by 1-methyl-4-phenyl-1,2,3,6-tetrahydropyridine (MPTP) and β-amyloid peptide.[99-102] The *in vitro* discovery led to findings that in rodent models of both permanent and transient cerebral focal ischemia, PARP inhibitors achieved substantial reduction of infarct volume.[84,88,103-107] Postischemia treatment with PARP inhibitors was also found to be highly effective.[84,107,108] Complementary to the pharmacological intervention studies, PARP knockout mice were highly resistant to neural damage after being subject to transient brain ischemia.[21,44] Together, these studies offered compelling evidence that PARP activation mediates neuronal death during ischemia and validated targeting PARP for neuroprotection. In one rodent model of MPTP-induced Parkinson's disease, dopamine neurons in the substantia nigra in PARP$^{-/-}$ mice could survive MPTP toxicity while most of those neurons in the wild-type littermates were degenerated.[109] Indeed, PARP inhibitors offered effective neuroprotection in MPTP-treated mice.[110] In other rodent models of traumatic brain injuries, PARP inhibition significantly ameliorated neural damage and PARP gene deletion offered significant functional benefit in motor skills and learning.[8,11] PARP activation mediated ischemia–reperfusion injury was also found in other organs, e.g., in heart and muscle, suggesting a general role for PARP in free radical–elicited necrotic cell death.[111,112] PARP$^{-/-}$ mice or hearts prepared from them were spared of myocardium and preserved myocardial functions after being subject to ischemia.[113-115] Although the *in vitro* cardioprotective effects of PARP inactivation could largely be attributed to preservation of cellular energy, part of the *in vivo* effects could be due to the additional role of PARP in mediating upregulation of P-selectin and ICAM-1, two key proteins essential for lymphocytes activation.[113,114] By suppressing neutrophil recruitment, PARP inhibition may provide strong anti-inflammatory effects. Indeed, inflammation was attenuated when the PARP knockout mice were subject to systemic multiple organ failure, and mucosal injury in colitis.[116,117]

In summary, experiments with the PARP knockout mice have yielded definitive evidence that PARP activation is responsible for tissue damage or degeneration in a broad spectrum of disease models. Taken together with pharmacological studies, the results have validated targeting PARP inhibition as a potentially effective means to ameliorate various diseases. In animal models, PARP inhibitors have already demonstrated remarkable efficacy in treating injuries due to ischemia–reperfusion or inflammation.

13.7 CONCLUSION

The biological roles of PARP have largely been deduced from the consequences of eliminating PARP activity on cellular functioning. Both genetic manipulation and pharmacological intervention have been instrumental in elucidating the involvement of PARP in maintaining genomic integrity, facilitating DNA repair, and participating in apoptosis. PARP knockout mice have contributed to the understanding of PARP activation in mediating a broad spectrum of diseases. The dual approaches have been essential in proving the principle of targeting PARP as a novel means to develop therapeutic treatment. Genetic manipulation such as expression of dominant-negative mutant or PARP antisense RNA may serve as a first step toward gene therapy to combat chronic diseases due to PARP activation. The remarkable protective effects that PARP inhibitors afford in animal models of diseases hold strong potentials for further drug development to treat ischemia–reperfusion tissue damage and inflammation-related injuries.

ACKNOWLEDGMENT

The authors thank V. Kalish and K. Tays for critically reviewing the manuscript.

REFERENCES

1. de Murcia, G. and Ménissier-de Murcia, J., Poly(ADP-ribose) polymerase: a molecular nick-sensor, *Trends Biochem. Sci.*, 19, 172, 1994.
2. Ueda, K. and Hayaishi, O., ADP-ribosylation, *Annu. Rev. Biochem.*, 54, 73, 1985.
3. Lazebnik, Y. A., Kaufmann, S. H., Desnoyers, S., Poirier, G. G., and Earnshaw, W. C., Cleavage of poly(ADP-ribose) polymerase by a proteinase with properties like ICE, *Nature*, 371, 346, 1994.
4. Banasik, M., Komura, H., Shimoyama, M., and Ueda, K., Specific inhibitors of poly(ADP-ribose) synthetase and mono(ADP-ribosyl)transferase, *J. Biol. Chem.*, 267, 1569, 1992.
5. Milam, K. M. and Cleaver, J. E., Inhibitors of poly(adenosine diphosphate-ribose) synthesis: effect on other metabolic processes, *Science*, 223, 589, 1984.
6. Ruf, A., Ménissier-de Murcia, J., de Murcia, G., and Schulz, G. E., Structure of the catalytic fragment of poly(AD-ribose) polymerase from chicken, *Proc. Natl. Acad. Sci. U.S.A.*, 93, 7481, 1996.
7. Zhang, J., PARP inhibition: a novel approach to treat ischaemia/reperfusion and inflammation-related injuries, in *Emerging Drugs: The Prospect for Improved Medicines*, Vol. 4, Fitzgerald, J. D., Bowman, W. C., and Taylor, J. B., Eds., Ashley Publication Ltd., London, 1999, chap. 10.
8. LaPlaca, M. C., Zhang, J., Li, J.-H., Raghupathi, R., Smith, F., Bareye, F. M., Graham, D. I., and McIntosh, T. K., Pharmacologic inhibition of poly(ADP-ribose) polymerase is neuroprotective following traumatic brain injury in rats, in *19th Annual Neurotrauma Society Symposium*, Miami, Florida, *J. Neurotrauma*, 16, 976, 1999 (abstract).
9. Zhang, J., Li, J.-H., Lautar, S., Zhou, Y., Mooney, M. L., and Williams, L. R., GPI 6150, a PARP inhibitor, exhibits strong anti-inflammatory effect in rat models of arthritis, in *6th International Meeting of Biology of Nitric Oxide*, Stockholm, Sweden, *Acta. Physiol. Scand.*, 167(Suppl.), 90, 1999 (abstract).
10. Szabó, C. and Dawson, V. L., Role of poly(ADP-ribose) synthetase in inflammation and ischaemia–reperfusion, *Trends Pharmacol. Sci.*, 19, 287, 1998.
11. Whalen, M. J., Clark, R. S. B., Dixon, C. E., Robinchaud, P., Marion, D. W., Vagni, V., Graham, S. H., Virág, L., Hasko, G., Stachlewitz, R., Szabó, C., and Kochanek, P. M., Reduction of cognitive and motor deficits after traumatic brain injury in mice dificient in poly(ADP-ribose) polymerase, *J. Cereb. Blood Flow Metab.*, 19, 835, 1999.
12. Kumari, S. R., Mendoza-Alvarez, H., and Alvarez-Gonzalez, R., Functional interactions of p53 with poly(ADP-ribose) polymerase (PARP) during apoptosis following DNA damage: covalent poly(ADP-ribosyl)ation of p53 by exogenous PARP and noncovalent binding of p53 to the Mr 85,000 proteolytic fragment, *Cancer Res.*, 58, 5075, 1998.
13. Simbulan-Rosenthal, C. M., Rosenthal, D. S., Luo, R., and Smulson, M. E., Poly(ADP-ribosyl)ation of p53 during apoptosis in human osteosarcoma cells, *Cancer Res.*, 59, 2190, 1999.
14. Uchida, K. and Miwa, M., Poly(ADP-ribose) polymerase: structural conservation among different classes of animals and its implications, *Mol. Cell. Biochem.*, 138, 25, 1994.
15. Tseng, A., Jr., Lee, W. M., Jakobovits, E. B., Kirsten, E., Hakam, A., McLick, J., Buki, K., and Kun, E., Prevention of tumorigenesis of oncogene-transformed rat fibroblasts with DNA site inhibitors of poly(ADP ribose) polymerase, *Proc. Natl. Acad. Sci. U.S.A.*, 84, 1107, 1987.

16. Simonin, F., Ménissier-de Murcia, J., Poch, O., Muller, S., Gradwohl, G., Molinete, M., Penning, C., Keith, G., and de Murcia, G., Expression and site-directed mutagenesis of the catalytic domain of human poly(ADP-ribose)polymerase in *Escherichia coli*. Lysine 893 is critical for activity, *J. Biol. Chem.*, 265, 19249, 1990.
17. Simonin, F., Poch, O., Delarue, M., and de Murcia, G., Identification of potential active-site residues in the human poly(ADP-ribose) polymerase, *J. Biol. Chem.*, 268, 8529, 1993.
18. Ruf, A., de Murcia, G., and Schulz, G. E., Inhibitor and NAD$^+$ binding to poly(ADP-ribose) polymerase as derived from crystal structures and homology modeling, *Biochemistry*, 37, 3893, 1998.
19. Berger, N. A., Poly(ADP-ribose) in the cellular response to DNA damage, *Radiat. Res.*, 101, 4, 1985.
20. Yamamoto, H. and Okamoto, H., Protection by picolinamide, a novel inhibitor of poly(ADP-ribose) synthetase, against both streptozotocin-induced depression of proinsulin synthesis and reduction of NAD content in pancreatic islets, *Biochem. Biophys. Res. Commun.*, 95, 474, 1980.
21. Endres, M., Wang, Z. Q., Namura, S., Waeber, C., and Moskowitz, M. A., Ischemic brain injury is mediated by the activation of poly(ADP-ribose)polymerase, *J. Cereb. Blood Flow Metab.*, 17, 1143, 1997.
22. Wielckens, K., Schmidt, A., George, E., Bredehorst, R., and Hilz, H., DNA fragmentation and NAD depletion. Their relation to the turnover of endogenous mono(ADP-ribosyl) and poly(ADP-ribosyl) proteins, *J. Biol. Chem.*, 257, 12872, 1983.
23. Ding, R., Pommier, Y., Kang, V. H., and Smulson, M. E., Depletion of poly(ADP-ribose) polymerase by antisense RNA expression results in a delay in DNA strand break rejoining, *J. Biol. Chem.*, 267, 12804, 1992.
24. Ding, R. and Smulson, M. E., Depletion of nuclear poly(ADP-ribose) polymerase by antisense RNA expression: influences on genomic stability, chromatin organization, and carcinogen cytotoxicity, *Cancer Res.*, 54, 4627, 1994.
25. Qu, Z., Fujimoto, S., and Taniguchi, T., Enhancement of interferon-gamma-induced major histocompatibility complex class II gene expression by expressing an antisense RNA of poly(ADP-ribose) synthetase, *J. Biol. Chem.*, 269, 5543, 1994.
26. Babiychuk, E., Cottrill, P. B., Storozhenko, S., Fuangthong, M., Chen, Y., O'Farrell, M. K., Van Montagu, M., Inze, D., and Kushnir, S., Higher plants possess two structurally different poly(ADP-ribose) polymerases, *Plant J.*, 15, 635, 1998.
27. Berghammer, H., Ebner, M., Marksteiner, R., and Auer, B., pADPRT-2: a novel mammalian polymerizing(ADP-ribosyl)transferase gene related to truncated pADPRT homologues in plants and *Caenorhabditis elegans*, *FEBS Lett.*, 449, 259, 1999.
28. Johansson, M., A human poly(ADP-ribose) polymerase gene family (ADPRTL): cDNA cloning of two novel poly(ADP-ribose) polymerase homologues, *Genomics*, 57, 442, 1999.
29. Ame, J. C., Rolli, V., Schreiber, V., Niedergang, C., Apiou, F., Decker, P., Muller, S., Hoger, T., Ménissier-de Murcia, J., and de Murcia, G., PARP-2, a novel mammalian DNA damage-dependent poly(ADP-ribose) polymerase, *J. Biol. Chem.*, 274, 17860, 1999.
30. Le Cam, E., Fack, F., Ménissier-de Murcia, J., Cognet, J. A., Barbin, A., Sarantoglou, V., Revet, B., Delain, E., and de Murcia, G., Conformational analysis of a 139 base-pair DNA fragment containing a single-stranded break and its interaction with human poly(ADP-ribose) polymerase, *J. Mol. Biol.*, 235, 1062, 1994.
31. Molinete, M., Vermeulen, W., Burkle, A., Ménissier-de Murcia, J., Kupper, J. H., Hoeijmakers, J. H., and de Murcia, G., Overproduction of the poly(ADP-ribose) polymerase DNA-binding domain blocks alkylation-induced DNA repair synthesis in mammalian cells, *EMBO J.*, 12, 2109, 1993.

32. Kupper, J. H., van Gool, L., and Burkle, A., Molecular genetic systems to study the role of poly(ADP-ribosyl)ation in the cellular response to DNA damage, *Biochimie*, 77, 450, 1995.
33. Kupper, J. H., Muller, M., Jacobson, M. K., Tatsumi-Miyajima, J., Coyle, D. L., Jacobson, E. L., and Burkle, A., *trans*-Dominant inhibition of poly(ADP-ribosyl)ation sensitizes cells against gamma-irradiation and N-methyl-N'-nitro-N-nitrosoguanidine but does not limit DNA replication of a polyomavirus replicon, *Mol. Cell. Biol.*, 15, 3154, 1995.
34. Schreiber, V., Hunting, D., Trucco, C., Gowans, B., Grunwald, D., de Murcia, G., and Ménissier-de Murcia, J., A dominant-negative mutant of human poly(ADP-ribose) polymerase affects cell recovery, apoptosis, and sister chromatid exchange following DNA damage, *Proc. Natl. Acad. Sci. U.S.A.*, 92, 4753, 1995.
35. Wang, Z. Q., Auer, B., Stingl, L., Berghammer, H., Haidacher, D., Schweiger, M., and Wagner, E. F., Mice lacking ADPRT and poly(ADP-ribosyl)ation develop normally but are susceptible to skin disease, *Genes Dev.*, 9, 509, 1995.
36. de Murcia, J. M., Niedergang, C., Trucco, C., Ricoul, M., Dutrillaux, B., Mark, M., Oliver, F. J., Masson, M., Dierich, A., LeMeur, M., Walztinger, C., Chambon, P., and de Murcia, G., Requirement of poly(ADP-ribose) polymerase in recovery from DNA damage in mice and in cells, *Proc. Natl. Acad. Sci. U.S.A.*, 94, 7303, 1997.
37. Masutani, M., Suzuki, H., Kamada, N., Watanabe, M., Ueda, O., Nozaki, T., Jishage, K., Watanabe, T., Sugimoto, T., Nakagama, H., Ochiya, T., and Sugimura, T., Poly(ADP-ribose) polymerase gene disruption conferred mice resistant to streptozotocin-induced diabetes, *Proc. Natl. Acad. Sci. U.S.A.*, 96, 2301, 1999.
38. Attardi, L. D. and Jacks, T., The role of p53 in tumour suppression: lessons from mouse models, *Cell. Mol. Life Sci.*, 55, 48, 1999.
39. Wang, Z. Q., Stingl, L., Morrison, C., Jantsch, M., Los, M., Schulze-Osthoff, K., and Wagner, E. F., PARP is important for genomic stability but dispensable in apoptosis, *Genes Dev.*, 11, 2347, 1997
40. Leist, M., Single, B., Kunstle, G., Volbracht, C., Hentze, H., and Nicotera, P., Apoptosis in the absence of poly-(ADP-ribose) polymerase, *Biochem. Biophys. Res. Commun.*, 233, 518, 1997.
41. Trucco, C., Oliver, F. J., de Murcia, G., and Ménissier-de Murcia, J., DNA repair defect in poly(ADP-ribose) polymerase-deficient cell lines, *Nucl. Acids Res.*, 26, 2644, 1998.
42. Masutani, M., Nozaki, T., Nishiyama, E., Shimokawa, T., Tachi, Y., Suzuki, H., Nakagama, H., Wakabayashi, K., and Sugimura, T., Function of poly(ADP-ribose) polymerase in response to DNA damage: gene-disruption study in mice, *Mol. Cell. Biochem.*, 193, 149, 1999.
43. Choi, D. W., At the scene of ischemic brain injury: is PARP a perp? *Nat. Med.*, 3, 1073, 1997.
44. Eliasson, M. J., Sampei, K., Mandir, A. S., Hurn, P. D., Traystman, R. J., Bao, J., Pieper, A., Wang, Z. Q., Dawson, T. M., Snyder, S. H., and Dawson, V. L., Poly(ADP-ribose) polymerase gene disruption renders mice resistant to cerebral ischemia, *Nat. Med.*, 3, 1089, 1997.
45. Burkart, V., Wang, Z. Q., Radons, J., Heller, B., Herceg, Z., Stingl, L., Wagner, E. F., and Kolb, H., Mice lacking the poly(ADP-ribose) polymerase gene are resistant to pancreatic beta-cell destruction and diabetes development induced by streptozotocin, *Nat. Med.*, 5, 314, 1999.
46. Pieper, A. A., Brat, D. J., Krug, D. K., Watkins, C. C., Gupta, A., Blackshaw, S., Verma, A., Wang, Z. Q., and Snyder, S. H., Poly(ADP-ribose) polymerase-deficient mice are protected from streptozotocin-induced diabetes, *Proc. Natl. Acad. Sci. U.S.A.*, 96, 3059, 1999.

47. Stone, P. R. and Shall, S., Poly(adenosine diphosphoribose) polymeraae in mammalian nuclei. Characterization of the activity in mouse fibroblasts (LS cells), *Eur. J. Biochem.,* 38, 146, 1973.
48. Purnell, M. R. and Whish, W. J. D., Novel inhibitors of poly(ADP-ribose) synthetase, *Biochem. J.,* 185, 775, 1980.
49. Durkacz, B. W., Omidiji, O., Gray, D. A., and Shall, S., (ADP-ribose)n participates in DNA excision repair, *Nature,* 283, 593, 1980.
50. Nishizuka,Y., Ueda, K., Nakazawa, K., Reeder, R. H., Honjo, T., and Hayaishi, O., Poly adenosine diphosphate ribose synthesis and nicotinamide adenine dinucleotide transglycosidases, *J. Vitaminol.,* 14, 143, 1968.
51. Clark, J. B., Ferns, G. M., and Pinder, S., Inhibition of nuclear NAD nucleosidase and poly ADP-ribose polymerase activity from rat liver by nicotinamide and 5-methyl nicotinamide, *Biochem. Biophys. Acta,* 238, 82, 1971.
52. Preiss, I., Schlseger, R., and Hitz, H., Specific inhibition of poly ADP ribose polymerase by thymidine and nicotinamide in HeLa cells, *FEBS Lett.,* 9, 244, 1971.
53. Uecia, K., Fukushima, M., Okaysma, H., and Havaishi, O., Nicotinamide adenine dinucleotide glycohydrolase from rat liver nuclei. Isolation and characterization of a new enzyme, *J. Biol. Chem.,* 250, 7541, 1975.
54. Claycomb, W. C., Poly(adenosine diphosphate ribose) polymerase activity and nicotinamide adenine dinucleotide in differentiating cardiac muscles, *Biochemistry,* 1154, 387, 1976.
55. Levi, V., Jacobson, E. L., and Jacobson, M. K., Inhibition of poly(ADP-ribose) polymerase by methylated xanthines and cytokinins, *FEBS Lett.,* 88, 144, 1978.
56. Terada, M., Fujiki, H., Marks, P. A., and Sugimura, T., Induction of erythroid differentiation of murine erythraleukernia cells by nicotinamide and related phosphate ribose) polymerase, *Proc. Natl. Acad. Sci. U.S.A.,* 76, 6411, 1979.
57. Yaniamoto, H. and Okanioto, H., Protection by picolinamide, a novel inhibitor of poly(ADP-ribose) synthetase, against both streptozotocin-induced depression of proinsulin synthesis and reduction of NAD content in pancreatic islets, *Biochem. Biophys. Res. Commun.,* 95, 474, 1980.
58. Oikawa, A., Tohua, H., Kanai, M., Miwa, M., and Sugimura, T., Inhibitors of poly(adenosine diphosphate ribose) polymerase induce sister chromatid exchanges, *Biochem. Biophys. Res. Commun.,* 97, 1311, 1980.
59. Buki, K. G., Bauer, P. L., Mendeleyev, J., Hakam, A., and Kun, E., Destabilization of Zn^{2+} coordination in ADP-ribose transferase (polymerizing) by 6-nitroso 1,2-benzopyrone coincidental with inactivation of the polymerase but not the DNA binding function, *FEBS Lett.,* 290 181, 1991.
60. Banasik, M. and Ueda, K., Inhibitors and activators of ADP-ribosylation reactions, *Mol. Cell. Biochem.,* 138, 185, 1994.
61. Sims, J. L. and Benjamin, R. C., Mechanism of ethanol stimulation of poly(ADP ribose) synthetase, in *ADP Ribosylation of Proteins,* Althaus, F. R., Htlz, H., and Shall, S., Eds., Springer-Verlag, Berlin, 1985, 124–128.
62. Banasik, M. and Ueda, K., Dual inhibitory effects of dimethyl sulfoxide on poly(ADP-ribose) synthetase, *J. Enzyme Inhibition,* 14, 239, 1999.
63. Rankin, P. W., Jacobson, E. L., Benjamin, R. C., Moss, I., and Jacobson, M. K., Quantitative studies of inhibitors of ADP-ribosylation *in vitro* and *in vivo, J. Biol. Chem.,* 264, 4312, 1989.
64. Borek, C., Morgan, W. F., Ong, A., and Cleaver, J. E., Inhibition of malignant transformation *in vitro* by inhibitors of poly(ADP-ribose) synthesis, *Proc. Natl. Acad. Sci. U.S.A.,* 81, 243, 1984.

65. Shinkwin, A. E., Whish, W. J. D., and Threadgill, M. D., Synthesis of thiophenecarboxamides, thieno[3,4-c]pyridin-4(5H)-ones and thieno[3,4-d]pyrimidin-4(3H)-ones and preliminary evaluation as inhibitors of poly(ADP-ribose)polymerase (PARP), *Bioorg. Med. Chem.*, 7, 297, 1999.
66. Sims, J. L., Sikorski, G. W., Catino, D. M., Berger, S. I., and Berger, N. A., Poly(adenosine diphosphoribose) polymerase inhibitors stimulate unscheduled deoxyribonucleic acid synthesis in normal human lymphocytes, *Biochemistry,* 21, 1813, 1982.
67. Watson, C. Y., Whish, W. I. D., and Threadgill, M. D., Synthesis of 3-substituted benzamides and 5-substituted isoquinolin-1(2H)-ones and preliminary evaluation as inhibitors of poly(ADP-ribose)polymerase (PARP), *Bioog. Med. Chem.*, 6, 721, 1998.
68. Griffin, R. J., Walk, L., Calvert, A. H., Curtin, N. J., Newell, D. R., and Golding, B. T., Benzamide analogs, useful as PARP (ADP-ribosyltransferase, ADPRT) DNA repair enzyme inhibitors, Int. Publication Number: WO/95/24379, 1995.
69. Suto, M. J., Turner, W. R., Arundel-Suto, C. K., Werbel, L. M., and Sebolt-Leopold, J. S., Dihydroisoquinolinones: the design and synthesis of a new series of potent inhibitors of poly(ADP-ribose)polymerase, *Anti-Cancer Drug Design,* 7, 101, 1991.
70. Banasik, M., Kornura, H., and Ueda, K., Inhibition of poly(ADP-ribose) synthetase by fatty acids, vitamins and vitamin-like substances, *FEBS Lett.,* 263, 222, 1990.
71. Cleaver, J. E., Bodell, W. I., Morgan, W. E., and Zelle, B., Differences in regulation by poly(ADP-ribose) of repair of DNA damage from alkylating agents and ultraviolet light according to cell type, *J. Biol. Chem.,* 258, 9059, 1983.
72. Cleaver, J. E., Differential toxicity of 3-aminobenzamide to wild-type and 6-thioguanine-resistant Chinese hamster cells by interference with pathways of purine biosynthesis, *Mutat. Res.,* 131, 123, 1984.
73. Milam, K. M. and Cleaver, J. E., Inhibitors of poly(adenosine diphosphate-ribose) synthesis effect on other metabolic processes, *Science,* 223, 589, 1984.
74. Banasik, M., Komura, H., Shimoyama, M., and Ueda, K., Specific inhibitors of poly(ADP-ribose) synthetase and mono(ADP-ribosyl)transferase, *J. Biol. Chem.*, 267, 1569, 1992.
75. Sestili, P., Spadons, G., Balsamini, C., Scovassi, I., Cattabeni, F., Duranti, E., Cantoni, O., Higgins, D., and Thomson, C., Structural requirements for inhibitor of poly(ADP-ribose) polymerase, *J. Cancer Res. Clin. Oncol.*, 116, 615, 1990.
76. Bowman, K. J., Curtin, N. J., Golding, B. T., Griffn, R. J., and White, A., Potentiation of anticancer agent cytotoxicity by the novel poly(ADP-ribose) polymerase inhibitors, NU1025 and NU1064, *Br. J. Cancer,* 78, 1269, 1998.
77. Griffin, R. J., Srinivasan, S., White, A. W., Bowman, K., Calvert, A. H., Curtin, N. J., Newell, D. R., and Golding, B. T., Novel benzimidazole and quinazolinone inhibitors of the DNA repair enzyme poly(ADP-ribose)polymerase, *Pharm. Sci.,* 2, 43, 1996.
78. Griffin, R. J., Srinivasan, S., Bowman, K., Calvert, A. H., Curtin, N. J., Newell, D. R., Pemberton, L. C., and Golding, B. T., Resistance-modifying agents. 5. Synthesis and biological properties of quinazolinone inhibitors of the DNA repair enzyme poly(ADP-ribose)polymerase (PARP), *J. Med. Chem.,* 41, 5247, 1998.
79. Showalter, H. D. H., Dihydro- and Tetrahydronaohthyridines, U.S. Patent 5,391,554, 1995.
80. Griffin, R. J., Pemberton, L. C., Rhodes, D., Blesadale, C., Bowman, K., Calvert, R. A. H., Cumin, N. J., Durkacz, B. W., Newell, D. B., Porteous, J. K., and Golding, B. T., Novel potent inhibitors of the DNA repair enzyme poly(ADP-ribose)polymerase (PARP), *Anti-Cancer Drug Design,* 10, 507, 1995.

81. Yoshida, S., Aoyagi, T., Harada, S., Matuda, N., Ikeda, T., Naganawa, H., Hamada, M., and Takeuchi, T., Production of 2-methyl-4[3H]quinazolinone, an inhibitor of poly(ADP-ribose) synthetase, by bacterium, *J. Antibiot.*, 44, 111, 1991.
82. Li, J-H., Tays, K. L., and Zhang, J., Oxo-substituted compounds, process of making, and compositions and methods for inhibiting PARP activity, Int. Publication Number: WO/99/11624, 1999.
83. Li, J-H., Zhang, J., Jackson, P. F., and Maclin, K. M., Poly(ADP-ribose) polymerase (PARP) inhibitors, methods and pharmaceutical compositions for treating neural or cardiovascular tissue damage, Int. Publication Number: WO/99/11645, 1999.
84. Zhang, J., Li, J-H., and Lautar, S., Post-ischemia protection by a potent PARP inhibitor in focal ischemia, *Soc. Neurosci. Abstr.*, 24, 1226, 1998.
85. Benjamin, R. C. and Gill, D. M., Poly(ADP-ribose)synthesis *in vitro* programmed by damaged DNA. A comparison of DNA molecules containing different types of strand breaks, *J. Biol. Chem.*, 255, 10502, 1980.
86. Pivazyan, A. D., Birks, E. M., Wood, T. G., Lin, T. S., and Prusoff, W. H., Inhibition of poly(ADP-ribose) polymerase activity by nucleoside analogs of thymidine, *Biochem. Pharmacol.*, 44, 947, 1992.
87. Kun, E., Methods for treating inflammation, inflammatory disease, arthritis and stroke using pADPT inhibitors, Int. Publication Number: WO/98/51307, 1998.
88. Endres, M., Scott, G. S., Salzman, A. L., Kun, E., Moskowitz, M. A., and Szabó, C., Protective effects of 5-iodo-6-amino-1,2-benzopyrone, and inhibitor of poly(ADP-ribose) synthetase against peroxynitrite-induced glial damage and stroke development, *Eur. J. Pharmacol.*, 351, 377, 1998.
89. Szabó, C., Virág, L., Cuzzocrea, S., Scott, G. S., Hake, P., O'Connor, M. P., Zingarelli, B., Salzman, A., and Kun, E., Protection against peroxynitrite-induced fibroblast injury and arthritis development by inhibition of poly(ADP-ribose) synthase, *Proc. Natl. Acad. Sci. U.S.A.*, 95, 3867, 1998.
90. Banasik, M., Komura, H., and Ueda, K., Inhibitions of poly(ADP-ribose) synthetase by unsaturated fatty acids, vitamins and vitamin-like substances, *FEBS Lett.*, 263, 222, 1990.
91. Vanni, A., Fiore, M., De Salvia, R., Cundari, E., Ricordy, R., Ceccarelli, R., and Degrassi, F., DNA damage and cytotoxicity induced by beta-lapachone: relation to poly(ADP-ribose) polymerase inhibition, *Mutat. Res.*, 401, 55, 1998.
92. Yager, J. W. and Wienke, J. K., Inhibition of poly(ADP-ribose) polymerase by arsenite *Mutat. Res.*, 386, 345, 1997.
93. Ueda, K., Banasik, M., Nakajima, S., Yook, H. Y., and Kido, T., Cell differentiation induced by poly(ADP-ribose) synthetase inhibitors, *Biochimie*, 77, 368, 1995.
94. Yoshida, S., Naganawa, H., Aoyagi, T., Takeuchi, T., and Umezawa, H., Benadrostin, new inhibitor of poly(ADP-ribose) synthetase, produced by actinomycetes. II. Structure determination, *J. Antibiot.*, 41, 1015, 1998.
95. Nduka, N., Skidmore, C. J., and Shall, S., The enhancement of cytotoxicity of *N*-methyl-*N*-nitrosourea and of gamma-radiation by inhibitors of poly(ADP-ribose) polymerase, *Eur. J. Biochem.*, 105, 525, 1980.
96. Ben-Hur, E., Utsumi, H., and Elkind, M. M., Inhibitors of poly(ADP-ribose) synthesis enhance X-ray killing of log-phase Chinese hamster cells, *Radiat. Res.*, 97, 546, 1984.
97. Yamamoto, H., Uchigata, Y., and Okamoto, H., Streptozotocin and alloxan induce DNA strand breaks and poly(ADP-ribose) synthetase in pancreatic islets, *Nature*, 294, 284, 1981.

98. Masiello, P., Cubeddu, T. L., Frosina, G., and Bergamini, E., Protective effect of 3-aminobenzamide, an inhibitor of poly(ADP-ribose) synthetase, against streptozotocin-induced diabetes, *Diabetologia*, 28, 683, 1985.
99. Wallis, R. A., Panizzon, K. L., Henry, D., and Wasterlain, C. G., Neuroprotection against nitric oxide injury with inhibitors of ADP-ribosylation, *Neuroreport*, 5, 245, 1999.
100. Zhang, J., Dawson, V. L., Dawson, T. M., and Snyder, S. H., Nitric oxide activation of poly(ADP-ribose) synthetase in neurotoxicity, *Science*, 263, 687, 1994.
101. Cosi, C., Suzuki, H., Milani, D., Facci, L., Menegazzi, M., Vantini, G., Kanai, Y., and Skaper, S. D., Poly(ADP-ribose) polymerase: early involvement in glutamate-induced neurotoxicity in cultured cerebellar granule cells, *J. Neurosci. Res.*, 39, 38, 1994.
102. Zhang, J., Pieper, A., and Snyder, S. H., Poly(ADP-ribose) synthetase activation: an early indicator of neurotoxic DNA damage, *J. Neurochem.*, 65,1411, 1995.
103. Takahashi, K., Greenberg, J. H., Jackson, P., Maclin, K., and Zhang, J., Neuroprotective effects of inhibiting poly(ADP-ribose) synthetase on focal cerebral ischemia in rats, *J. Cereb. Blood Flow Metab.*, 17, 1137, 1997.
104. Lo, E. H., Bosque-Hamilton, P., and Meng, W., Inhibition of poly(ADP-ribose) polymerase reduction of ischemic injury and attenuation of N-methyl-D-aspartate-induced neurotransmitter dysregulation, *Stroke*, 29, 830, 1997.
105. Ayoub, I. A., Ogilvy, C. S., and Maynard, K. I., Nicotinamide reduces infarct in a permanent model of focal cerebral ischemia in Wistar rats, *Soc. Neurosci. Abstr.*, 24, 214, 1998.
106. Tokime, T., Nozaki, K., Sugino, T., Kikuchi, H., Hashimoto, N., and Ueda, K., Enhanced poly(ADP-ribosyl)ation after focal ischemia in rat brain, *J. Cereb. Blood Flow Metab.*, 18, 991, 1998.
107. Sun, A.-Y. and Cheng, J.-S., Neuroprotective effects of poly(ADP-ribose) polymerase inhibitors in transient focal cerebral ischemia of rats, *Acta Pharmacol. Sin.*, 19, 104, 1998.
108. Takahashi, K., Pieper, A. A., Croul, S. E., Zhang, J., Snyder, S. H., and Greenberg, J. H., Post-treatment with an inhibitor of poly(ADP-ribose) polymerase attenuates cerebral damage in focal ischemia, *Brain Res.*, 829, 46, 1999.
109. Mandir, A. S., Przedborski, S., Jackson-Lewis, V., Wang, Z. Q., Simbulan-Rosenthal, C. M., Smulson, M. E., Hoffman, B. E., Guastella, D. B., Dawson, V. L., and Dawson, T. M., Poly(ADP-ribose) polymerase activation mediates 1-methyl-4-phenyl-1,2,3,6-tetrahydropyridine (MPTP)-induced parkinsonism, *Proc. Natl. Acad. Sci. U.S.A.*, 96, 5774, 1999.
110. Cosi, C. and Marien, M., Decreases in mouse brain NAD$^+$ and ATP induced by 1-methyl-4-phenyl-1, 2,3,6-tetrahydropyridine (MPTP): prevention by the poly(ADP-ribose) polymerase inhibitor, benzamide, *Brain Res.*, 809, 58, 1998.
111. Thiemermann, C., Bowes, J., Myint, F. P., and Vane, J. R., Inhibition of the activity of poly(ADP-ribose) synthetase reduces ischemia–reperfusion injury in the heart and skeletal muscle, *Proc. Natl. Acad. Sci. U.S.A.*, 94, 679, 1997.
112. Zingarelli, B., Cuzzocrea, S., Zsengeller, Z., Salzman, A. L., and Szabó, C., Protection against myocardial ischemia and reperfusion injury by 3-aminobenzamide, an inhibitor of poly(ADP-ribose) synthetase, *Cardiovasc. Res.*, 36, 205, 1997.
113. Zingarelli, B., Salzman, A. L., and Szabó, C., Genetic disruption of poly(ADP-ribose) synthetase inhibits the expression of P-selectin and intercellular adhesion molecule-1 in myocardial ischemia/reperfusion injury, *Circ. Res.*, 83, 85, 1998.

114. Grupp, I. L., Jackson, T. M., Hake, P., Grupp, G., and Szabó, C., Protection against hypoxia-reoxygenation in the absence of poly (ADP-ribose) synthetase in isolated working hearts, *J. Mol. Cell. Cardiol.,* 31, 297, 1999.
115. Walles, T., Wang, P., Pieper, A., Li, J.-H., Zhang, J., Snyder, S. H., and Zweier, J. L., Mice lacking poly(ADP-ribose) polymerase gene show attenuated cellular energy depletion and improved recovery of myocardial function following global ischemia, *Circ. Suppl.,* 98, I-260, 1998.
116. Szabó, C., Lim, L. H., Cuzzocrea, S., Getting, S. J., Zingarelli, B., Flower, R. J., Salzman, A. L., and Perretti, M., Inhibition of poly(ADP-ribose) synthetase attenuates neutrophil recruitment and exerts anti-inflammatory effects, *J. Exp. Med.,* 186, 1041, 1997.
117. Zingarelli, B., Szabó, C., and Salzman, A. L., Blockade of poly(ADP-ribose) synthetase inhibits neutrophil recruitment, oxidant generation, and mucosal injury in murine colitis, *Gastroenterology,* 116, 335, 1999.
118. Suto, M. J., Turner, W. R., and Werbel, L. M., Substituted Dihydroisoquinolinones and Related Compounds as Potentiators of the Lethal Effects of Radiation and Certain Chemotherapeutic Agents; Selected Compounds, Analogs and Process, U.S. Patent 5,177,075, 1993.
119. Griffin, R. J., Curtin, N. J., Newell, D. R., and Golding, B. T., Quinazolinone Compounds as Chemotherapeutic Agents, European Patent Application EP 0897915, 1999.
120. Griffin, R. J., Curtin, N. J., Newell, D. R., and Golding, B. T., Benzamide Analogues, Useful as PARP (ADP-Ribosyltransferas, ADPRT) DNA Repair Enzyme Inhibitors, European Patent Application EP 0879820, 1998.

14 Determination of Poly(ADP-Ribose) Polymerase Activation and Cleavage during Cell Death

*El Bachir Affar, Stéphane Gobeil,
Marc Germain, Eric Winstall, Rashmi Shah,
Sylvie Bourassa, Jérôme St-Cyr, Claudia Boucher,
and Guy G. Poirier*

CONTENTS

14.1 Introduction 306
14.2 Procedures 307
    14.2.1 Determination of NAD Level 307
        14.2.1.1 Extraction and Purification of NAD 308
        14.2.1.2 Determination of NAD by Cycling Assay 308
        14.2.1.3 Determination of NAD by HPLC 308
    14.2.2 Determination of pADPr 309
        14.2.2.1 Isolation of pADPr Following Cellular
                Radiolabeling 309
        14.2.2.2 Isolation of pADPr from Cells and Tissue for
                Nonisotopic Detection Methods 310
        14.2.2.3 Purification of pADPr by DHBB 311
        14.2.2.4 Determination of pADPr by HPLC 311
        14.2.2.5 Determination of pADPr by the Immunodot Blot
                Method 312
        14.2.2.6 Determination of pADPr by Enzyme-Linked
                Immunosorbent Assay 313
        14.2.2.7 High-Resolution Gel Electrophoresis and
                Immunodetection of pADPr 313
        14.2.2.8 Silver Stain of pADPr 315
        14.2.2.9 Detection of pADPr by Immunocytochemistry and
                Flow Cytometry 315

14.2.3 Detection of PARP Cleavage during Cell Death 316
        14.2.3.1 Western Blotting Detection of PARP 316
        14.2.3.2 Activity Western Blot for PARP Detection 317
Abbreviations 318
References 319

14.1 INTRODUCTION

Until recently, poly(ADP-ribose) (pADPr) was known to be synthesized by the nuclear enzyme poly(ADP-ribose) polymerase (PARP). Studies in the last 2 years have reported the presence of several PARP-like proteins in mammalian cells, suggesting the extent and complexity of poly(ADP-ribosyl)ation reactions.[1-5] Three members of this new family of enzymes have been shown to carry out pADPr synthesis. A 140-kDa human enzyme named tankyrase (due to its similarity to ankyrins and the catalytic domain of PARP) is suggested to be associated with telomeres.[1] A 62-kDa enzyme with homology to the catalytic domain of PARP has been described to synthesize pADPr in the presence of DNA strand breaks.[4] We have also isolated a 55-kDa enzyme that synthesizes pADPr in a manner independent of DNA strand breaks.[5]

The DNA nick sensor protein PARP catalyzes the synthesis of pADPr from β-NAD in response to DNA strand breaks. This 113-kDa protein is organized into three functional domains. The 42-kDa N-terminal DNA binding domain (DBD) contains the two zinc fingers responsible for the binding function of the DNA breaks. The 16-kDa automodification domain contains amino acid residues to which pADPr is covalently linked. The 55-kDa catalytic C-terminal domain is responsible for the initiation, elongation, and branching of pADPr.[6,7] On the other hand, pADPr catabolism is achieved by poly(ADP-ribose) glycohydrolase (PARG). This enzyme was recently found to be a 110-kDa protein[8,9] displaying a predominantly cytoplasmic distribution.[9,10]

Many laboratories have made significant efforts to establish procedures that permit studies on the involvement of pADPr in different features of cell functions such as DNA repair, recombination, transcription, and cell death. In the last two decades, we have been studying pADPr metabolism in *in vitro* models and cellular systems. Our laboratory has been actively involved in the establishment and improvement of several methods to study pADPr synthesis and catabolism. Among these are purification of PARG[11] and PARP;[12,13] detection of PARP by activity Western blot[14] and enzyme-linked immunosorbent assay (ELISA);[13] and detection of PARG by activity zymogram.[9,15] Recently, we established methods for the detection and characterization of pADPr using specific antibodies: determination of the cellular level of pADPr by immunodot-blot and ELISA;[16,17] analysis of pADPr size using high-resolution gel electrophoresis followed by immunoblotting;[17] and detection of pADPr using flow cytometry analysis.[17]

This chapter summarizes the main methods used to study pADPr metabolism and PARP detection with particular emphasis on common techniques that could be adopted to analyze pADPr synthesis and PARP cleavage during cell death (Figure 14.1).

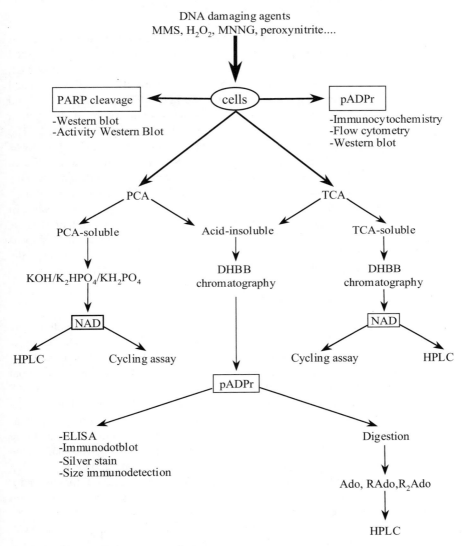

FIGURE 14.1 Schematic representation of the different techniques used for the determination of PARP activation and cleavage during cell death.

14.2 PROCEDURES

14.2.1 Determination of NAD Level

NAD, the substrate of PARP, is largely decreased during excessive DNA damage. The depletion of this molecule was suggested to induce ATP depletion and cell death[18] (see Introduction). Thus, NAD level determination could give valuable information regarding the role of PARP in cell death during excessive DNA damage.

14.2.1.1 Extraction and Purification of NAD

NAD extraction and purification were done as described in the original method[19] as modified by us.[20] After drug treatment, the cells (2 to 10×10^6) were rapidly washed with cold HEPES-saline buffer (10 mM HEPES, pH 7.4; 140 mM NaCl; 7 mM KCl; 6 mM glucose) and were kept on ice for 15 min with 1 ml of 20% TCA. Attached cells were harvested by scraping, rinsed with another milliliter of 20% TCA, and pooled with the previous fraction. The TCA pellet was collected by centrifugation at 1800 g for 10 min at 4°C and processed for pAPDr analysis (Section 14.2.2). The TCA soluble fraction was diluted with 0.25 M ammonium acetate (pH 9.0) and was adjusted to pH 9.0 with ammonium hydroxide. The NAD purification was performed using the dihydroxyboronyl Bio-Rex (DHBB) anion exchange chromatography. This affinity resin was synthesized according to Wielckens et al.[21] as modified by us.[20] The purification of NAD was done by using 1 ml of resin (0.5 ml packed resin) in an Econocolumn (Bio-Rad). The resin was washed twice with 5 ml of 0.25 M ammonium acetate (pH 9.0). After loading of the samples (two times), the resin was washed twice with 10 ml of 0.25 M ammonium acetate (pH 9.0). The elution of the NAD was done using 4 ml of water prewarmed at 37°C (4 × 1 ml).

14.2.1.2 Determination of NAD by Cycling Assay

The assay was done according to Bernofsky and Swan,[22] which was adapted to be used as a microassay.[20] Samples or known amounts of NAD (50 µl) (standard curve, 0 to 30 pmol/well) were dispensed in each well of a polystyrene, flat-bottomed, 96-well microplate. Freshly made reagent mixture (100 µl) containing at the final concentrations, 100 mM bicine (pH 7.8), 500 mM ethanol, 4.17 mM EDTA, 0.8 mg/ml bovine serum albumin (BSA), 0.42 mM phenylmethylsulfonylfluoride, 3-[4,5-dimethylthiazol-2-*yl*]-2,5 diphenyltetrazolium bromide (MTT), 1.66 mM phenazine ethosulfate (PES), was added to the wells. The reaction was initiated by addition of 20 µl (2 U) of alcohol dehydrogenase (EC 1.1.1.1) prepared in 100 mM bicine (pH 7.8). The incubation was done for 30 min at 30°C in the dark and the absorbance was read at 490 nm with a microplate reader (Model EL 340, Bio-Tek Instruments). A correlation coefficient of 0.99 was usually obtained for the standard curve, which was used for NAD determination. The absorbance could also be read at 550 nm as described by Shah et al.[20] The NAD could also be determined directly after PCA precipitation, without its purification on DHBB resin, according to the method described by Jacobson.[23] Briefly, the PCA soluble fraction (3 ml) was adjusted to pH 7.5 by adding 1.5 ml of 1 M KOH, 0.33 M K_2HPO_4–KH_2PO_4 (pH 7.5), and kept on ice for 15 min. The $KClO_4$ precipitate was removed by centrifugation at 2000 g for 15 min at 4°C. The final supernatant was frozen until use.

14.2.1.3 Determination of NAD by HPLC

An anion exchange high-pressure liquid chromatography (HPLC) was used to measure the cellular amount of NAD according to Alvarez-Gonzalez et al.[24] NAD was resolved from the other residual products of the DHBB eluate using a Partisil SAX column (10 µm, 250 × 4.6 mm) preceded by a guard column (75 × 2.1 mm) containing the same resin. The NAD was eluted isocratically with 50 mM K_2HPO_4–KH_2PO_4 (pH 4.7) at the flow rate of 1.0 ml/min. The absorbance was read at 254 nm and the peak area was integrated to calculate the amount of NAD using a standard curve of known quantities of NAD.

14.2.2 Determination of pADPr

Analysis of pADPr provides direct proof of its presence in cell and tissue samples. The level of pADPr in untreated cells is usually very low, approximately 2 to 10 pmol/mg of DNA.[25] This basal level of pADPr could be ascribed to PARP as well as to other pADPr-synthesizing enzymes. In contrast to wild-type cells, PARP-deficient cells exhibit a feeble synthesis of pADPr in response to DNA damage.[26] This suggests that PARP is the main enzyme responsible for pADPr synthesis in response to DNA strand breaks caused by DNA-damaging agents. In the case of the excessive DNA damage associated with pathological conditions, pADPr synthesis would be an accurate marker for the detection of PARP activation (see Introduction).

The main problem with the estimation of pADPr levels in cells is its occurrence in low amounts. In fact, picomolar amounts of pADPr are purified from micromolar amount of other nucleic acids. Thus, investigations on pADPr metabolism have been restricted by the scarcity of methods available to determine pADPr levels and its size in cells and tissues. Poly(ADP-ribosyl)ation has been extensively investigated with isolated nuclei or permeabilized cells. The main drawback of these approaches is the loss of cellular integrity. In fact, these nonphysiological conditions give very high amounts of pADPr and nonspecific poly(ADP-ribosyl)ation of proteins.[27,28] For example, permeabilized keratinocytes displayed a 400-fold higher pADPr amount than intact cells,[28] probably due to the release of PARG and also to DNA breaks introduced during permeabilization.

To overcome these difficulties, other methods that permit the qualitative and/or quantitative determination of pADPr have been established. Among these methods, radiolabeling of the cellular NAD pool with ^{3}H-adenine[19] and conversion of pADPr to fluorescent derivatives[29] have been described. Following PCA or TCA precipitation of poly(ADP-ribosyl)ated proteins, the acid insoluble fraction was further used to purify the pADPr, which was digested to adenosine (Ado), ribosyladenosine (RAdo), and diribosyladenosine (R_2Ado). These nucleosides were resolved by HPLC and the radioactivity or the fluorescence was determined in each fraction to estimate the amount of pADPr. Although these methods give an accurate estimation of pADPr, they are difficult to perform for routine determination. Recently, we established nonisotopic and simple techniques for pADPr detection in intact cells.[16,17] These methods allow the quantification of pADPr and could be used for its determination in a large number of samples.

14.2.2.1 Isolation of pADPr Following Cellular Radiolabeling

This procedure described by Aboul-Ela et al.[19] was modified as reported by Shah et al.[20] As the cells are not permeable to NAD, tritiated adenine that is incorporated in the intracellular pool of NAD was used for pADPr radiolabeling. Cells (2 to 10 × 10^6) were incubated for 16 h in 5 ml of fresh complete medium containing 20 MCi/ml of ^{3}H-adenine (^{3}H-Ade; 30 Ci/mmol). After drug treatments in fresh complete medium, the TCA-insoluble fraction was obtained as described (Section 14.2.1.1). The protein-bound ^{3}H-pADPr in the TCA pellet was dissolved in 0.2 ml of formic acid and diluted with 3.7 ml of water. BSA (1 mg) and cold pADPr (1 nmol) were added before precipitation with 1 ml of 100% TCA (4°C, 15 min) and centrifugation at 1800 g for 10 min at 4°C. Addition of BSA increases the precipitation of proteins. The exogenous pADPr allows detection of its digestion products (absorbance at 254 nm) and calculation of the yield. The TCA pellet was further resuspended in 1 ml of 250 mM ammonium acetate, 6 M guanidine hydrochloride, 10 mM EDTA (pH 6.0) (AAGE6). ^{3}H-pADPr was separated from the proteins by incubation at 37°C for 2 h with 1 ml of 1 M KOH, 50 mM EDTA with periodic vortexing. Incubation at this temperature prevents possible ^{3}H exchange, which could result in underestimation of pADPr amounts. The resulting sample was completed to 10 ml with 250 mM ammonium acetate, 6 M guanidine hydrochloride, 10 mM EDTA (pH 9.0) (AAGE9), and adjusted to pH 9.0 with HCl. The pADPr was then purified by DHBB chromatography (Section 14.2.2.3). The TCA supernatant was used for the purification of NAD (Section 14.2.1.1). The total amount of NAD was determined by cycling assay (Section 14.2.1.2), and the ^{3}H-NAD was resolved and determined by HPLC (Section 14.2.1.3). The radioactivity of NAD was used to estimate the specific activity of NAD, which was then used to calculate the amount of pADPr.

14.2.2.2 Isolation of pADPr from Cells and Tissue for Nonisotopic Detection Methods

The pADPr was isolated from the cells essentially according to the radiolabeling method (Section 14.2.2.1), except that the radiolabeling and formic acid precipitation step (addition of BSA and cold-pADPr) were omitted. After drug treatments and TCA precipitation, the pellet was washed twice with ethanol and resuspended in 1 ml of AAGE6. The pADPr was separated from the proteins by incubation at 60°C for 1 h with 1 ml of 1 M KOH, 50 mM EDTA. The fraction obtained was completed to 10 ml by AAGE9 and purified by DHBB chromatography (Section 14.2.2.3).

The polymer was isolated from tissues by using a modified method[20] previously described by Jacobson et al.[29] The tissue ($\cong$ 300 mg) was powdered in liquid nitrogen using a mortar and pestle. The resulting material was transferred to a polypropylene tube and homogenized in 20 ml of 1 N PCA using a polytron homogenizer at maximum speed for 30 s (twice). The PCA insoluble material was precipitated for 1 h at 4°C and recovered by centrifugation at 1000 g for 15 min at 4°C. The PCA soluble fraction (5 ml) was neutralized with 2.5 ml of 2 N KOH; 0.33 M K_2HPO_4–KH_2PO_4 (pH 7.5). After removing of $KClO_4$ precipitate by centrifugation at 2000 g for 15 min at 4°C, the supernatant was frozen at –20°C until NAD analysis (Section 14.2.1.2). The PCA insoluble pellet containing poly(ADP-ribosyl)ated proteins was washed twice in ethanol (–20°C) and diethyl ether (–20°C). Dispersion with polytron during the washes allowed the removal of residual PCA. The pellet was dried and stored at –70°C until use. The pADPr was released from protein by alkali digestion in 10 ml of 1 M KOH, 50 mM EDTA at 60°C for 90 min with periodic vortexing and completed with 14 ml of AAGE 9 (the pH was adjusted to 9.0 with HCl). Complete dissolution of the pellet before alkali treatment is required for efficient pADPr detachment from proteins. The fraction obtained was used for DHBB purification of pADPr (Section 14.2.2.3).

14.2.2.3 Purification of pADPr by DHBB

DHBB resin (0.5 ml of packed resin) was poured into a 10 ml Econocolumn and sequentially washed with 5 ml AAGE9, 10 ml water, and 10 ml AAGE9. The samples containing pADPr were passed twice through the DHBB resin. The column was washed with 8 ml of AAGE9 (three times) followed by 5 ml of 1 M ammonium acetate (pH 9.0) (two times). The pADPr was eluted with 4 ml of water prewarmed at 37°C (4 × 1 ml). The DHBB eluate was lyophilized and frozen until use. For tissue samples, some modifications of the purification procedure were made to increase the yield of pADPr.[20]

A known amount of ^{32}P-automodified PARP synthesized according to Brochu et al.[30] can be added to separate samples of acid-insoluble fraction before alkali treatment. This operation allows the calculation of the percentage of pADPr recovered after alkali treatment, DHBB purification, and the subsequent analysis steps.

14.2.2.4 Determination of pADPr by HPLC

Unlabeled and ^{3}H-pADPr eluted from DHBB were digested to nucleosides which were quantified by HPLC as originally described[29] and modified by us.[20] Briefly, the polymer samples were incubated at 37°C for 2 h with 25 U of RNase A (EC 3.1.27.5) prepared in 10 mM Tris (pH 7.4), 1 mM EDTA, to digest any contaminating RNA. The presence of RNA could affect the efficiency of the digestion of pADPr by the enzymes used at the following step. pADPr was then digested at 37°C overnight with 0.25 U of snake venom phosphodiesterase (EC 3.1.4.1) and 2.5 U of bacterial alkaline phosphatase (EC 3.1.3.1) to form Ado, RAdo, and R_2Ado. These nucleosides were resolved isocratically by HPLC on a C_{18} reverse-phase column (5 µm, 4.6 × 250 mm) and eluted at room temperature using 7 mM ammonium formate (pH 5.8), 6% methanol at flow rate of 1 ml/min. RAdo represents the main digestion product of pADPr. Its isolation and determination permits quantification of pADPr. The endogenous RAdo amount was estimated from radioactivity present in the ^{3}H-RAdo peak and from the specific activity of the ^{3}H-NAD pool in the cells.

For the pADPr isolated without radiolabeling, HPLC detection was done following the formation of fluorescent derivatives of the obtained nucleosides as described by Jacobson et al.[29] Briefly, the nucleoside RAdo was incubated in 0.5 ml of 1 M sodium citrate (pH 4.5), 1 M chloroacetaldehyde at 60°C for 4 h in the dark before HPLC separation, which is coupled to fluorescence detection.

Figure 14.2 represents an HPLC isolation of ^{3}H-RAdo derived from ^{3}H-pADPr synthesized in C3H10T1/2 cells treated with heat shock, which induces pADPr accumulation in the cells. The peak of Ado is derived from the pADPr termini and possibly from contaminating RNA. Considering the amount of the exogenous pADPr recovered, the yield of cellular pADPr is approximately 70%.

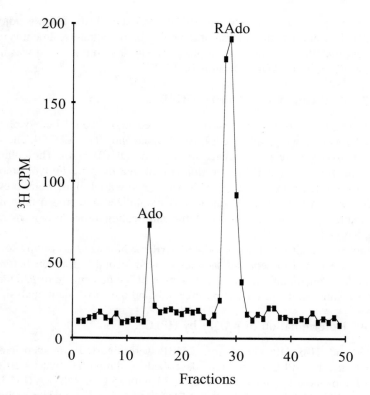

FIGURE 14.2 HPLC separation of ^{3}H-RAdo derived from ^{3}H-pADPr synthesized in C3H10T1/2 cells treated with heat shock at 42°C for 30 min.

14.2.2.5 Determination of pADPr by the Immunodot Blot Method

This nonisotopic immunodetection method of pADPr was recently developed by us.[16] The procedure is based on the fixation of pADPr on a positively charged membrane followed by its specific detection with anti-pADPr antibodies. The DHBB-purified pADPr was diluted in 0.4 M NaOH containing 10 mM EDTA and loaded on the Hybond N+ nylon membrane (Amersham Life Science) using a dot blot manifold system (Life Technologies). The membrane was then washed once with 0.4 M NaOH, removed from the manifold, and kept in water for further processing. The membrane was saturated in PBS-MT (phosphate-buffered saline, pH 7.4, containing 5% nonfat dried milk and 0.1% Tween 20) then incubated with the first anti-pADPr antibody. The membrane was washed with PBS-MT and incubated with a peroxidase-conjugated antimouse or antirabbit IgG secondary antibody (Jackson Immunoresearch). The blot was again washed with PBS-MT followed by washes in PBS prior to analysis by chemiluminescence using the Chemiluminescence Reagent Plus kit (Dupont NEN) or Super Signal Ultra (Pierce). As a large amount of long and branched pADPr is obtained during its synthesis by purified PARP, the pADPr used for the standard curve was prepared from C3H10T1/2 cells treated with N-methyl-N'-nitro-N-nitrosoguanidine (MNNG) after labeling with ^{3}H-Ade. Because of the action of PARG in the cells, the size distribution of this pADPr population would be similar to that obtained in other cell types. Due to the heterogeneity of pADPr, its concentration is expressed as ADP-residue equivalents.

Polymer synthesized by purified PARP or in cultured cells was purified on DHBB resin and detected by anti-pADPr antibodies (Figure 14.3). It could be seen that the chemiluminescence signal was obtained with femtomolar concentrations of pADPr (Panel A). Panel B shows the pADPr synthesized in C3H10T1/2 cells in various conditions. Heat-shock treatment increases pADPr accumulation by inhibiting PARG.[31,32] The DNA-damaging agent MNNG induces pADPr synthesis following the formation of DNA strand breaks. By using ^{3}H-pADPr for a standard curve, the cellular level of pADPr was estimated in C3H10T1/2 following treatments with MNNG and menadione (Panel C). Short pADPr less than six residues has a weak binding capacity for the membrane and is thus lost during the immunodetection steps.

14.2.2.6 Determination of pADPr by Enzyme-Linked Immunosorbent Assay

The assay was done as described by us.[17] Microplates (Falcon 3912) were coated overnight at 4°C with 1 µg/ml of poly-L-lysine made in PBS-T$_2$ (PBS, pH 7.4, containing 0.05% Tween 20). The plates were washed with PBS before the addition of pADPr prepared in 50 mM carbonate/bicarbonate buffer (pH 7.5) and incubated overnight at 37°C. The wells were washed with PBS-T$_2$, saturated for 1 h with PBS-MT$_2$ (PBS-T$_2$ containing 1% nonfat powdered milk) and incubated with the first anti-pADPr antibody. After several washes with PBS-T$_2$, the immune complex was incubated with a peroxidase-conjugated secondary antibody for 30 min. The wells were washed with PBS-T$_2$; then PBS, and detection was done with a 2,2'-azino-bis(3-ethylbenzthiazoline-6-sulfonic acid) (ABTS)/H$_2$O$_2$ solution. The absorbance was read at 405 nm using a microplate reader. The pADPr used for the standard curve was prepared as described (Section 14.2.2.5).

14.2.2.7 High-Resolution Gel Electrophoresis and Immunodetection of pADPr

Vertical electrophoresis of pADPr was carried out according to Alvarez-Gonzalez and Jacobson.[33] A 20% acrylamide gel (20 × 20 × 0.15 cm; acrylamide:bis-acrylamide (19:1) in 80 mM Tris-borate (pH 8.0) containing 2 mM EDTA, 0.07% ammonium persulfate, and 0.05% TEMED) was prerun in 42 mM Tris-borate (pH 8.3), 1.2 mM EDTA for 1 to 2 h at 400 V before loading the samples. Polymer samples were dried in a speed vac and resuspended in 10 µl of the loading buffer (50% urea, 25 mM NaCl, 4 mM EDTA, 0.02% xylene cyanol, and 0.02% bromophenol blue).

Electrophoretic resolution of pADPr was followed by its transfer onto a hybond N+ membrane in 35 mM Tris-borate (pH 8.3) and 1 mM EDTA using a transblot electrophoresis transfer apparatus (Bio-Rad). The transfer was performed at 0.4 A for 1.5 h at room temperature. After the transfer, the membrane was dried and pADPr was cross-linked on the membrane for 5 min using an ultraviolet (312 nm) transilluminator apparatus (Fisher Scientific). After pADPr fixation, the membrane was

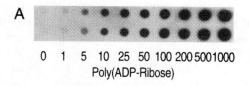
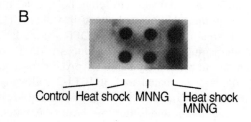
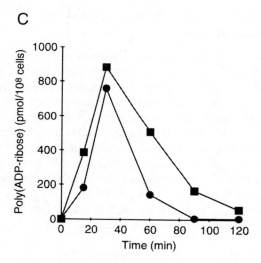

FIGURE 14.3 Detection of pADPr by the dot-immunoblot technique, using anti-pADPr antibody. (A) Various femtomolar amounts of pADPr synthesized *in vitro* were loaded onto a positively charged membrane and detected with LP96-10 antibody (dilution, 1/5000). (B) Polymer synthesis from 4×10^4 C3H10T1/2 cells subjected to heat shock at 45°C for 30 min and/or treated with 64 μM MNNG for 30 min was detected with the LP96-10 antibody (dilution, 1/5000). (C) Quantification of pADPr levels in C3H10T1/2 cells treated with 100 μM MNNG (squares) or 100 μM menadione (circles). (From Affar, E. B. et al., *Anal. Biochem.*, 259, 280, 1998.)

saturated with PBS-MT for 1 h and processed for immunodetection as described (Section 14.2.2.5).

The results presented in Figure 14.4 show the immunodetection of pADPr synthesized with purified PARP (Panel A) or following treatment of C3H10T1/2 cells with MNNG (Panel B). Most pADPr sizes were detected except short pADPr below 8 mers. Due to its weak binding to the membrane, this class of pADPr is lost

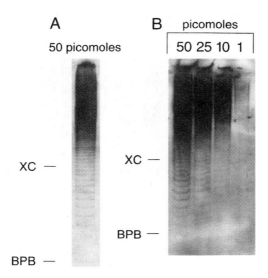

FIGURE 14.4 Immunodetermination of pADPr sizes by polyacrylamide gel electrophosis and immunoblotting. Poly(ADP-ribose) synthesized *in vitro* (A) or from intact cultured cells treated with MNNG (B) was resolved by polyacrylamide gel electrophoresis and transferred onto a positively charged membrane before blotting with the anti-pADPr antibody, LP96-10 (dilution 1/5000). (From Affar, E. B. et al., *Biochim. Biophys. Acta*, 1428, 137, 1999. With permission from Elsevier Science.)

during washing of the membrane for immunodetection. Using the LP96-10 anti-pADPr antibody, quantities of 1 pmol of pADPr isolated from C3H10T1/2 were immunodetected. We noticed that with this technique the 10H anti-pADPr antibody failed to detect pADPr below 25 mers.

14.2.2.8 Silver Stain of pADPr

Determination of pADPr chain length by silver staining has been described by Hacham and Ben-Ishai.[34] This method has recently been improved by Malanga et al. for pADPr size quantification.[28] The pADPr separated by high-resolution electrophoresis is stained directly on the gel using a GELCODE color Silver Stain kit (Pierce). Despite the fact that this procedure permits the detection of polymers up to 3 mers, it is less sensitive than the immunodetection method. The results presented in Figure 14.5 represent a silver stain coloration of various concentrations of pADPr synthesized with purified PARP.

14.2.2.9 Detection of pADPr by Immunocytochemistry and Flow Cytometry

For immunocytochemistry, cells grown on coverslips were washed following treatments with cold PBS, fixed on ice in 3.7% formaldehyde for 15 min, washed with PBS and permeabilized in PBS, containing 0.2% NP-40 for 15 min at room tem-

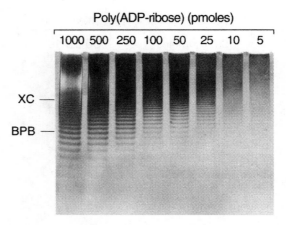

FIGURE 14.5 Determination of pADPr sizes by silver staining. pADPr synthesized *in vitro* was resolved by polyacrylamide gel electrophoresis and revealed by silver staining of the gel.

perature. After another washing with PBS, coverslips were saturated with PBS-MT for 1 h and incubated overnight at 4°C with anti-pADPr antibody. Cells were washed with PBS-MT prior to incubation for 30 min with fluorescein isothiocyanate (FITC)-conjugated second IgG (Jackson ImmunoResearch). Coverslips were washed with PBS and water after which they were dried. The mounting medium *p*-phenylenediamine diluted in glycerol (0.1% final) was used to prevent fluorescence fading.

Analysis of pADPr by flow cytometry was recently described by us.[17] The cells were washed in cold PBS and fixed for 10 min using a cold mixture (–20°C) of acetone–methanol (0.3/0.7 v/v). Immunodetection was performed according to the protocol described for immunocytochemistry, except that between each step cells were spun in a tabletop microcentrifuge for 20 s. The cells were analyzed on an Epics Elite ESP cytometer (Coulter Electronics).

14.2.3 Detection of PARP Cleavage during Cell Death

PARP is cleaved during apoptosis into a 89-kDa and a 24-kDa fragment[35-37] (see Chapter 10 by Germain et al.). PARP cleavage is concomitant with the internucleosomal degradation of DNA and was observed in almost all cases of apoptosis. Due to its high abundance, early cleavage, and the availability of specific antibodies, detection of PARP cleavage is widely used as a specific marker for apoptosis. PARP is also cleaved during necrosis into 50-, 40-, and 35-kDa fragments.[38] Thus, the differential cleavage of PARP during apoptosis and necrosis could be used to determine the mode of cell death.

14.2.3.1 Western Blotting Detection of PARP

Preparation of cell extracts was done as described previously.[20] After drug treatments, cells were washed with ice-cold HEPES, lysed in reducing loading buffer (62.5 m*M*

Tris-HCl, pH 6.8; 4 M urea; 10% glycerol; 2% SDS; 0.3% bromophenol blue; 5% β-mercaptoethanol) and sonicated to break the DNA. Samples were incubated at 65°C for 15 min before electrophoresis.

Tissue samples were prepared according to Shah et al.[20] Briefly, tissues were homogenized in ice-cold extraction buffer containing 50 mM glucose, 25 mM Tris-HCl (pH 8.0), 10 mM EDTA, 1 mM phenylmethylsulfonyl fluoride (PMSF), and 0.5 μg/ml each of leupeptin, aprotinin, and antipain using a glass homogenizer with a Teflon pestle. Fractions of the tissue extract were diluted in reducing loading buffer, sonicated, and heated before electrophoresis.

For electrophoresis, a sample volume equivalent to 1 to 2×10^5 cells or 20 μg of protein was loaded and resolved on an 8% polyacrylamide gel according to Laemmli.[39] The electrophoresis was performed in running buffer (25 mM Tris, 192 mM glycine, 0.1% SDS) at 200 V for 30 min. Proteins were then transferred onto a nitrocellulose membrane in transfer buffer (25 mM Tris, 192 mM glycine, 20% methanol) with constant stirring at 125 V for 60 min. After the transfer, the membrane was stained with Ponceau S dye (0.1% Ponceau S, w/v; 5% acetic acid, v/v) for 1 min and washed in deionized water. Molecular weight standards were then identified. The nitrocellulose membrane was saturated in PBS-MT for 60 min and incubated with the monoclonal $C_{II}10$ anti-PARP antibody in PBS-MT for time ranging from 2 h to overnight. The membrane was washed in PBS-MT, incubated with the peroxidase-conjugated second antibody for 30 min, and washed again with PBS-MT. The blot was then washed in PBS, and PARP was revealed by a chemiluminescence kit.

Figure 14.6 shows an immunodetection of PARP in HL-60 cells treated with etoposide, an inhibitor of topoisomerase II. It could be seen that a time-dependent cleavage of PARP is obtained following the treatment. Similar results were obtained for other cell types and treatments.

14.2.3.2 Activity Western Blot for PARP Detection

This technique allows the immunodetection of poly(ADP-ribose) synthesized following electrophoresis and renaturation of PARP on the membrane.[14] This procedure

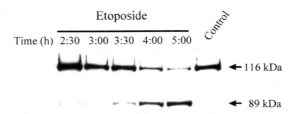

FIGURE 14.6 Western blot detection of PARP and its 89-kDa apoptotic fragment. HL-60 cells were treated with apoptosis inducer, etoposide, for the indicated times. Following treatment, 10 μg of cell lysates were resolved using 8% SDS-PAGE, transferred onto a nitrocellulose membrane, and subsequently analyzed by immunoblotting for PARP cleavage using the monoclonal anti-PARP $C_{II}10$ antibody (dilution 1/10,000).

can be used for the detection of PARP in many species, particularly when anti-PARP antibodies are not available.

The preparation of samples and electrophoresis were carried out as for a regular Western blot (Section 14.2.3.1). After electrophoresis, the gel was incubated at 37°C for 1 h in running buffer containing freshly added 0.7 M β-mercaptoethanol. This treatment is necessary to avoid cross-linking between proteins. Proteins were then transferred onto a nitrocellulose membrane as described (Section 14.2.3.1). Prestained or biotinylated molecular weight standards were used to avoid staining under acidic conditions, which could destroy the polymerase activity. The membrane was incubated for 1 h at room temperature in renaturation buffer (50 mM Tris-HCl, pH 8.0; 100 mM NaCl; 1 mM DTT; 0.3% Tween 20; 2 mM MgCl$_2$; and 20 µM Zn (II) acetate). After this step, the membrane was incubated in renaturation buffer containing 100 µM of NAD. After pADPr synthesis step, the blot was washed in renaturation buffer without MgCl$_2$, Zn (II) acetate, and DNA. To remove pADPr noncovalently, but strongly attached to others proteins, the membrane was washed in SDS buffer (50 mM Tris-HCl, pH 8.0; 100 mM NaCl; 1 mM DTT; 2% SDS) as described.[14] To detect automodified PARP, Western blotting was done with anti-polymer antibodies as described (Section 14.2.3.1).

Membranes were eventually erased to be reprobed with the same or other antibodies according to the erasable Western blot technique described by Kaufmann et al.[40] Briefly, membranes were washed in PBS-T (PBS containing 0.1% Tween 20) and incubated for 30 min at 65°C in stripping buffer (62.5 mM Tris-HCl, pH 6.8; 2% SDS; 100 mM β-mercaptoethanol freshly added). The membranes were washed several times in PBS-T and processed as for a regular Western blot.

Figure 14.7 shows a detection of PARP and its 89-kDa apoptotic fragment by Western and activity-Western blotting. Blot A represents the immunodetection of PARP purified from bovine thymus or PARP from HL-60 cells treated with etoposide using C$_{II}$10 antibody. The same samples were used for activity blot in the absence (Figure 14.7B) or in the presence of activated DNA (Figure 14.7C). Because the addition of activated DNA results in increased automodification of the full-length enzyme but not the 89-kDa fragment, it could be omitted if comparison between PARP and its fragments is required.

ABBREVIATIONS

The abbreviations used are AAGE6, 250 mM ammonium acetate, 6 M guanidine hydrochloride, 10 mM EDTA (pH 6.0); AAGE9, 250 mM ammonium acetate, 6 M guanidine hydrochloride, 10 mM EDTA (pH 9.0); ABTS, 2,2′-azino-bis(3-ethylbenzthiazoline-6-sulfonic acid); Ado, adenosine; BPB, bromophenol blue; BSA, bovine serum albumin; DBD, DNA binding domain; DHBB, dihydroxyboronyl Bio-Rex; DTT, dithiothreitol; ELISA, enzyme-linked immunosorbent assay; HEPES, N-[2-hydroxyethyl]piperazine-N'-[2-ethanesulfonic acid]; HPLC, high-performance liquid chromatography; PES, phenazine ethosulfate; PMSF, phenylmethylsulfonyl fluoride; MNNG, 1-methyl-3-nitro-1-nitrosoguanidine; MTT, phenylmethylsulfonylfluoride, 3-[4,5-dimethylthiazol-2-yl]-2,5-diphenyltetrazolium bromide; NAD,

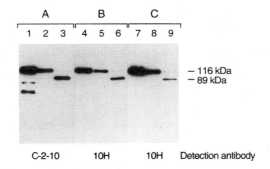

FIGURE 14.7 Activity western blot detection of PARP. Purified PARP (120 ng) (lane 1) and cells extracts (1.5×10^4 cells) from control (lane 2) or apoptotic HL-60 cells (lane 3) were electrophoresed and transferred onto a nitrocellulose membrane. PARP was revealed using the monoclonal anti-PARP antibody $C_{II}10$ (A) or by activity Western blot using the 10H anti-pADPr antibody. pADPr synthesis was done either in the presence of NAD alone (B) or in the presence of NAD and activated DNA (C). (From Shah, G. M. et al., *Anal. Biochem.*, 232, 251, 1995.)

nicotinamide adenine dinucleotide; PARP, poly(ADP-ribose) polymerase; PARG, poly(ADP-ribose) glycohydrolase; pADPr, poly(ADP-ribose); PBS, phosphate-buffered saline; PBS-MT, PBS (pH 7.4) containing 5% nonfat dried milk and 0.1% Tween 20; PBS-T$_2$, PBS (pH 7.4) containing 0.05% Tween 20; PBS-MT$_2$, PBS-T$_2$ containing 1% nonfat powdered milk; PBS-T, PBS containing 0.1% Tween 20; PCA, perchloric acid; RAdo, ribosyl adenosine; R$_2$Ado, diribosyl adenosine; TCA, trichloroacetic acid; XC, xylene cyanol.

REFERENCES

1. Smith, S., Giriat, I., Schmitt, A. and de Lange, T., Tankyrase, a poly(ADP-ribose) polymerase at human telomeres, *Science*, 282, 1484, 1998.
2. Johansson, M., A human poly(ADP-ribose) polymerase gene family (ADPRTL): cDNA cloning of two novel poly(ADP-ribose) polymerase homologues, *Genomics*, 57, 442, 1999.
3. Berghammer, H., Ebner, M., Marksteiner, R., and Auer, B., pADPRT-2: a novel mammalian polymerizing (ADP-ribosyl)transferase gene related to truncated pADPRT homologues in plants and *Caenorhabditis elegans*, *FEBS Lett.*, 449, 259, 1999.
4. Amé, J. C., Rolli, V., Schreiber, V., Niedergang, C., Apiou, F., Decker, P., Muller, S., Hoger, T., de Murcia, J. M., and de Murcia, G., PARP-2, A novel mammalian DNA damage-dependent poly(ADP-ribose) polymerase, *J. Biol. Chem.*, 274, 17860, 1999.
5. Sallmann, F. R., Vodenicharov, M. D., Wang, Z. Q., and Poirier, G. G., Characterization of sPARP, a novel poly(ADP-ribose) polymerase with DNA strand breaks independent activity, *J. Biol. Chem.*, in press.
6. Lautier, D., Lagueux, J., Thibodeau, J., Ménard, L., and Poirier, G. G., Molecular and biochemical features of poly(ADP-ribose) metabolism, *Mol. Cell. Biochem.*, 122, 171, 1993.

7. D'Amours, D., Desnoyers, S., D'Silva, I., and Poirier, G. G., Poly(ADP-ribosyl)ation reactions in the regulation of nuclear functions, *Biochem. J.*, 342, 249, 1999.
8. Lin, W., Amé, J. C., Aboul-Ela, N., Jacobson, E. L., and Jacobson, M. K., Isolation and characterization of the cDNA endcoding bovine poly(ADP-ribose) glycohydrolase, *J. Biol. Chem.*, 272, 11895, 1997.
9. Winstall, E., Affar, E. B., Shah, R., Bourassa, S., Scovassi, A. I., and Poirier, G. G., Poly(ADP-ribose) glycohydrolase is present and active in mammalian cells as a 110-kDa protein, *Exp. Cell Res.*, 246, 395, 1999.
10. Winstall, E., Affar, E. B., Shah, R., Bourassa, S., Scovassi, A. I., and Poirier, G. G., Preferential perinuclear localization of poly(ADP-ribose) glycohydrolase, *Exp. Cell Res.*, 251, 372, 1999.
11. Thomassin, H., Jacobson, M. K., Guay, J., Verreault, A., Aboul-Ela, N., Ménard, L., and Poirier, G. G., An affinity matrix for the purification of poly(ADP-ribose) glycohydrolase, *Nucl. Acids Res.*, 18, 4691, 1990.
12. D'Amours, D., Duriez, P. J., Orth, K., Shah, R. G., Dixit, V. M., Earnshaw, W. C., Alnemri, E. S., and Poirier, G. G., Purification of the death substrate poly(ADP-ribose) polymerase, *Anal. Biochem.*, 249, 106, 1997.
13. Sallmann, F. R., Plancke, Y. D., and Poirier, G. G., Rapid detection of poly(ADP-ribose) polymerase by enzyme-linked immunosorbent assay during its purification and improvement of its purification, *Mol. Cell. Biochem.*, 185, 199, 1998.
14. Shah, G. M., Kaufmann, S. H., and Poirier, G. G., Detection of poly(ADP-ribose) polymerase and its apoptosis-specific fragment by a nonisotopic activity-Western blot technique, *Anal. Biochem.*, 232, 251, 1995.
15. Brochu, G., Shah, G. M., and Poirier, G. G., Purification of poly(ADP-ribose) glycohydrolase and detection of its isoforms by a zymogram following one- or two-dimensional electrophoresis, *Anal. Biochem.*, 218, 265, 1994.
16. Affar, E. B., Duriez, P. J., Shah, R. G., Sallmann, F. R., Bourassa, S., Küpper, J. H., Bürkle, A., and Poirier, G. G., Immunodot blot method for the detection of poly(ADP-ribose) synthesized *in vitro* and *in vivo*, *Anal. Biochem.*, 259, 280, 1998.
17. Affar, E. B., Duriez, P. I., Shah, R. G., Winstall, E., Germain, M., Boucher, C., Bourassa, S., Kirkland, J. B., and Poirier, G. G., Immunological determination and size characterization of poly(ADP-ribose) synthesized *in vitro* and *in vivo*, *Biochim. Biophys. Acta*, 1428, 137, 1999.
18. Berger, N. A., Poly(ADP-ribose) in the cellular response to DNA damage, *Radiat. Res.*, 101, 4, 1985.
19. Aboul-Ela, N., Jacobson, E. L., and Jacobson, M. K., Labeling methods for the study of poly- and mono(ADP-ribose) metabolism in cultured cells, *Anal. Biochem.*, 174, 239–250, 1988.
20. Shah, G. M., Poirier, D., Duchaine, C., Brochu, G., Desnoyers, S., Lagueux, J., Verreault, A., Hoflack, J. C., Kirkland, J. B., and Poirier, G. G., Methods for biochemical study of poly(ADP-ribose) metabolism *in vitro* and *in vivo*, *Anal. Biochem.*, 227, 1, 1995.
21. Wielckens, K., Bredehorst, R., and Hilz, H., Quantification of protein-bound ADP-ribosyl and (ADP-ribosyl) in residues, *Methods Enzymol.*, 106, 472, 1984.
22. Bernofsky, C., and Swan, M., An improved cycling assay for nicotinamide adenine dinucleotide, *Anal. Biochem.*, 53, 452, 1973.
23. Jacobson, E. L. and Jacobson, M. K., Pyridine nucleotide levels as a function of growth in normal and transformed 3T3 cells, *Arch. Biochem. Biophys.*, 175, 627, 1976.

24. Alvarez-Gonzalez, R., Eichenberger, R., Loetscher, P., and Althaus, F. R., A new highly selective physicochemical assay to measure NAD^+ in intact cells, *Anal. Biochem.*, 156, 473, 1986.
25. Affar, E. B., Shah, R. G., and Poirier, G. G., Poly(ADP-ribose) turnover in quail myoblast cells: relation between the polymer level and its catabolism by glycohydrolase, *Mol. Cell. Biochem.*, 193, 127, 1999.
26. Shieh, W. M., Ame, J. C., Wilson, M. V., Wang, Z. Q., Koh, D. W., Jacobson, M. K., and Jacobson, E. L., Poly(ADP-ribose) polymerase null mouse cells synthesize ADP-ribose polymers, *J. Biol. Chem.*, 273, 30069, 1998.
27. Tanuma, S., Kawashima, K., and Endo, H., Comparison of ADP-ribosylation of chromosomal proteins between intact and broken cells, *Biochem. Biophys. Res. Commun.*, 127, 896, 1985.
28. Malanga, M., Bachmann, S., Panzeter, P. L., Zweifel, B., and Althaus, F. R., Poly(ADP-ribose) quantification at the femtomole level in mammalian cells, *Anal. Biochem.*, 228, 245, 1995.
29. Jacobson, M. K., Payne, D. M., Alvarez-Gonzalez, R., Juarez-Salinas, H., Sims, J. L., and Jacobson, E. L., Determination of *in vivo* levels of polymeric and monomeric ADP-ribose by fluorescence methods, *Methods Enzymol.*, 106, 483, 1984.
30. Brochu, G., Shah, G. M., and Poirier, G. G., Purification of poly(ADP-ribose) glycohydrolase and detection of its isoforms by a zymogram following one- or two-dimensional electrophoresis, *Anal. Biochem.*, 218, 265, 1994.
31. Jonsson, G. G., Ménard, L., Jacobson, E. L., Poirier, G. G., and Jacobson, M. K., Effect of hyperthermia on poly(adenosine diphosphate-ribose) glycohydrolase, *Cancer Res.*, 48, 4240, 1988.
32. Jonsson, G. G., Jacobson, E. L., and Jacobson, M. K., Mechanism of alteration of poly(adenosine diphosphate-ribose) metabolism by hyperthermia, *Cancer Res.*, 48, 4233, 1988.
33. Alvarez-Gonzalez, R. and Jacobson, M. K., Characterization of polymer of ADP-ribose generated *in vitro* and *in vivo*, *Biochemistry*, 26, 3218, 1987.
34. Hacham, H. and Ben-Ishai, R., Determination of poly(ADP-ribose) chain length distribution on polyacrylamide gels by silver staining, *Anal. Biochem.*, 184, 83, 1990.
35. Kaufmann, S. H., Desnoyers, S., Ottaviano, Y., Davidson, N. E., and Poirier, G. G., Specific proteolytic cleavage of poly(ADP-ribose) polymerase: an early marker of chemotherapy-induced apoptosis, *Cancer Res.*, 53, 3976, 1993.
36. Lazebnik, Y. A., Kaufmann, S. H., Desnoyers, S., Poirier, G. G., and Earnshaw, W. C., Cleavage of poly(ADP-ribose) polymerase by a proteinase with properties like ICE, *Nature*, 371, 346, 1994.
37. Tewari, M., Quan, L. T., O'Rourke, K., Desnoyers, S., Zeng, Z., Beidler, D. R., Poirier, G. G., Salvesen, G. S., and Dixit, V. M., Yama/CPP32 beta, a mammalian homologue of CED-3, is a CrmA-inhibitable protease that cleaves the death substrate poly(ADP-ribose) polymerase, *Cell*, 81, 801, 1995.
38. Shah, G. M., Shah, R. G., and Poirier, G. G., Different cleavage pattern for poly(ADP-ribose) polymerase during necrosis and apoptosis in HL-60 cells, *Biochem. Biophys. Res. Commun.*, 229, 838, 1996.
39. Laemmli, U. K., Cleavage of structural proteins during the assembly of the head of bacteriophage T4, *Nature*, 227, 680, 1970.
40. Kaufmann, S. H., Ewing, C. M., and Shaper, J. H., The erasable Western blot, *Anal. Biochem.*, 161, 89, 1987.

15 Isoforms of Poly(ADP-Ribose) Polymerase: Potential Roles in Cell Death

Elaine L. Jacobson and Myron K. Jacobson

CONTENTS

15.1 Introduction 323
15.2 PARP-1 323
15.3 Tankyrase 325
15.4 PARP-2 327
15.5 Vault-PARP 327
15.6 Final Remarks 327
Acknowledgments 328
References 328

15.1 INTRODUCTION

Poly(ADP-ribose) polymerase 1 (PARP-1), discovered more than 30 years ago, was thought until recently to be solely responsible for the synthesis of ADP-ribose polymers in animal cells. The recent identification of multiple proteins with PARP activity[1] dictates careful reevaluation of the roles of PARP-1 in cell death. Further, the newly discovered members of the PARP family need to be considered for potential involvement in cell death signaling pathways. All of the members of the PARP family have a catalytic domain that contains highly conserved residues referred to as a PARP signature sequence.[2] Table 15.1 compares some of the other properties of the emerging family of PARP proteins.

15.2 PARP-1

The function of PARP-1 has long been closely linked to cell death responses to genotoxic stress. Early studies demonstrated that PARP-1 and poly(ADP-ribose) glycohydrolase (PARG) catalyze opposing arms of ADP-ribose polymer cycles with kinetics that result in a very transient existence of individual polymer molecules.[3]

TABLE 15.1
Comparative Properties of Different Proteins with PARP Activity

Protein	Size, kDa	Activated by DNA Damage	Catalyzes Automodification	Substrates	Cellular Location
PARP-1[a]	113	Yes	Yes	Histones Other DNA binding proteins	Nuclear
Tankyrase[b]	142	No	Yes	Telomere-specific protein TRF1	Chromosome telomeres, peri-nuclear
PARP-2[c]	62	Yes	Yes	Unknown	Nuclear
Vault-PARP[d]	193	No	Yes	MVP	Cytoplasmic, nuclear

[a] Reference 2.
[b] Reference 11.
[c] Reference 22.
[d] Reference 25.

Isoforms of PARP: Potential Roles in Cell Death

The kinetics of ADP-ribose polymer cycles suggested that they participate in stress signaling functions. The ability of PARP-1 to deplete a large fraction of the cellular NAD pool provided a possible mechanism by which energy metabolism regulates cell death pathways.[4,5] From structural studies, we have learned that the common factor of genotoxic stress that activates PARP-1 is the recognition of DNA strand breaks by the double-zinc-finger DNA-binding domain of PARP-1, which in some way leads to rapid activation of the catalytic domain of the enzyme.[2] Studies demonstrating that PARP-1 inhibitors or molecular genetic approaches to inactivate PARP-1 could modulate cell death following genotoxic stress have led to the postulated roles of PARP-1 in the cell death pathways of apoptosis and necrosis.[6]

The availability of animals and cells in which the PARP-1 gene has been disrupted has greatly facilitated understanding of PARP-1 function in cell death.[7,8] The picture emerging from PARP-1 knockout (KO) cells and animals is that PARP-1 can serve as either a survival factor or as an effector of cell death following genotoxic stress.[6] For example, PARP-1 KO animals and cells derived from them show decreased survival following treatment with alkylating agents and γ irradiation.[8] Under these conditions, PARP-1, functioning in concert with other DNA damage response checkpoint proteins, appears to promote cell survival by halting cell cycle progression and activating DNA damage repair pathways.[8] In contrast to the increased sensitivity to alkylating agents and γ irradiation, PARP-1 KO animals are highly resistant to the damaging effects of oxidative stresses.[6] The involvement of PARP-1 in the promotion of cell death following oxidative stresses can occur by at least two possible mechanisms. In one case, excessive DNA damage results in hyperactivation of PARP-1, resulting in depletion of the cellular NAD pool to the extent that impaired energy metabolism leads to cellular necrosis. In the second mechanism, PARP-1 functions in signaling pathways that promote cell death by apoptosis. Most studies have attributed the oxidative stress resistance of PARP KO animals to the absence of PARP-1-mediated necrosis. A number of studies have demonstrated that PARP-1 mediates some pathways of apoptosis.[9] Therefore, the role of PARP-1 in effecting apoptosis vs. necrosis following oxidative stress needs to be examined. The function of PARP-1 as an effector of cell death has both positive and negative consequences for the organism. PARP-1 functions to maintain the genomic integrity of the organism in cases where cell death is promoted in cells that might undergo malignant transformation. On the other hand, in conditions such as ischemia–reperfusion injury, the PARP-1 death effector function threatens the life of the organism.

15.3 TANKYRASE

The availability of PARP-1 KO animals has contributed to a major paradigm shift in understanding of ADP-ribose polymer metabolism. The demonstration that cells with a disrupted PARP-1 gene retain the ability to synthesize ADP-ribose polymers[10] suggested the presence of additional enzyme(s) with PARP activity. Indeed, this has proved to be the case. The discovery that chromosome termini (telomeres) containing a unique PARP activity brings new perspective for the involvement of ADP-ribose polymer cycles in modulating cell cycle regulation and cell death.[11] Tankyrase is a protein that has a catalytic domain homologous to PARP-1; however, its overall domain structure is vastly different. In sharp contrast to PARP-1, tankyrase does not appear to require DNA for activity. Thus, the discovery of tankyrase provided the first example for involvement of ADP-ribose polymer cycles in functions unrelated to genotoxic stress responses. Telomeres are the terminal regions of chromosomes that contain unique repetitive DNA sequences and G-rich single-stranded overhangs that are stabilized by the telomere specific proteins TRF1 and TRF2.[12] Telomere maintenance is closely linked to pathways of cell death. Because of the single-strand overhangs, telomeres cannot be replicated completely by the replication machinery responsible for overall chromosome replication. Telomere length can be extended by action of a cellular reverse transcriptase termed telomerase, but most normal human tissues do not contain this activity.[13-16] Thus, in most cells, chromosome telomeres shorten with each round of replication, resulting in telomere erosion.[14] Telomere erosion leads to cellular senescence and induction of cell death when telomeres shorten to the point that a stable telomere structure cannot be maintained. Senescence and induction of cell death are attributed to the detection of eroded telomeres by DNA damage checkpoint proteins like p53,[14] which leads to prolonged cell cycle blocks and induction of apoptosis.

Until recently no clear vision of how telomeres escape detection by DNA damage checkpoint proteins had emerged. Recently, a model for a telomere loop structure[12] has been proposed in which TRF1 binds to and aligns double-stranded telomere DNA[17] into a structure (t-loop) that allows the G-rich overhangs to invade and form a displacement loop (D-loop) with double-stranded telomere DNA. The formation and/or stabilization of the D-loop is promoted by TRF2. In contrast to normal cells, most cancer cells contain high levels of telomerase and thus maintain unlimited proliferation potential by avoiding the cell death that results from telomere erosion. When telomerase is present, the telomere loop structure also likely needs to be disassembled and reassembled to allow telomere length maintenance in cancer cells. Tankyrase provides a potentially interesting link to telomere maintenance and thus to mechanisms of cell death in both normal and abnormal cells. *In vitro*, tankyrase interacts with and catalyzes ADP-ribose polymer modification of the telomere-specific protein TRF1. The telomere loop model suggests a role for tankyrase and PARG in disassembly and reassembly of the loop structure to allow for telomere replication and/or telomere length maintenance in cells containing telomerase.[1] Tankyrase-catalyzed addition of negatively charged polymers to TRF1 could allow disassociation of TRF1 from telomere DNA and disassembly of the loop. Likewise, PARG-catalyzed degradation of TRF1-associated ADP-ribose polymers could result in reassociation of TRF-1 and reassembly of the telomere loop. Limiting tankyrase or PARG activity may result in cell cycle blocks since TRF-1 may not be removed under these circumstances. While there is limited information on telomere replication in higher organisms, it appears to be a late S-phase event. Several studies of PARP inhibitors have shown them to cause late S and G_2 cell cycle blocks.[18-21] The effects of PARP inhibitors and substrate restriction on cell cycle progression need to be reexamined in terms of possible effects on tankyrase. Further, studies of cell cycle progression in PARP-1-disrupted cells and the construction of cell and animal models with tankyrase disruption and genetic crosses between PARP-1 and tankyrase KO

Isoforms of PARP: Potential Roles in Cell Death

animals should provide interesting insight into the role of tankyrase in cell cycle regulation and cell death.

15.4 PARP-2

A second link of ADP-ribose polymer metabolism to genotoxic stress has been established by the discovery of a protein termed PARP-2.[22-24] Like PARP-1, PARP-2 is activated by DNA breaks, but it detects DNA breaks by a DNA-binding domain very different from that of PARP-1.[22] A human cDNA with high sequence homology to PARP-2 also has been reported,[24] but the protein encoded by this cDNA has not yet been shown to have PARP catalytic activity. Specific function(s) for PARP-2 are not yet known, but its nuclear location and activation by DNA damage makes a role in cell death a likely possibility. PARP-2 activation by genotoxic stress may represent a signaling function distinct from that catalyzed by PARP-1. For example, PARP-2 may function in regions of chromatin (e.g., telomeres) where activation of PARP-1 must be avoided.

15.5 VAULT-PARP

A second link of ADP-ribose polymer metabolism to processes unrelated to genotoxic stress has been provided by the discovery of a PARP activity in association with cell structures known as vaults.[25] Vaults are large ribonucleoprotein complexes of unknown function located primarily in the cytoplasmic compartment;[26] thus the discovery of Vault-PARP suggests that ADP-ribose polymer cycles are not restricted to the nucleus. Like tankyrase, Vault-PARP does not require DNA for activity, but it contains a domain that interacts with and modifies the major protein component of vaults (MVP). A third protein component of vaults is TEP-1, which is also a component of the telomerase complex. Functions of Vault-PARP are not known, but its association with telomerase-associated proteins and its additional localization in mitotic spindles are consistent with functions related to cell death–signaling pathways.

15.6 FINAL REMARKS

The presence of multiple PARPs dictates a careful reassessment of the modulation of cell death by compounds that inhibit the activity of PARP-1. Most compounds developed against PARP-1 act at or near the nicotinamide region of the NAD-binding site of the enzyme; thus it is not surprising that commonly used PARP-1 inhibitors also inhibit the other known PARPs.[2,11,22,25] Indeed, some of the differences observed between the effects of PARP inhibitors and those observed in PARP-1 KO cells and animals may reflect biological effects due to inhibition of one or more of the other PARPs. The remarkable resistance of PARP-1 KO animals to myocardial infarction, stroke, shock, diabetes, and neurodegeneration strongly reinforces the notion that PARP-1 is an important player in cell death pathways. As seen in other contributions to this book, PARP-1 has emerged as an important potential therapeutic target for the modulation of cell death pathways. The discovery of multiple proteins with PARP activity now makes it likely that the successful therapeutic targeting of PARP-1 for the modulation of cell death will require the development of compounds that can selectively inhibit PARP-1.

ACKNOWLEDGMENTS

The authors are supported by research grants from the National Institutes of Health (CA43894, CA65579, NS38496) and Niadyne, Inc. (The authors are principals in Niadyne Inc., whose sponsored research is managed in accordance with University of Kentucky conflict-of-interest policies.)

REFERENCES

1. Jacobson, M. K. and Jacobson, E. L., Discovering new ADP-ribose polymer cycles: protecting the genome and more, *Trends Biochem. Sci.*, 24, 415–417, 1999.
2. de Murcia, G. and Ménissier-de Murcia, J., Poly(ADP-ribose) polymerase: a molecular nick-sensor, *Trends Biochem. Sci.*, 19, 172–176, 1994.
3. Juarez-Salinas, H., Sims, J. L., and Jacobson, M. K., Poly(ADP-ribose) levels in carcinogen-treated cells, *Nature*, 282, 740–741, 1979.
4. Jacobson, M. K., Levi, V., Juarez, S. H., Barton, R. A., and Jacobson, E. L., Effect of carcinogenic N-alkyl-N-nitroso compounds on nicotinamide adenine dinucleotide metabolism, *Cancer Res.*, 40, 1797–1802, 1980.
5. Berger, N. A., Sims, J. L., Catino, D. M., and Berger, S. J., Poly(ADP-ribose) polymerase mediates the suicide response to massive DNA damage: studies in normal and DNA-repair defective cells, in *ADP-Ribosylation, DNA Repair and Cancer*, Miwa, M., Ed., Japan Scientific Societies Press, Tokyo, 1983, 219–226.
6. Pieper, A. A., Verma, A., Zhang, J., and Snyder, S. H., Poly(ADP-ribose) polymerase, nitric oxide and cell death, *Trends Pharmacol. Sci.*, 20, 171–181, 1999.
7. Wang, Z. Q., Auer, B., Stingl, L., Berghammer, H., Haidacher, D., Schweiger, M., and Wagner, E. F., Mice lacking ADPRT and poly(ADP-ribosyl)ation develop normally but are susceptible to skin disease, *Genes Dev.*, 9, 509–520, 1995.
8. Ménissier-de Murcia, J., Niedergang, C., Trucco, C., Ricoul, M., Dutrillaux, B., Mark, M., Oliver, F. J., Masson, M., Dierich, A., LeMeur, M., Walztinger, C., Chambon, P., and de Murcia, G., Requirement of poly(ADP-ribose) polymerase in recovery from DNA damage in mice and in cells, *Proc. Natl. Acad. Sci. U.S.A.*, 94, 7303–7307, 1997.
9. Simbulan-Rosenthal, C. M., Rosenthal, D. S., Iyer, S., Boulares, A. H., and Smulson, M. E., Transient poly(ADP-ribosyl)ation of nuclear proteins and role of poly(ADP-ribose) polymerase in the early stages of apoptosis, *J. Biol. Chem.*, 273, 13703–13712, 1998.
10. Shieh, W. M., Ame, J. C., Wilson, M. V., Wang, Z. Q., Koh, D. W., Jacobson, M. K., and Jacobson, E. L., Poly(ADP-ribose) polymerase null mouse cells synthesize ADP-ribose polymers, *J. Biol. Chem.*, 273, 30069–30072, 1998.
11. Smith, S., Giriat, I., Schmitt, A., and de Lange, T., Tankyrase, a poly(ADP-ribose) polymerase at human telomeres, *Science*, 282, 1484–1487, 1998.
12. Griffith, J. D., Comeau, L., Rosenfield, S., Stansel, R. M., Bianchi, A., Moss, H., and de Lange, T., Mammalian telomeres end in a large duplex loop, *Cell*, 97, 503–514, 1999.

13. Hahn, W. C., Counter, C. M., Lundberg, A. S., Beijersbergen, R. L., Brooks, M. W., and Weinberg, R. A., Creation of human tumour cells with defined genetic elements, *Nature,* 400, 464–488, 1999.
14. Hahn, W. C., Stewart, S. A., Brooks, M. W., York, S. G., Eaton, E., Kurachi, A., Beijersbergen, R. L., Knoll, J. H., Meyerson, M., and Weinberg, R. A., Inhibition of telomerase limits the growth of human cancer cells, *Nat. Med.,* 5, 1164–1170, 1999.
15. Meyerson, M., Counter, C. M., Eaton, E. N., Ellisen, L. W., Steiner, P., Caddle, S. D., Ziaugra, L., Beijersbergen, R. L., Davidoff, M. J., Liu, Q., Bacchetti, S., Haber, D. A., and Weinberg, R. A., hEST2, the putative human telomerase catalytic subunit gene, is up-regulated in tumor cells and during immortalization, *Cell,* 90, 785–795, 1997.
16. Zhu, J., Wang, H., Bishop, J. M., and Blackburn, E. H., Telomerase extends the lifespan of virus-transformed human cells without net telomere lengthening, *Proc. Natl. Acad. Sci. U.S.A.,* 96, 3723–3728, 1999.
17. Griffith, J., Bianchi, A., and de Lange, T., TRF1 promotes parallel pairing of telomeric tracts *in vitro, J. Mol. Biol.,* 278, 79–88, 1998.
18. Hori, T., High incidence of sister chromatid exchanges and chromatid interchanges in the conditions of lowered activity of poly(ADP-ribose)polymerase, *Biochem. Biophys. Res. Commun.,* 102, 38–45, 1981.
19. Oikawa, A., Tohda, H., Kanai, M., Miwa, M., and Sugimura, T., Inhibitors of poly(adenosine diphosphate ribose) polymerase induce sister chromatid exchanges, *Biochem. Biophys. Res. Commun.,* 97, 1311–1316, 1980.
20. Schwartz, J. L., Morgan, W. F., Kapp, L. N., and Wolff, S., Effects of 3-aminobenzamide on DNA synthesis and cell cycle progression in Chinese hamster ovary cells, *Exp. Cell Res.,* 143, 377–382, 1983.
21. Utakoji, T., Hosoda, K., Umezawa, K., Sawamura, M., Matsushima, T., Miwa, M., and Sugimura, T., Induction of sister chromatid exchanges by nicotinamide in Chinese hamster lung fibroblasts and human lymphoblastoid cells, *Biochem. Biophys. Res. Commun.,* 90, 1147–1152, 1979.
22. Amé, J. C., Rolli, V., Schreiber, V., Niedergang, C., Apiou, F., Decker, P., Muller, S., Hoger, T., Ménissier-de Murcia, J., and de Murcia, G., PARP-2, a novel mammalian DNA damage-dependent poly(ADP-ribose) polymerase, *J. Biol. Chem.,* 274, 17860–17868, 1999.
23. Berghammer, H., Ebner, M., Marksteiner, R., and Auer, B., pADPRT-2: a novel mammalian polymerizing(ADP-ribosyl)transferase gene related to truncated pADPRT homologues in plants and *Caenorhabditis elegans, FEBS Lett.,* 449, 259–263, 1999.
24. Johansson, M., A human poly(ADP-ribose) polymerase gene family (ADPRTL): cDNA cloning of two novel poly(ADP-ribose) polymerase homologues, *Genomics,* 57, 442–445, 1999.
25. Kickhoefer, V. A., Siva, A. C., Kedersha, N. L., Inman, E. M., Ruland, C., Streuli, M., and Rome, L. H., The 193-kD vault protein, VPARP, is a novel poly(ADP-ribose) polymerase, *J. Cell Biol.,* 146, 917–928, 1999.
26. Kickhoefer, V. A., Rajavel, K. S., Scheffer, G. L., Dalton, W. S., Scheper, R. J., and Rome, L. H., Vaults are up-regulated in multidrug-resistant cancer cell lines, *J. Biol. Chem.,* 273, 8971–8974, 1998.

Index

A

Acetylcholine, 84, 88, 91
Aconitase
    inactivation by nitric oxide, 115
Activation of PARP, *see also* PARP
    measurements of, 307–315
Activator-dependent transcription, 265–270
Activity zymogram
    determination of PARG activity by, 306–307
Adenosine, *see* purines
Adhesion molecules, 97,153
    intracellular adhesion molecule-1(ICAM-1), 29, 51–55, 64–67, 72, 148–149, 295
    P-selectin, 51–55
Adriamycin, 171–177
Adult respiratory distress syndrome, 62
AIDS, 169
Alkaline phosphatase, 93
Alkylating agents, 175, 213–215, 218, 283, 294
Alpha-amino-3-hydroxy-5-methylisoxazole-4-propionic acid (AMPA), 8
Alzheimer's disease, 7, 23, 32–34
Amiphostine, 174
AMPA, *see* Alpha-amino-3-hydroxy-5-methylisoxazole-4-propionic acid
Annexin, 140–142
Antibodies
    against PARP, 316–319
    against poly(ADP-ribose), 313–315
Antisense depletion of PARP, 283–284
Antitumor agents, 167–177
Antiviral agents, 167–177
Aminobenzamide derivatives, *see* PARP, inhibitors of
Amyloid protein, 32–33
AP-1 complex, 64–65, 72
Aphidicolin, 265
Apoptosis, 3, 24–25, 62, 168, 195–197, 209–219, 227–242, *see also* Caspases
    detection of cell death by, 140–142
    in diabetes, 114–117
    induced by cisplatin, 176
    morphological features of, 24–25, 210
    role of bcl proteins, 146–147
    switch from necrosis to, 139–142, 197, 216
    time course of, 227–239
Apoptotic bodies, 24
AP-2, 267–270
Arrhythmia, 41
Arthritis, 55, 65, 68–73
    MHC-II, role of, 148–149
    lymphocytes, role of, 152
Atherosclerosis, 62
ATM, 213
ATP
    cellular depletion, 1–3, 16, 24–25, 27, 31, 42–46, 64–67, 82, 196, 228, 235, 306
Auto-poly(ADP-ribosyl)ation, *see* PARP, domains of

B

Bacterial lipopolysaccharide, *see* Endotoxin
Base excision repair, 175, 184, 188, 190–193, 284, *see also* DNA, repair of
Bcl protein family, 146, 235, 238, 256–257
Bcl-2, 146–147, 195
Berger, Nathan, 1
Beta cells, *see* Islet cells
BGP-15, *see* PARP, inhibitors of
Biobreeding rat
    as a diabetes model, 105
Bleomycin
    lung injury induced by, 172
Blood flow, 27–28, 45, 67
Blood pressure, 87
Blood vessels
    cerebral artery, 7–17
    mesenteric artery, 51–54, 88, 91
B-lymphocyte
    development of, 151–153
Body weight loss
    in cisplatin intoxication, 173
    in colitis, 67–69
Bone marrow cells, 191
Brain trauma 2, *see also* Neurotrauma
BRCT motif, 186

C

Calcium, intracellular
  as a cytotoxic factor, 12, 16, 26–29, 48, 54, 89, 281
  interactions with mitochondrial alterations, 144–146
Calcium-magnesium-dependent endonuclease, see Endonuclease
Calmodulin
  as a cofactor in activating NOS, 12
Canavanine, 84
Carcinogenesis, 13, 73, 151
Cardiomyopathy, 49
Carrageenan, 51, 65–66
Caspases, 210–212, 215, 227–239; see also Apoptosis
  S-nitrosylation of, 139
  Caspase-1, 138
  Caspase-3
    role in apoptotic cell death, 138–148, 195–197, 210–212, 227–239, 254
    role in islet injury, 116
    role in neuroinjury, 25
CD95, 197
Cell cycle, 184–199, 235–240, 255–258, 265–270, 326
Ceramide, 147
c-fos, 54, 93
Chambon, 184
Chemokines, 65, 148–149, 153
Chromatin architecture, 184, 259
Chromosomal damage, 190–193
Cisplatin, 171–177
Citrulline
  staining in the central nervous system, 10
c-jun, 267
Cleavage of PARP, see PARP, cleavage of
Clinical trials
  nicotinamide intervention trials in diabetes, 118–119
Cochrane, Charles, 1
Colitis, 2, 52, 62, 65–70
Collagenase, 54, 148–149
Comparative genomic hybridization, 240
Complement, 65, 72, 90, 153
Creatinine phosphokinase, 45
CREB, 267
Cyclooxygenase, 65
Cyclophosphamide
  as a diabetes coinducer, 111
Cytochrome C, see Apoptosis

D

Dendritic cells, 132
Dexamethasone
  apoptosis, induced by, 141–142, 147, 214, 231–232, 282
Diabetes mellitus, 1, 2, 43, 73, 103–120, 198, 228, 294
  insulin secretion, loss of, 104
  MHC-II, role of, 148–149
  stages of development, 105
  type I (insulin dependent), 104–105
Diphteria toxin, 187
Dopamine, 15, 30–32
DNA
  alkylation of, 168
  cleavage, see Apoptosis
  mitochondrial, 169
  repair of, 1, 16, 184–199, 212, 218–219, 236, 282–283
  strand breaks, see also DNA, repair of; Radiation, ionizing
    in diabetes, 107–109, 116–119
    in inflammation, 63, 72
    in myocardial infarction, 42, 43, 55
    in neuroinjury, 1, 12–13, 28, 30, 31
    in shock, 82–83, 94–97
    induced by bleomycin, 172
DNA-binding domain of PARP, see PARP, domains of
DNA helicase, 260
DNA ligase, 186, 194, 260
DNA methyltransferase, 93
DNA-PK, 151, 196, 213, 217, 258
DNA polymerases, 193–194, 235, 260–270, 280
DNA primase, 260, 264
DNA synthesome, 259–264
  components of, 259–260

E

8-hydroxydeoxyguanosine, 43
Electron leakage, see Mitochondria
ELISA
  determination of PARP activity by, 306–307, 313
Endonuclease, 259, 280
Endothelial nitric oxide synthase, see Nitric oxide synthases
Endothelium, 51, 52, 62, 89, 91
Endothelium-derived relaxing factor, see Nitric oxide

Index

Endotoxin, 54, 84–85, 93; *see also* Shock, endotoxic
Epithelium
  in colitis, 52, 64, 177
  in Peyer's plaques, 138–139
  in shock, 86, 88
E2F-1, 264–270
Etoposide, 317
Excitotoxicity, *see* Glutamate

F

Fas ligand
  apoptosis, induced by, 141–142, 211–212, 214, 227–239, 252, 254, 256, 284
Flow cytometry
  detection of cell death by, 140–142
  detection of chromosomal alterations by, 239–242
  detection of PARP activation by, 315–316
Fluid extravasation, 90

G

Gel electrophoresis
  determination of PARP activity by, 306–307, 313–314
  gene amplification, 188
Gene expression, modulation by PARP, *see* PARP, modulation of gene expression by
Genetic background
  of knockout animals, 14
Genomic integrity, 188, 227–242, 239–242
Glia, 31, 33
Glucocorticoids, 71
Glutamate, 7–17, 23–34, 294
  measurement of by microdyalisis, 7
Glutathione, 72, 91, 172
GMP, cyclic, 8–9, 84
  levels in the brain, 8–9
GPI-650, *see* PARP, inhibitors of
G-proteins, 8, 10
Guanidinoethyl disulfide, 84
Guanylyl cyclase, soluble, 10

H

Heart, *see* Myocardium
Heat labile enterotoxin, 187
HeLa, 212, 252
Histamine, 63
Histones, 184, 212, 235, 253–255, 268, 280

HIV, 168
HL-60, 147, 212, 231, 317
HMG, 184
Human studies, *see also* Clinical trials
  increased PARP activation in the brain, 15–16
Huntington's disease, 7
Hydrogen peroxide, 1, 2, 49, 62, 83, 86, 96
Hydroxyl radical, 83, 84, 89, 93–95
Hypochlorous acid, 83
Hypoxanthine, *see* Purines

I

Immunodot method
  determination of PARP activity by, 306–307, 312–313
Inducible nitric oxide synthase, *see* Nitric oxide synthases
Inflammation
  role of PARP, 61–73
Inflammatory bowel disease, *see* Colitis
Infection, 73, 94
INH2BP, *see* PARP, inhibitors of
Inosine, *see* Purines
Insulin secretion
  impairment of, 104–112
Insulinoma cell line, 111–112
Interferon-gamma
  effect on cells, 136, 282
  in diabetes, 106–107, 111–112
  in shock, 93–94
Interleukin-1
  in diabetes, 106–107, 111–112, 115–117
  in inflammation, 54, 65
Interleukin-6, 65
Interleukin-10, 54
Intestinal barrier
  in colitis, 64–65
Intracellular adhesion molecule-1 (ICAM-1), *see* Adhesion molecules
Intracellular calcium, *see* Calcium, intracellular
Intravital microscopy, 51
Ionizing radiation, *see* Radiation, ionizing
Iron chelation, 172
Ischemia-reperfusion injury
  cerebral, *see* Stroke
  myocardial, *see* Myocardial ischemia-reperfusion
  skeletal muscle, *see* Skeletal muscle ischemia-reperfusion
  splanchnic, *see* Splanchnic ischemia-reperfusion

Islet cells
   death of, 1, 73, 96, 104–106, 177, 228
   immune response against, 104–105
Isoforms of nitric oxide synthase, see Nitric oxide synthases, isoforms of
Isoforms of PARP, see PARP, isoforms of
Isolated perfused heart, see Myocardium
Isoquinolinones, see PARP, inhibitors of
Isotype switch, 152

J

Joint disease, see Arthritis

K

Keratinocytes, 229, 252
Knockout animals
   of NO synthase, see Nitric oxide synthases, knockout animals
   of PARP, see PARP, knockout animals

L

Lactams, see PARP, inhibitors of
Laryngeal injury, 45–46
Leukocyte, see Neutrophil granulocyte
Lipid peroxidation, 16, 83
Lipopolysaccharide, see Endotoxin
Liver, 2–3, 86, 90
Long patch repair, 193
Lung, 90
Lymphocytes, see T-lymphocyte

M

Macrophages
   in diabetes, 112
   in inflammation, 62
   in neurodegeneration, 33
   in shock, 83, 86, 94
   interactions with T-lymphocytes, 136
Malignant transformation, see Carcinogenesis
Malondialdehyde, 67
MAO-B, 31
MAP kinase, 64–65, 72, 93
Measurements of poly(ADP-ribose), see Poly(ADP-ribose), measurements of; see also PARP
Methylmethanesulfonate, 188
MHC-II, 148–149, 282

Middle cerebral artery, see Blood vessels
Mitochondria
   dysfunction of
      in diabetes, 107
      in myocardial reperfusion, 42, 46–49
      in neuroinjury, 24, 26, 31
      in shock, 86
      in thymocyte death, 144–146
   electron leakage, 48, 144–146
   respiratory chain, 48, 83, 86, 144–146, 169–171
   transmembrane potential, 49, 144–146
MK801, 27
Mortality
   after radiation injury, 189–190
   in cisplatin intoxication, 173
   in colitis, 67–70
   in myocardial reperfusion, 45
   in shock, 86–87, 95–97
MPTP, 8, 15, 30–32
Multiple organ dysfunction, 86; see also Shock
Multiprotein DNA replication complex, see DNA synthesome
Myeloperoxidase
   as an indicator of neutrophil infiltration
      in colitis, 66–68
      in myocardial reperfusion, 44, 52
      in shock, 88
Myocardial ischemia-reperfusion, 17, 29, 41–56, 94, 169, 173, 177, 280, 295
   role of protein kinase C, 146
Myocardium, 2, 41–56, 87
   isolated perfused heart, 45, 169, 173, 295
   left ventricular pressure, 45–47

N

NAD, depletion of; see also ATP; Necrosis
   in cell death, 10, 14, 16, 24, 27, 31, 139–142, 219, 227–239, 255, 281
   in diabetes, 104, 108, 111–117
   in inflammation, 64
   in myocardial infarction, 44–46
   in shock, 96–97
   induced by cisplatin, 176
   measurements of, 307–309
      cycling assay, 308
      extraction and purification, 308
      HPLC method, 308–309
   nuclear, 227
NADPH diaphorase
   as an indicator of iNOS, 132–138
NADPH oxidase, 43, 72
Naphtalidimes, see PARP, inhibitors of

Index

Necrosis, 1–3, 198
 detection by flow cytometry, 140–142
 in diabetes, 115–116
 in inflammation, 64–65, 72
 in myocardial reperfusion, 42, 45
 in neuroinjury, 24–25
 in shock, 82
 morphological features of, 24–25
 of thymocytes, 139–144
Nephrotoxicity
 induced by cisplatin, 174
Neuritic plaques, 32
Neuronal nitric oxide synthase, *see* Nitric oxide synthases
Neurotoxicity, 1–3, 7–17, 176
Neurotrauma, 30, 88, 260, 295
Neutrophil granulocyte
 in inflammation, 62, 65–66, 70
 in myocardial reperfusion, 43, 45–46, 51–54
 in neuroinjury, 29
 in shock, 83, 89–95
Nicotinamide, *see* PARP, inhibitors of
Nitric oxide (NO)
 as a trigger of PARP activation, 1–3, 132
 in diabetic islet destruction, 106, 112, 117–119
 in inflammation, 61–64, 68
 in myocardial reperfusion injury, 42, 49, 54
 in neurodegeneration, 26–34
 in shock, 81–97
 in stroke, 8, 10
 endothelium-derived relaxing factor, 8, 62, 85
 neurotransmitter roles of, 8, 10–12, 62
Nitric oxide synthases (NOS)
 endothelial, 62, 81
 inducible
  detection by NADPH diaphorase, 133–138
  expression of, 148–149
  in colitis, 62, 65, 72
  in diabetes, 106, 112
  in lymphoid organs, 133–138
  in myocardial reperfusion, 54
  in neuroinjury, 30–33
  in shock, 81–89
 inhibitors of, 87, 91
 knockout animals, 9, 32, 136
 neuronal, 7–17, 26, 32, 62
  alternatively spiced forms of, 9
Nitrotyrosine
 as a marker of peroxynitrite generation
  in circulatory shock, 82, 88, 91
  in inflammation, 63, 67
  in lymphoid organs, 134–138
  in myocardial reperfusion, 43
  in stroke, 10

Nitroxonium, 83
N-methyl-D-aspartate (NMDA)
 as a neurotoxin, 8–14, 26–30
Nonobese diabetic mice
 as a model of diabetes, 105–110
Nuclear factor kappa B
 activation of
  in inflammation, 64, 72
  in myocardial injury, 54
  in shock, 89, 94–97, 199
 poly(ADP-ribosyl)ation of, 268–270

O

Obesity
 in PARP-deficient animals, 284
Oct-1, 186, 199
Okamoto, 107, 108
Osteosarcoma cells, 227–239, 253, 257
Oxygen-free radicals
 in diabetes, 106
 in neuroinjury, 7, 12, 26, 28, 31, 33, 281
Oxygen-glucose deprivation, 14

P

Parkinson's disease, 2, 7, 23, 30–32, 43, 228, 280
PARP
 cleavage of
  measurement of, 316–318
  role in apoptosis, 1, 3, 24–25, 139–143, 185, 195–197, 215–218, 228
 crystal structure of, 187, 281
 domains of, 185–187, 230–231, 281
 inhibitors of, 11, 29, 279–295
  aminobenzamide derivatives, 11, 12, 29, 32, 48, 49, 51–55, 66, 82, 85–97, 108–110, 140, 188, 212, 216, 255, 262, 280
  BGP-15, 169–177
  clinical uses of, 16–17, 72–73, 94, 117–119, 171, 199, 282, 294–295
  Consensus structural requirements, 293–294
  GPI-650, 14–15, 44, 48, 292
  INH2BP, 29, 49, 66, 70, 86–87, 93, 140, 293
  isoquinolinones, 13–14, 44, 48, 86, 187, 287–294
  lactams, 288–294
  naphtalidimes, 288–294
  nicotinamide, 14, 44, 48, 52, 70, 73, 86, 108–110, 118–119, 153, 285–288
 interactions with transcription factors, 193–195, 199, 252–270, 285–288
 isoforms of, 13–14, 197–199, 253, 323–328

knockout animals
  apoptosis in cells from, 197, 213–217, 232–234
  E2F-1 expression in cells from, 265
  exposed to whole body irradiation, 189–190
  generation of, 112–113, 189, 214, 252, 283–285
  genomic instability of, 190–193, 239–242
  in diabetes, 112–115, 228
  in immune-mediated diseases, 153
  in inflammation, 65–70, 198
  in myocardial reperfusion, 43, 45, 51
  in neuroinjury, 27, 30–32, 228
  in shock, 89–91, 95–97
  in stroke, 12–17
  thymocytes from, 140–146
  modulation of gene expression by, 1, 148–151
  in inflammation, 72
  in myocardial infarction, 54
  in shock, 92–97
  molecular organization of, 42, 63–64, 184–188, 209, 281
  noncleavable mutant of, 195–197, 219, 227–239
  physical association of, 259–270
PARP-2, *see* PARP, isoforms of
Paw edema, 51, 66
Peritonitis, 51
Peroxidation, of lipids, *see* Lipid peroxidation
Peroxynitrite
  cytotoxic effects
    in diabetes, 107
    in inflammation, 63
    in myocardial injury, 49, 53
    in neurons, 10, 26–27, 30, 31
    in shock, 82–84, 89, 94–97
    in thymocytes, 132–153, 214
  generation of
    in inflammation, 42, 44, 63, 132
    in shock, 83
    in the central nervous system, 10
    in the heart, 26–27
Pertussis toxin, 187
Peyer plaques, *see* Epithelium
P53
  apoptosis, role of, 148, 212–213, 235–239, 255–258, 264, 267–270, 280, 284
Phospholipase A2, 28, 48
Pleurisy, 51
Poly(ADP-ribose), *see also* PARP
  immunohistochemical staining for, 15, 33, 44
  measurements of, 309–315
    by cellular radiolabeling, 309–310
    by ELISA, *see* ELISA

    by flow cytometry, *see* Flow cytometry
    by gel electrophoresis, *see* Gel electrophoresis
    by HPLC, 311
    by immunodot, *see* Immunodot
    by nonisotopic methods, 310–316
    purification of, 311
Poly(ADP-ribose) glycohydrolase, 64, 184, 230, 256, 281, 306, 323
Poly(ADP-ribosyl)ation, *see also* PARP
  of nuclear proteins, 252–268
Postsynaptic density protein (PSD), 9, 12
Potassium
  neuroinjury, role of, 27
  withdrawal of, apoptosis induced by, 147
Procaspase, 231–232, 254, *see also* Caspases
Proliferation, 191
Propydium iodide, 140–142
Prostaglandins, 66–68
Proteases, 90
Protein kinase C, 54, 93, 146, 215
P-selectin, *see* Adhesion molecules
P21-activated kinase, 196, 217
Purines
  as inhibitors of PARP, 145–146, 292
Pyknosis
  in diabetes, 114
  in neuroinjury, 25
Pyramidal cells, 30

R

Radiation, ionizing, 1, 11, 14, 83, 116, 148, 184, 188–190, 215, 239, 252, 294
Random mutagenesis, 188
Reactive oxygen species, *see* Superoxide
Recombination, 151–154
Repair, *see* DNA, repair of
Reperfusion, *see* Ischemia-reperfusion injury
Replication forks, 185
Respiratory chain, *see* Mitochondria
Retinoids, 150, 153
Retinoid X receptor, 149
RXR-alpha, 185, 199

S

SCID, 151, 190
Sepsis, *see* Shock, endotoxic
Serum starvation, 265
Shock, 81–97, 280
  endotoxic, 62, 65, 81–87, 136, 198

Index

hemorrhagic, 62, 87–88
lymphocyte function and, 137
Short patch repair, 193
Silver stain of PARP, 315
Sister chromatid exchange, 14, 73, 239, 253
S-nitroso-N-acetyl-DL-penicillamine (SNAP)
  as a nitric oxide donor, 49, 85
Skeletal muscle ischemia-reperfusion, 48
Skin lesions
  in PARP-deficient animals, 284
Splanchnic ischemia-reperfusion, 46, 48, 65, 88–89
SP-1, 267
Staphylococcal enterotoxin B
  effects on thymus, 133, 136
Stem cells, 191
Streptozocin
  as a diabetes inducer, 105–111, 113–115
  DNA alkylation by, 108
Stroke, 1–3, 7–17, 25–30, 48, 198, 280, 294
Sulfur mustard, 229
Superoxide anion
  in diabetes, 107, 111, 115–119
  in inflammation, 62–64, 132
  in myocardial reperfusion, 48, 49
  in neuroinjury, 26, 31–32
  in shock, 81–82, 89, 94–97
  in thymocyte death, 143–145
  production of by antiviral agents, 168–171
Superoxide dismutase
  activity of, in the central nervous system, 26
  activity of, in the heart, 48
  activity of, in shock, 82, 84, 86

T

Tankyrase, 198, 306, 324–327, *see also* PARP, isoforms of
  and islet cell death, 117
TEF-1, 185, 267
Telomerase, 168
Tetraploidity, 240
TFIIF, 267
3-aminobenzamide, *see* Aminobenzamide derivatives
3′-azido-3′-deoxythymidine (AZT), 168–171
Thymocytes
  mechanism of cell death, 52–53
  nitric oxide and peroxynitrite toxicity, 132–153
  selection, 133, 142–144, 210
Tissue plasminogen activator (TPA)
  and stroke treatment, 16

T-lymphocytes
  development of, 151–153
  in diabetes, 106
  interactions with macrophages, 136
  proliferation, gene expression, 137–138
Topoisomerase I
  inhibition of, 190, 214, 235, 253, 258–264
Topoisomerase II, 168, 235, 253, 258–264
TPEN, 144
Transgultaminase, 150–151
Traumatic brain injury, *see* Neurotrauma
Traumatic spinal cord injury, *see* Neurotrauma
Trinitrobenzenesulfonic acid (TNBS)
  colitis, induced by, 6, 66–68
Tumor growth delay, 174
Tumor necrosis factor alpha
  apoptosis induced by, 147, 196, 211–214, 218, 284
  in colitis, 65, 72
  in diabetes, 106–107, 111
  in myocardial reperfusion, 51–55
  in neuroinjury, 25
  in shock, 94
  receptor family, 133, 195
2′3′-dideoxycytidine, 169–171

X

Xanthine oxidase
  in colitis, 62–63, 72
  in diabetes, 107, 111
  in myocardial reperfusion, 43
  in shock, 83
XRCC1, 193–195

Y

YY1, 186, 199, 267

U

Ubiquitin-conjugating enzyme, 194
Ultrapotent inhibitors of PARP, *see* PARP, inhibitors of

W

Western blot
  determination of PARP activity by, 306–307, 319
  determination of PARP cleavage, 316–318

V

Vascular reactivity, 82, 91, 96
Vasoconstriction, 81, 91
Vasorelaxation, 84, 91
Vault-PARP, 327; *see also* PARP, isoforms of
V(D)J recombination, 151–153
Venules, 52

Z

Zidovudine, 49
Zinc
    as a co-factor in PARP related assays, 318
    intracellular, role in cell death, 144–146
Zinc fingers of PARP, *see* PARP, domains of
Zymosan
    inflammation, induced by, 51–52, 63, 89–91